DIGITAL ELECTRONICS
With Engineering Applications

PRENTICE-HALL SERIES IN COMPUTER APPLICATIONS IN ELECTRICAL ENGINEERING

Frank Kuo, *editor*

DAVIS
Computer Data Displays
JENSEN & LIEBERMAN
IBM Circuit Analysis: Program: Techniques and Applications
KUO & MAGNUSON
Computer Oriented Circuit Design
SIFFERLEN & VARTANIAN
Digital Electronics with Engineering Applications

DIGITAL ELECTRONICS

With Engineering Applications

Thomas P. Sifferlen

Manager, Sonar Systems Department
Submarine Signal Division
Raytheon Company

Vartan Vartanian

Manager, Underwater Technology Department
Ocean Systems Division
Sanders Associates, Incorporated; and
Adjunct Associate Professor of Engineering
Brown University

Prentice-Hall, Inc. Englewood Cliffs, N. J.

Current printing (last digit):

10 9 8 7 6 5 4 3 2 1

13-214304-6
Library of Congress Catalog Card Number: 79-108608
Printed in the United States of America

PRENTICE-HALL INTERNATIONAL, INC., *London*
PRENTICE-HALL OF AUSTRALIA, PTY. LTD., *Sydney*
PRENTICE-HALL OF CANADA LTD., *Toronto*
PRENTICE-HALL OF INDIA PRIVATE LTD., *New Delhi*
PRENTICE-HALL OF JAPAN, INC., Tokyo

PREFACE

This book is intended to provide a unified treatment of the theory, design techniques, components, and applications for digital electronic systems. The design examples are not limited to specific types of components but utilize the modular approach to logic functions that can be implemented with discrete components, integrated circuits, and large-scale integration of semiconductor circuits. It is believed that the design approaches taken will not become obsolete, but rather will become more enhanced as further developments are made in integrated-circuit technology.

The book may be used as a text for a one-semester course in an undergraduate engineering curriculum. The course, to be more effective, should include laboratory work where the student can design, build, and test representative circuits. In curricula where an active-circuits course with pulse and wave shaping is included, Chapter 5 may be omitted. By expanding on this material and including a discussion of Z-transforms and digital filters, the instructor may use the text for a two-semester course.

Practicing electronics engineers, who require an updating of their technical skills, will find the book an effective means for self study. The vast majority of the engineers who completed their formal education before the past decade did not take a comprehensive course in digital electronics. When attempting self-study in the advances in digital techniques, the engineers find that most books on the subject are either large-computer oriented, or are limited to the theoretical aspects of Boolean algebra and sequential machines. Equally unsatisfactory is the periodical literature that provides the handbook approach to digital circuits without relating adequately to the theoretical fundamentals; an approach that is aimed more for technicians rather than for professional engineers.

An attempt has been made to provide a balanced treatment of theory and practice. The design examples given, and subjects discussed, are relevant to the electronics engineer in industry. The basic theory given, coupled with applications, should provide the reader with sufficient background to enable a sound understanding of the important subject of digital electronics.

T. P. SIFFERLEN, V. VARTANIAN

CONTENTS

* References and problems follow each chapter.

DIGITAL ELECTRONICS
With Engineering Applications

CHAPTER 1

SWITCHING CIRCUIT DESIGN

Switching circuits form the fundamental building blocks of digital computers and electronic data processing equipment. These circuits operate in one or the other of two states that are represented by two nonoverlapping ranges of values. Switching circuits can be designed with good noise immunity and high operational reliability. Many electrical devices exhibit bistable operating characteristics; e.g., switches and relays are normally open or closed, nonconducting or conducting. Also, active devices such as transistors can be operated in one of two stable states, i.e., either cut off or fully conducting.

The properties of complex digital processing systems can be completely described by the operation of their individual building blocks and the interrelationships of their bistable states. To be able to describe the logical functions of these switching circuits a mathematics known as *Boolean algebra* is used.

Boolean algebra was first developed by the mathematician George Boole[1]* (1815–1864) in connection with the study of logic. His treatise, *An Investigation of the Laws of Thought on Which Are Founded the Mathematical Theories of Logic and Probabilities*, presented the first practical application of logic in an algebraic form. Claude E. Shannon[2] developed the application of Boolean algebra in the design of telephone switching circuits and published a paper in 1938 entitled "A Symbolic Analysis of Relay and Switching Circuits."

1-1 BOOLEAN ALGEBRA

Boolean algebra was originally developed to produce a means for representing and solving problems in logic. The variables in Boolean algebra

*Superscript numbers designate references at the end of the chapter.

are assigned only two sets of values, TRUE and FALSE; i.e., a variable representing a logical statement is either true or false. For example, let us consider a switch that can be in either the *closed* or *open* position, as shown in Fig. 1-1.

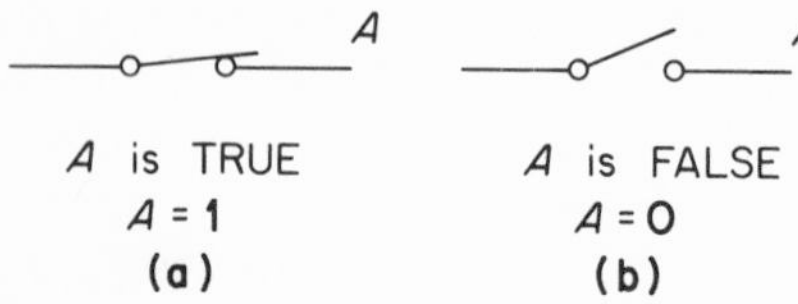

A is TRUE
$A = 1$
(a)

A is FALSE
$A = 0$
(b)

FIGURE 1-1. The Two Logical Switch Positions.

A logical statement concerning this switch is "The switch is closed"; and we shall represent this statement by the symbol A. The symbol A, representing the switch, now becomes a variable that can take on either of two values, TRUE or FALSE. This is illustrated in Fig. 1-1. When the switch is closed, the variable A is TRUE; conversely, when the switch is open, the variable A is FALSE.

As we have seen, the Boolean variable is two-valued (also called *binary*); therefore, when it represents the state of a switch or relay, the two values are CLOSED and OPEN or ON and OFF. Actually, any convenient choice of words, including HIGH-LOW, TRUE-FALSE, **1-0**, etc., can be used to describe the state of a binary, or two-valued device; but symbolically the Boolean variable is usually represented by **1** and **0**. These symbols are also shown in Fig. 1-1. Therefore, $A = \mathbf{1}$ or else $A = \mathbf{0}$. There is no numerical significance to the **1** and **0**; there is only a logical significance.

Since a binary variable has two values, the rules of Boolean algebra are quite different from those of ordinary algebra. The primary objective in Boolean algebra is the determination of the truth of propositions or algebraic functions and, when dealing with combinations of propositions, the truth value of the combination. Boolean algebra as originally defined used three fundamental *connectives* or so-called *propositional operations*: NOT, AND, and OR. The NOT operation is a *negation* or *complementation* function. The AND and OR are a form of Boolean *multiplication* and *addition*, respectively; however, the rules are different from ordinary algebra. It is easier to think of these connectives literally as AND and OR operations. The symbols used most commonly for the Boolean AND and OR operations are the dot $(\cdot)$ and plus sign $(+)$, respectively.

Boole defined TRUE and FALSE statements as **1** and **0**, respectively. This definition is still used and has been extended to binary digital circuits. It was noted above that an open relay contact or switch may be defined as being in the **0** state and a closed contact in the **1** state. Similarly, a transistor in saturation can be defined as representing a **0** and cut off as a **1**. Also, the **0** and **1** are defined in terms of voltages (or ranges of voltages); i.e., **0** may be defined as -3 V or 0 V or $+4.5$ to $+5.0$ V in a given system. The definition of states is usually left to the designer of the particular equipment and even can be different in parts of the same equipment.

There are two classes of logic circuits, *combinational* and *sequential*. The combinational circuit has a specific output for a given set of inputs, whereas the output of a sequential circuit depends on the time sequence of inputs. Combinational circuits are discussed in this chapter, and sequential circuits are introduced in Chap. 3.

Fundamental Operations

The NOT Connective. The NOT connective is a propositional operation that negates or inverts. As defined, a Boolean variable A has two possible logical values, **0** and **1**. If A is *not* equal to **0**, then it must equal **1**. Similarly, if A is *not* equal to **1**, then it must equal **0**. The negation or complementation of A is read as NOT A and is usually written as $\bar{A}$ or A'.

Negation or inversion of an electrical voltage level is conveniently obtained in an electronic amplifier, e.g., a common emitter transistor amplifier. The amplifier output waveform is the inverse of the input waveform. The circuit symbol for negation is shown in Fig. 1-2(a). It is an operational amplifier with a small circle at the output to indicate inversion. The input to the amplifier is a two-valued voltage, designated by the Boolean variable A. The two values of A are **0** and **1**, where, e.g., **0** may represent a voltage of 0 V and **1** may represent a voltage of 5 V. If the input to the inversion amplifier is 5 V(a **1**), then the output will be 0 V(a **0**). Similarly, if the amplifier input is a **0**, then the output will be a **1**. A table listing all the possible states of a circuit, showing the relationship among the input and output variables, is called a *truth table*. A truth table for the NOT connective is shown in Fig. 1-2(b).

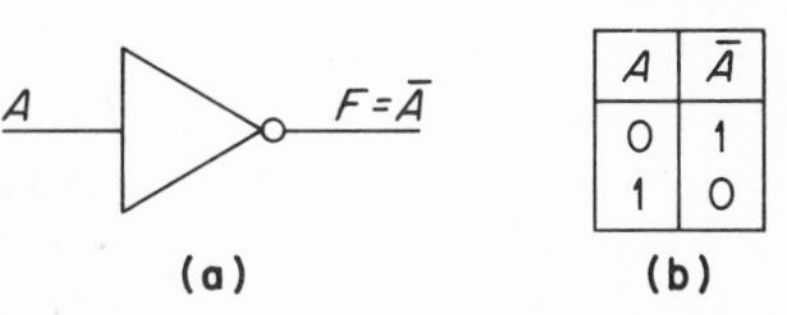

A	$\bar{A}$
0	1
1	0

FIGURE 1-2. (a) NOT Circuit Symbol. (b) Truth Table for the NOT Connective.

Amplifiers in digital circuits are used for negation, as drivers to provide inputs to other circuits, and for restoring voltage levels. Digital amplifiers are quite different from analog amplifiers because they are not used for linear signal amplification. Circuit design details for digital amplifiers are given in Chap. 2.

The AND Connective. The AND connective is read as A AND B, and is generally written $A \cdot B$ or simply AB. A Boolean function, $F = A \cdot B$, means that both the *A and B* variables must be TRUE for F to be TRUE; i.e., $F = \mathbf{1}$

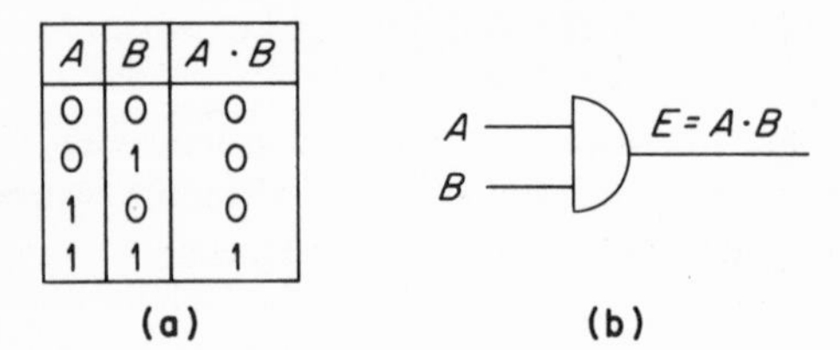

A	B	$A \cdot B$
0	0	0
0	1	0
1	0	0
1	1	1

FIGURE 1-3. (a) Truth Table for AND Connective. (b) The AND Circuit Symbol.

when, and only when, $A = B = \mathbf{1}$. The truth table for the AND operation, shown in Fig. 1-3(a), lists all the possible combinations for the input variables for a two-variable case. The AND gate symbol is illustrated in Fig. 1-3(b). The AND function, in general, can involve more than two variables. An AND gate with n inputs will have a TRUE or **1** output only when *all* the inputs are **1**'s.

Considerable insight into the physical significance of the AND connective is gained when the circuit is implemented with relay or switch contacts. An AND circuit using switch contacts is shown in Fig. 1-4. Note that in switching circuits, as shown in Fig. 1-4, we assume that a logic level **1** persists

Input "1" — A — B — C — Output = $A \cdot B \cdot C$

FIGURE 1-4. Switch Contact Representation of an AND Circuit with Three Input Variables.

at the input; a **1** will exist at the output when the switches are in the proper state as indicated by the Boolean function. Therefore, it is readily seen that the output is a **1** when all switches A AND B AND C are in the **1** state (closed).

The OR Connective. The OR connective is read as A OR B and is written $A + B$. A Boolean function, $F = A + B$, means that either *A or B* must be TRUE for F to be TRUE; i.e., $F = \mathbf{1}$ when *A or B or* both equals **1**. This is illustrated for the two-variable case with a truth table for the OR circuit in Fig. 1-5(a). The symbol for the OR gate is shown in Fig. 1-5(b).

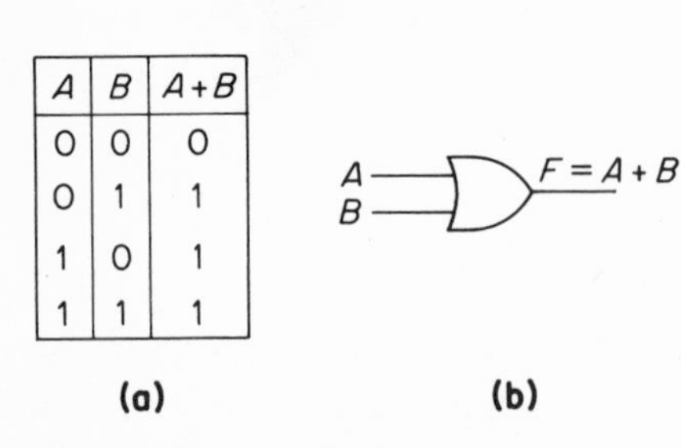

A	B	$A+B$
0	0	0
0	1	1
1	0	1
1	1	1

FIGURE 1-5. (a) Truth Table for OR Connective. (b) The OR Circuit Symbol.

A switch contact representation of a three-variable OR circuit is shown in Fig. 1-6. We see that the output is a **1** when either switch A or B or C, or any combination thereof, is in the closed or **1** position. The OR circuit, of course, also can be expanded to contain n input variables.

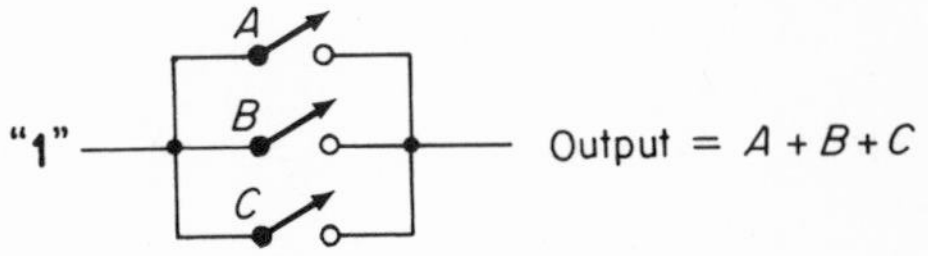

FIGURE 1-6. Switch Contact Representation of an OR Circuit with Three Input Variables.

Postulates

The following postulates are adopted for the three fundamental operations NOT, AND, and OR:

NOT	AND	OR
$\bar{0} = 1$	$0 \cdot 0 = 0$	$0 + 0 = 0$
$\bar{1} = 0$	$0 \cdot 1 = 0$	$0 + 1 = 1$
	$1 \cdot 0 = 0$	$1 + 0 = 1$
	$1 \cdot 1 = 1$	$1 + 1 = 1$

The above NOT relations are basic and are consistent with the definition of a Boolean variable as being two-valued; i.e., if the variable A is not **1**, then it must be **0**, and vice versa. The AND and OR functions are consistent with the switch contact representation, where the AND connective is represented by contacts in series and the OR connective by contacts in parallel. (The **1** and **0**, of course, represent closed and open switches, respectively.) All Boolean relations can be derived from these postulates. (Actually, as we will see later, only two of the above postulates are fundamental, the NOT and either the AND or the OR.)

It frequently is desirable to manipulate Boolean expressions for simplification of the functions and arrangement of the terms and connectives in a form that is easily implemented with standard circuit modules. The following theorems are useful in the manipulation of Boolean algebraic expressions; they are easily derived from the above postulates and can be verified with the switch contact representation:

1. ONE and ZERO rules	$0 + A = A, \quad 0 \cdot A = 0$
	$1 + A = 1, \quad 1 \cdot A = A$
2. Commutative laws	$A + B = B + A, \quad AB = BA$
3. Associative laws	$A + (B + C) = (A + B) + C$
	$A(BC) = (AB)C$

4. Distributive laws $A + BC = (A + B)(A + C)$
 $A(B + C) = AB + AC$
5. Idempotence laws $A + A = A, \quad A \cdot A = A$
6. Complementary laws $A + \bar{A} = 1, \quad A \cdot \bar{A} = 0$
7. Absorption laws $A + AB = A$
 $A(A + B) = A$
 $A + \bar{A}B = A + B$
8. Involution $\overline{(\bar{A})} = A$
9. Inversion laws (DeMorgan's theorems) $\overline{(A + B)} = \bar{A} \cdot \bar{B}$
 $\overline{A \cdot B} = \bar{A} + \bar{B}$

There is a natural tendency to make an analogy between Boolean algebra and ordinary algebra, which compares the AND operation to ordinary multiplication and the OR operation to addition. The Boolean variable, of course, by definition only has two values, and the AND multiplication comparison appears valid. However, extreme caution must be taken in comparing the OR operation with ordinary addition, as indicated in the distributive law, law number 4. The switch contact representation of this relation is shown in Fig. 1-7, illustrating the truth of the Boolean equality.

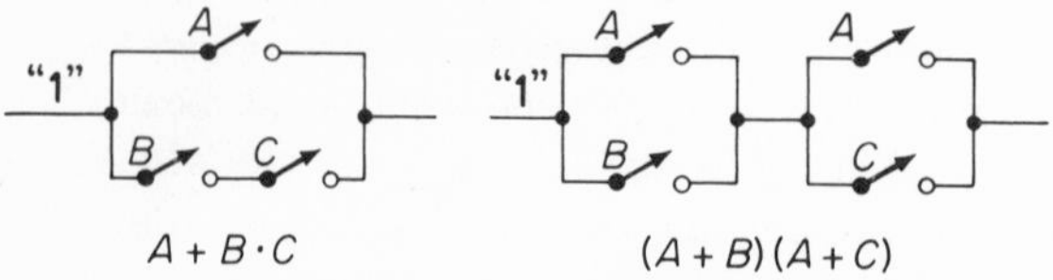

FIGURE 1-7. Switch Contacts Illustrating $A + BC = (A + B)(A + C)$.

DeMorgan's Theorems

DeMorgan's theorems, law number 9, are used when it is necessary to form the complement of a Boolean function. These theorems are proven in Sec. 1-3 by the use of *Venn diagrams*. To find the complement of a function, the following steps are taken:

1. Negate the entire Boolean function
2. Change all AND's to OR's and all OR's to AND's
3. Negate each variable.

We will illustrate these rules with the following example. To find an equivalent form for the expression $F = \bar{A} + B + \bar{C}$, first,

$$\bar{F} = \overline{\bar{A} + B + \bar{C}} \qquad \text{(Step 1)}$$

then, changing the OR's to AND's,

$$\bar{A} \cdot B \cdot \bar{C} \qquad \text{(Step 2)}$$

and finally negating each variable,

$$\bar{F} = A \cdot \bar{B} \cdot C \qquad \text{(Step 3)}$$

Negating the entire equation gives

$$F = \overline{A \cdot \bar{B} \cdot C}$$

which is an equivalent expression.

DeMorgan's theorems convert an expression that is in the form of a *sum of product* to one that is a *product of sums* and vice versa. This is illustrated with the following example:

$$F = A\bar{B}C + \bar{A}B + A\bar{B}\bar{C} \qquad \text{(1-1a)}$$

Then

$$\bar{F} = (\bar{A} + B + \bar{C}) \cdot (A + \bar{B}) \cdot (\bar{A} + B + C) \qquad \text{(1-1b)}$$

DeMorgan's theorems are used in manipulating complex Boolean expressions. Frequently, successive application of the theorems facilitates the simplification of these expressions. Also, manipulation of Boolean functions can lead to forms that may be easier to implement with a given type of integrated circuit component.

DeMorgan's theorems point out the relationship between the AND and OR operations through the use of the NOT operation. Negating an AND function results in an OR function and, conversely, negating an OR function results in an AND function. Because of this property an entire switching algebra can be postulated using only two basic operations, the AND and NOT connectives or the OR and NOT connectives. This leads to NAND and NOR logic elements, which are combinations of NOT and AND and of NOT and OR circuits.

NAND And NOR Logic

The NOT, AND, and OR connectives were given as the basic operations in Boolean algebra. However, only the NOT and either the AND or the OR connectives are required to synthesize any logic function. One advantage of using NAND or NOR logic is that it minimizes the number of different circuits used and these circuits can be easily fabricated using integrated circuit modules.

The NAND Operation. The NAND operation is implemented with an AND gate followed by an inverting amplifier, which forms the output driving circuit. A truth table for a two-input NAND function is shown in Fig. 1-8(a).

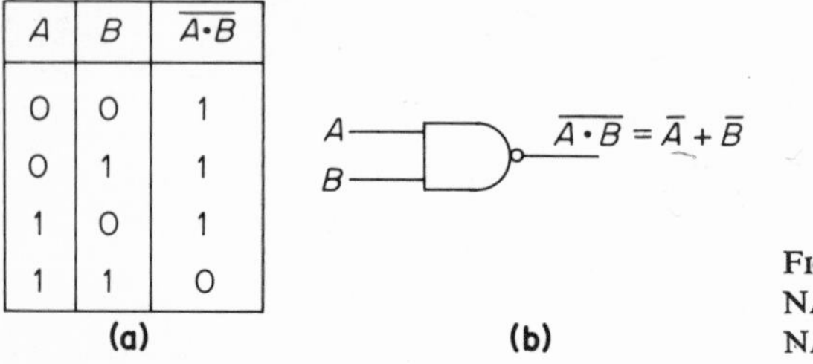

A	B	$\overline{A \cdot B}$
0	0	1
0	1	1
1	0	1
1	1	0

FIGURE 1-8. (a) Truth Table for NAND Operation. (b) The NAND Gate Symbol.

The symbol for this function is shown in Fig. 1-8(b), which is the AND gate symbol followed by a circle which denotes inversion.

If any one of the input variables to a NAND gate is **0**, the output will be in the **1** state. Also, if *all* input variables are in the **1** state, then the output will be in the **0** state. Integrated circuit NAND gates are available that accept up to eight input variables, with three and four-input gates being more common. These gates can be expanded to accommodate even larger numbers of inputs.

The NOR Operation. The NOR operation is performed by inverting the output of an OR gate. The truth table for the NOR function and the symbol for a two-input NOR gate is shown in Fig. 1-9. *All* inputs to the NOR gate must be in the **0** state for the output to be a **1**; otherwise the output will be in the **0** state.

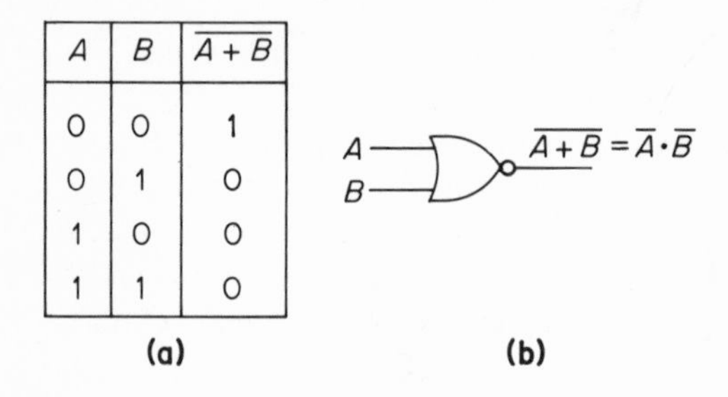

A	B	$\overline{A + B}$
0	0	1
0	1	0
1	0	0
1	1	0

FIGURE 1-9. (a) Truth Table for NOR Operation. (b) The NOR Gate Symbol.

The Exclusive-OR Operation

The exclusive-OR is a function that appears frequently in digital circuitry and, for convenience, it has been assigned a special symbol. However, it is *not* an independent operation and therefore can be expressed in terms of the basic connectives. The expression *A exclusively OR B* is written $A \oplus B$. This function is TRUE ($A \oplus B = 1$) when A or B is TRUE *and*, at the same time, both are *not* TRUE. This statement can be expressed symbolically as

$$A \oplus B = (A + B)\cdot\overline{AB}$$

Using DeMorgan's theorem,

$$A \oplus B = (A + B)(\bar{A} + \bar{B}) = A\bar{A} + A\bar{B} + \bar{A}B + B\bar{B} = A\bar{B} + \bar{A}B$$

The truth table for the exclusive-OR function is given in Fig. 1-10(a).

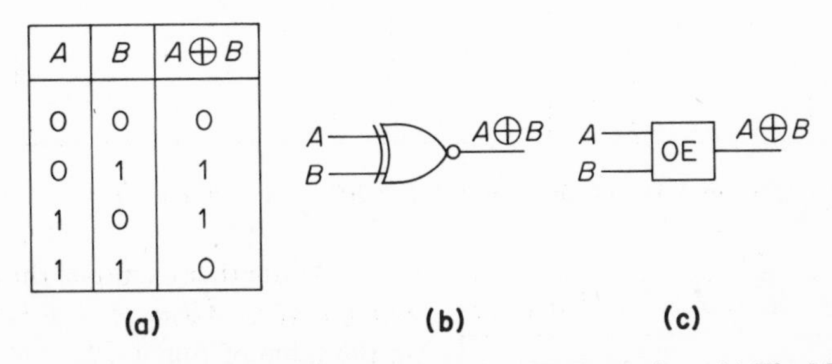

A	B	A⊕B
0	0	0
0	1	1
1	0	1
1	1	0

FIGURE 1-10. (a) Truth Table for Exclusive-OR Operation. (b) The MIL Standard Exclusive-OR Symbol. (c) An Alternate Symbol.

The standard circuit symbol used in government documentation for military equipment is shown in Fig. 1-10(b). Many authors use the alternate symbol shown in Fig. 1-10(c).

1-2 TRUTH TABLE OF COMBINATIONS

Analysis

A table listing all the possible values of a Boolean function is called a *truth table*. Since a Boolean variable can have either of two values, a function with n variables will have 2^n listings in the table to cover all possible permutations. The last column in the table represents the value of the function for the particular values of the variables in each row. These tables are used in the previous section to provide insight and understanding of the basic Boolean operations. In complex Boolean functions, additional columns may be added to facilitate determining the value of the complete expression. Also, Boolean equations and identities may be verified by using a truth table. If two functions are equal for all possible values of their variables, then the two functions are equal. This constitutes a mathematical proof by *induction*. As an example, the relation $A + BC = (A + B)(A + C)$, which is the *distributive law* for the OR operation, is verified in the table of Fig. 1-11.

A	B	C	BC	$A+BC$	$A+B$	$A+C$	$(A+B)(A+C)$
0	0	0	0	0	0	0	0
0	0	1	0	0	0	1	0
0	1	0	0	0	1	0	0
0	1	1	1	1	1	1	1
1	0	0	0	1	1	1	1
1	0	1	0	1	1	1	1
1	1	0	0	1	1	1	1
1	1	1	1	1	1	1	1

FIGURE 1-11. Truth Table for $A + BC = (A + B)(A + C)$.

A	B	$A+B$	$\bar{A}$	$\bar{A}B$	$A+\bar{A}B$
0	0	0	1	0	0
0	1	1	1	1	1
1	0	1	0	0	1
1	1	1	0	0	1

FIGURE 1-12. Truth Table for $A + \bar{A}B = A + B$.

As another example, the *absorption law* $A + \bar{A}B = A + B$ is verified in the table of Fig. 1-12.

The two possible values of a binary variable, **0** and **1**, represent a state or condition (e.g., *false* or *true*, *open* or *closed*, etc.) and not a numerical quantity. In a table of combinations of possible states of binary variables, it is convenient to list the combinations in a sequential order corresponding to numbers in the binary number system. This could be considered a method of *bookkeeping*. The association of combinations of binary variables with the binary number system is made frequently in logic circuit design and analysis. A binary number, of course, also can represent a quantity, e.g., a voltage measurement converted into a binary number. Each digit of this binary number is treated as an independent binary variable when its voltage representation is applied to logic circuits. The association of binary numbers with combinations of binary variables is natural. Some authors, however, use a different face type to distinguish a binary number from a binary variable.

The proof of any Boolean expression may be obtained using a truth table. However, it is frequently easier to obtain a proof by manipulation of the algebraic expressions using the Boolean laws given in Sec. 1-1. This is certainly true for expressions containing many variables where the construction of a truth table is a lengthy procedure.

The *absorption law*, $A + AB = A$, can be verified easily using a truth table, but a simple manipulation using the *distributive law* for the AND operation verifies the result. The left side of this equation is

$$\begin{aligned} A + AB &= A(1 + B) && \text{(by distributive law)} \\ &= A(1) && \text{(by ONE and ZERO rule)} \\ &= A \end{aligned}$$

All the Boolean relations are useful in the manipulation of Boolean expressions. The absorption laws and the distributive law for the OR operation are quite different from ordinary algebra, and therefore they should be memorized to facilitate their use in the manipulation of these expressions. The following relation is used as an example:

$$\begin{aligned} A + B &= A + ABC + BC\bar{A} + \bar{A}B + DA + \bar{D}A \\ &= A + BC(A + \bar{A}) + \bar{A}B + A(D + \bar{D}) && \text{(by distributive law)} \\ &= A + BC + \bar{A}B + A && \text{(by complementary law)} \\ &= A + BC + \bar{A}B && \text{(by idempotence law)} \\ &= (A + \bar{A}B) + BC && \text{(by associative law)} \\ &= A + B + BC && \text{(by absorption law)} \\ &= A + (B + BC) && \text{(by associative law)} \\ &= A + B && \text{(by absorption law)} \end{aligned}$$

The steps taken in the above example are not unique in arriving at the solution. Other distributive groupings can be made and the laws applied in a different sequence to obtain the same result. Reduction of Boolean expressions by algebraic manipulation is not a systematic procedure, and other methods have been devised. Reduction using mapping techniques is discussed in Sec. 1-3.

Synthesis

If a table of combinations is given that defines the output for *all* possible combinations of the input variables, the corresponding Boolean function can be determined. A process, which is an inverse operation of the analysis procedure described above, is used to synthesize the function. If there are n variables, as before, there are 2^n permutations entered into the table. It is preferable to enter these permutations in a binary sequence. The desired Boolean function is obtained by *summing* (OR operation) all input AND expressions for which the function is TRUE, a **1**. Alternately, the function may be written as the sum of the complements of all terms for which the function is **0**.

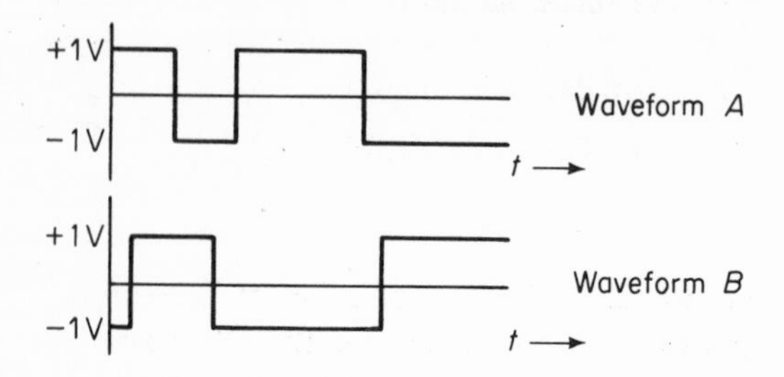

FIGURE 1-13. Binary Waveforms.

As an example, it is desired to design a circuit that will multiply the waveforms of the two binary signals shown in Fig. 1-13. The waveforms are normalized to values of ±1V. A table of permutations of all possible values of the product of the two waveforms is given in Fig. 1-14. The waveforms are

Waveform A	Waveform B	Product A x B
−1	−1	+1
−1	+1	−1
+1	−1	−1
+1	+1	+1

FIGURE 1-14. Table of Combinations of Product Waveforms.

coded for a binary representation as follows:

0 denotes −1 V

1 denotes +1 V

A	B	f(A, B)
0	0	1
0	1	0
1	0	0
1	1	1

FIGURE 1-15. Truth Table for $f(A, B)$.

The table in Fig. 1-14 is rewritten as a truth table for the binary variables A and B, representing the waveforms A and B, respectively, in Fig. 1-15. It is desired to determine the Boolean function, $f(A, B)$. The function is TRUE when A is NOT TRUE *and* B is NOT TRUE *or* A is TRUE *and* B is TRUE. Symbolically, this is written

$$f(A, B) = \bar{A}\cdot\bar{B} + A\cdot B$$

Using DeMorgan's theorem,

$$\begin{aligned}\overline{f(A, B)} &= \overline{\bar{A}\cdot\bar{B} + A\cdot B} \\ &= (A + B)\cdot(\bar{A} + \bar{B}) \\ &= A\bar{A} + A\bar{B} + B\bar{A} + B\bar{B}\end{aligned}$$

$$= A\bar{B} + B\bar{A}$$
$$= A \oplus B$$

Therefore,

$$f(A, B) = \overline{A \oplus B}$$

Alternately, the function may be synthesized by using the terms where the $f(A, B)$ is **0**. Eq. (1-2).

$$f(A, B) = \overline{\bar{A} \cdot B + A \cdot \bar{B}} \tag{1-2}$$

Manipulation of Eq. (1-2) gives the same result and is left as an exercise.

A circuit diagram using logic symbols to implement this function is shown in Fig. 1-16.

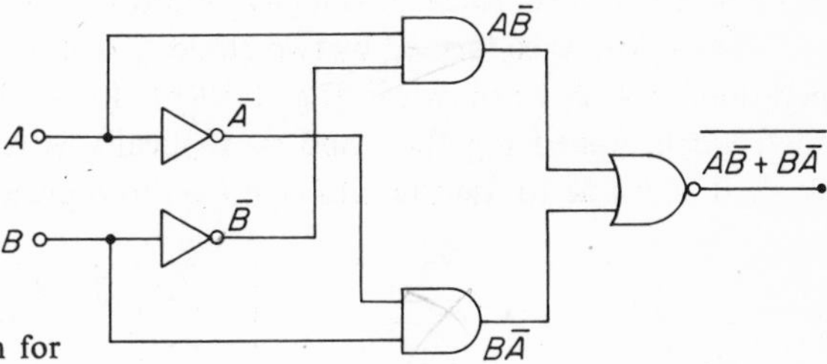

FIGURE 1-16. Logic Diagram for $\overline{A \oplus B} = \overline{A\bar{B} + B\bar{A}}$

1-3 GRAPHICAL REPRESENTATION OF SWITCHING VARIABLES AND FUNCTIONS

Venn Diagrams

Boolean algebra can be interpreted as an algebra of *sets*. This enables one to visualize, in terms of areas in closed curves, the truth value of propositions by determining whether or not a point is in a set. A switching variable is represented as a circle in a diagram called a Venn diagram, after the nineteenth-century British mathematician.

A single-variable Venn diagram is shown in Fig. 1-17. Everything, i.e., all points, within the circle represents the variable A and everything outside the circle is $\bar{A}$. The entire rectangle represents a **1** or all possible TRUE conditions. With this interpretation it is evident in Fig. 1-17 that $A + \bar{A} = \mathbf{1}$.

A two-variable Venn diagram is shown in Fig. 1-18. Everything within the left-hand circle is represented by the switching variable A, and everything outside that circle is $\bar{A}$. Similarly, the area within the right-hand circle represents the variable B, and everything outside that circle is $\bar{B}$. Where the two

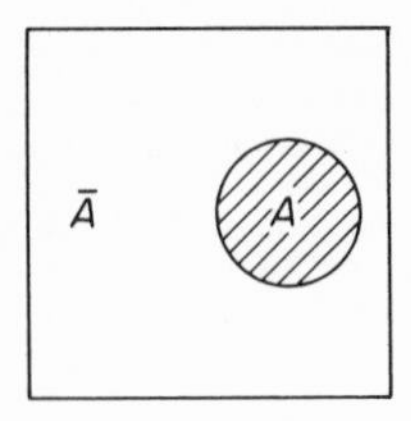

FIGURE 1-17. Venn Diagram for One Variable.

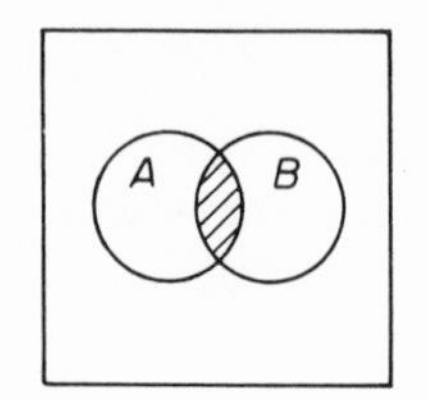

FIGURE 1-18. Two-Variable Venn Diagram.

circles intersect, the overlapping area represents $A \cdot B$, as shown by the shaded area in Fig. 1-18 and also in Fig. 1-19(a).

In the literature on set theory and logic, the AND operation is called *intersection* and symbolically is represented by a cap ($\cap$). We shall continue to use the dot ($\cdot$) to represent this logical multiplication.

The entire area formed by two circles joined or united represents the OR operation, $A + B$, as shown in Fig. 1-19(b). In set theory the OR operation is called *union*, joined together, and symbolically is represented by a cup ($\cup$). We shall continue to use the plus sign ($+$) to represent this logical addition.

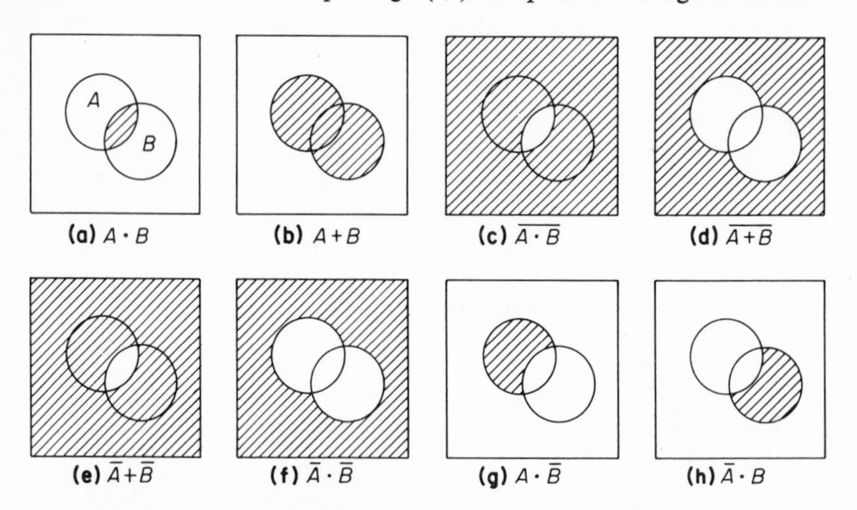

FIGURE 1-19. Venn Diagrams with Cross-Hatched Areas Representing Various Switching Functions.

The cross-hatched area in Fig. 1-19(c) represents NOT ($A \cdot B$), i.e., $\overline{A \cdot B}$. This is clearly everything outside the area indicated by $A \cdot B$ in Fig. 1-19(a). Similarly, Fig. 1-19(d) represents NOT ($A + B$) and should be compared with Fig. 1-19(b).

The cross-hatched area in Fig. 1-19(e) represents NOT A OR NOT B. This is equivalent to Fig. 1-19(c). Therefore,

$$\overline{A \cdot B} = \bar{A} + \bar{B}$$

which constitutes a proof of DeMorgan's theorem. The intersection of NOT A AND NOT B is shown in Fig. 1-19(f). Comparison of this with Fig. 1-19(d) proves the other DeMorgan relation:

$$\overline{A + B} = \bar{A} \cdot \bar{B}$$

Figure 1-19(g) shows a representation of A AND $\bar{B}$, which is the intersection of these two areas. Similarly, Fig. 1-19(h) represents $\bar{A}$ AND B.

The Boolean relations can be verified by use of Venn diagrams. As another example, Fig. 1-19(a) and (g) shows that

$$A \cdot B + A \cdot \bar{B} = A$$

Also, the area represented in Fig. 1-19(b) is the logical sum of the areas in Fig. 1-19(a), (g) and (h); therefore,

$$A + B = A \cdot B + A \cdot \bar{B} + \bar{A} \cdot B$$

A three-variable Venn diagram is shown in Fig. 1-20. The cross-hatched area shows $A \cdot B \cdot C$. In a manner that is similar to that used with the two-variable Venn diagrams, each separate area can be represented as a function of the Boolean variables A, B, and C. Boolean relations can be obtained by equating areas using the following rules:

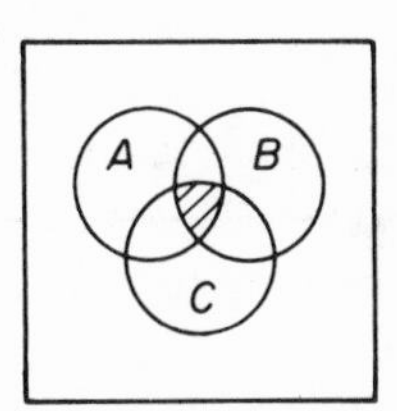

FIGURE 1-20. Venn Diagram Showing $A \cdot B \cdot C$

1. The AND operation is represented by an *intersection*, a coincidence, or overlapping of the variable areas.
2. The OR operation is represented by the total area formed by the *union* or joining of the variable areas. (Overlapping areas within the total area representing the function do not alter the logical sum.)
3. The *negation* of a variable or function is represented by everything outside the area representing the variable or function.

In the above Venn diagrams the variables are represented as circles. These diagrams are limited to three variables, but beyond that the circles must be distorted in order to form all possible combinations of overlapping areas. If rectangles are used to represent variables instead of circles, the overlapping areas are arranged in a tabular or matrix form and are referred to as a *map.* A two-variable map or Venn diagram is shown in Fig. 1-21.

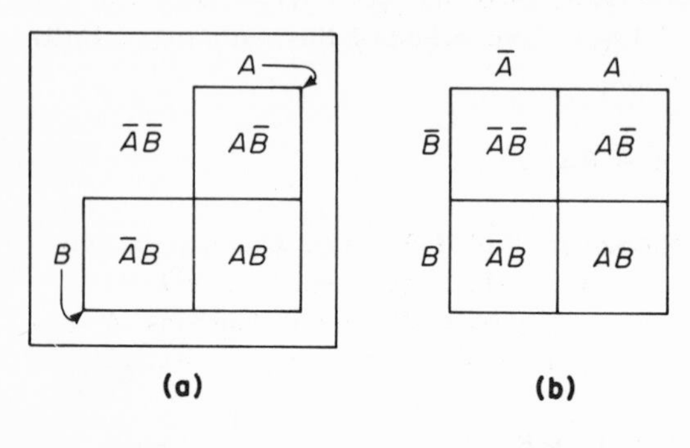

FIGURE 1-21. (a) Two-Variable Venn Diagram Using Rectangles to Represent the Variables. (b) Equivalent Map Drawn in Matrix Form.

In Fig. 1-21(a) the large rectangles within the square represent the variables A and B. The various intersecting areas also are shown. This map is redrawn in Fig. 1-21(b) in matrix form, and all variables and functions of the variables are easily identified. The vertical rectangles are $\bar{A}$ and A, and the horizontal rectangles are $\bar{B}$ and B. The four intersecting areas are shown as squares on the map.

Maps arranged in matrix form are particularly useful for simplification of Boolean functions. It is simpler to group the areas on the maps representing the functions than it is to manipulate the Boolean expressions using the theorems. Considerable work on applying mapping techniques for the representation of Boolean expressions was done by M. Karnaugh.[3] A four-variable Karnaugh map is shown in Fig. 1-22.

	$\bar{C}$	C		$\bar{C}$	
$\bar{A}$	0000	0100	1100	1000	$\bar{B}$
A	0001	0101	1101	1001	
	0011	0111	1111	1011	B
$\bar{A}$	0010	0110	1110	1010	
	$\bar{D}$		D		

FIGURE 1-22. Four-Variable Karnaugh Map.

In Fig. 1-22 the rectangles representing the four variables are depicted by braces. Each square represents a Boolean product or *intersection* of the variables. The numbers in the squares are a *binary representation* of the Boolean product. For example, the function $D\bar{C}BA$ is represented by 1011. This is a method of notation, and it is very useful in mapping techniques for simplification of Boolean functions. Using this notation an unbarred variable is assigned a **1** and a barred, or negated variable, a **0**; and the binary number corresponds to

$$N = A2^0 + B2^1 + C2^2, \text{ etc.}$$

where A is the least significant digit. For example, the Boolean function, which is a *sum-of-product terms*, Eq. (1-3),

$$\begin{aligned} F &= DCBA + DCB\bar{A}\bar{A} + D\bar{C}BA \quad (1\text{-}3) \\ &= m_{1111} + m_{1100} + m_{1011} \\ &= m_{15} + m_{12} + m_{11} \\ &= \sum(11, 12, 15) \end{aligned}$$

is represented as indicated by m-terms with the binary or equivalent decimal subscripts. The terms in a Boolean equation, like Eq. (1-3), are usually written in alphabetical sequence, e.g., $AB\bar{C}D$; however, the A factor still denotes the least significant digit in the binary number representation.

Sum of Products and Product of Sums

Two basic forms of Boolean expressions are the *sum of products* and the *product of sums*. A function in the form of a product of sums can be converted to a sum of products by expansion. For example,

$$\begin{aligned} F &= (A + B)(C + B), \text{ a product of sums} \\ &= AC + AB + BC + B, \text{ a sum of products} \quad (1\text{-}4) \end{aligned}$$

Eq. (1-4) contains redundant terms and therefore can be simplified. The representation of a Boolean function on a Karnaugh map provides insight that facilitates simplification of the function more easily than by direct algebraic manipulation.

Minterms and Maxterms. The AND products of all the Boolean variables, or their complements, are called *minterms.* Three variables form eight possible minterms: $\bar{A}\bar{B}\bar{C}$, $A\bar{B}\bar{C}$, $\bar{A}B\bar{C}$, $AB\bar{C}$, $\bar{A}BC$, $A\bar{B}C$, $\bar{A}\bar{B}C$, and ABC. Similarly, four variables form 16 minterms. The expression arises from the fact that these product terms represent a minimum area on a Karnaugh map, i.e., one elemental area. For example, the minterm $\bar{A}\bar{B}\bar{C}\bar{D}$ is represented by the square 0000 in Fig. 1-22. The minterm is designated by the lowercase letter m with a numerical subscript to indicate the variables. As indicated above with Eq. (1-3), $DCBA = m_{15}$, $DCB\bar{A} = m_{12}$, etc.

In a similar but converse manner, the logical sum of the Boolean variables, or their complements, represents a maximum area on the map and are called *maxterms.* By maximum area we mean all but one elemental square of the map. For example, the maxterm $A + B + C + D$ is represented by the areas including all the squares but 0000 in Fig. 1-22. This is, of course, consistent with DeMorgan's theorem, since

$$\overline{\bar{A}\bar{B}\bar{C}\bar{D}} = A + B + C + D$$

With four variables there are also 16 maxterms. The maxterm is designated by the capital letter M with the appropriate subscript to indicate the variables. For example, $(\bar{D} + C + \bar{B} + A) = M_5$, $(\bar{D} + \bar{C} + \bar{B} + \bar{A}) = M_0$, etc.

A sum-of-products Boolean expression contains a sum of minterms. Similarly, a product-of-sums expression is a product of maxterms.

1-4 MINIMIZATION USING MAPPING TECHNIQUES

Referring to the Karnaugh map of Fig. 1-22, we see that adjacent elemental squares or cells differ by only one variable. This is true for both a horizontal and vertical movement from one cell to an adjacent cell. This property also holds true in a movement from one end cell to the other end of the same row; therefore, end cells are also adjacent. Similarly, top and bottom cells of the same column are adjacent. It is this property of adjacent cells in the Karnaugh map that aids in simplifying Boolean functions. As an example, the function $F = m_5 + m_{13}$ is plotted on the Karnaugh map in Fig. 1-23. The four-variable map in Fig. 1-23 is identical to that in Fig. 1-22, only a different notation is used to represent the cells by the rows and columns. The function is represented on the map of Fig. 1-23 by a **1** in the m_5 and m_{13} cells. Sometimes a diagonal line is used in the selected cell for this purpose. We see that the area represented by these adjacent cells is independent of the variable D; i.e., D is **0** in one cell and **1** in the adjacent cell; therefore, the variable D can be eliminated from the function. Again referring to the map, we see that this area is represented by $A\bar{B}C$. Therefore,

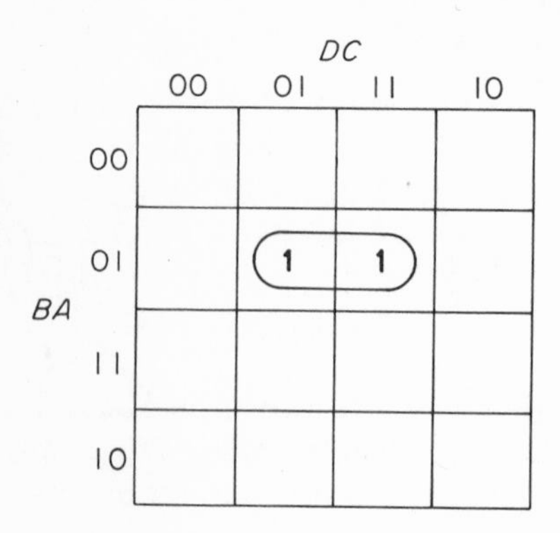

FIGURE 1-23. A Karnaugh Map of Function $F = m_5 + m_{13}$.

$$F = m_5 + m_{13} = A\bar{B}C\bar{D} + A\bar{B}CD = A\bar{B}C$$

which, of course, agrees with the algebraic postulate. This example illustrates how adjacent cells form a *couple* to eliminate one variable in the function.

The method of ordering the rows and columns of the map with binary numbers, as shown in Fig. 1-23, was devised by Karnaugh. Another method, where the row and column binary indications increase in numerical sequence, was developed by E. W. Vietch.[4] The Karnaugh map is more convenient to use for Boolean function simplification because adjacent cells form couples. A uniform method of variable placement on a Karnaugh map was presented by M. V. Mahoney.[5] This method expands a map to any number of variables by a folding technique or forming mirror images. The method is illustrated by taking the single variable map shown in Fig. 1-24(a) and folding it downward to form the two-variable map in Fig. 1-24(b). The folding is indicated by the axis on Fig. 1-24(b). The top row is called $\bar{B}$ and the bottom row B.

The numbers representing the new cells are obtained by adding *two* to the original cells. The three-variable map is obtained by folding the two-variable map to the right and forming a mirror image as indicated by the axis in Fig. 1-24(c). Note that the A and $\bar{A}$ columns are represented as mirror images. The side to the left of the axis represents $\bar{C}$ and the right side C. The numerical representation of the new cells is obtained by adding *four* to the corresponding image cells. This, of course, is adding another significant digit of weighted value four. Using this folding method the adjacent cells in couples differ in their numerical designation by powers of two (i.e., 1, 2, 4, etc.).

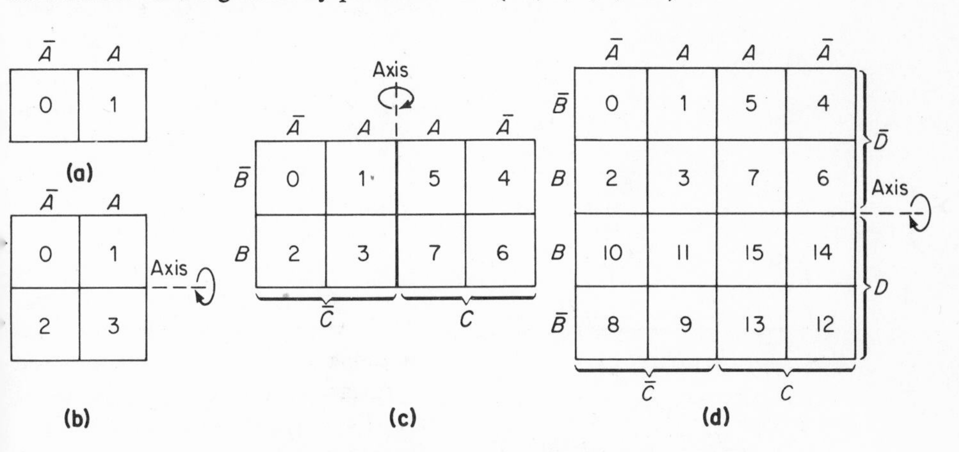

FIGURE 1-24. (a) Single-Variable Map. (b) Two-Variable Map. (c) Three-Variable Map. (d) Four-Variable Map.

Four-Variable Map. In a similar manner as described above, the four-variable map is obtained by folding the three-variable map downward and is shown in Fig. 1-24(d). The original cells are represented by $\bar{D}$ and the new cells by D. Since the new cells correspond to the fourth most significant digit in the binary representation, the new cells are numbered by adding *eight* to the corresponding image cells.

Higher order maps, i.e., with four or more variables, are very useful in the simplification of Boolean functions. By simplification of functions we mean that we desire to find the expression that uses the minimum number of gates to implement the function in the sum-of-products or product-of-sums form. Functions with three or less variables usually can be simplified easily by direct algebraic means, although mapping techniques can be used. Simplification using the mapping technique also lets you know when you are finished with the process, whereas using algebraic manipulation it is not apparent that the function cannot be further simplified.

Boolean expressions can be simplified, i.e., terms combined to form a new term that contains fewer variables, when the terms form couples on the map. This is illustrated with the following examples.

Simplify the following function:

$$\begin{aligned} F &= \bar{A}\bar{B}\bar{C}\bar{D} + \bar{A}\bar{B}C\bar{D} + AB\bar{C}D + ABCD \\ &= \sum(0, 4, 11, 15) \end{aligned}$$

This function is plotted in Fig. 1-25. Two sets of couples are formed, and the

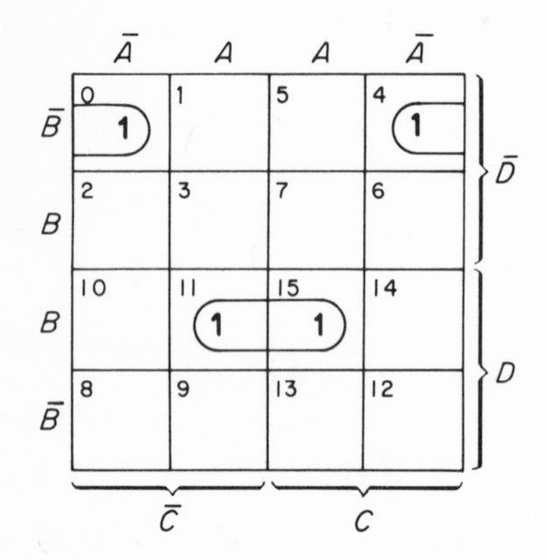

FIGURE 1-25. Map of the Function $F = \sum(0, 4, 11, 15)$.

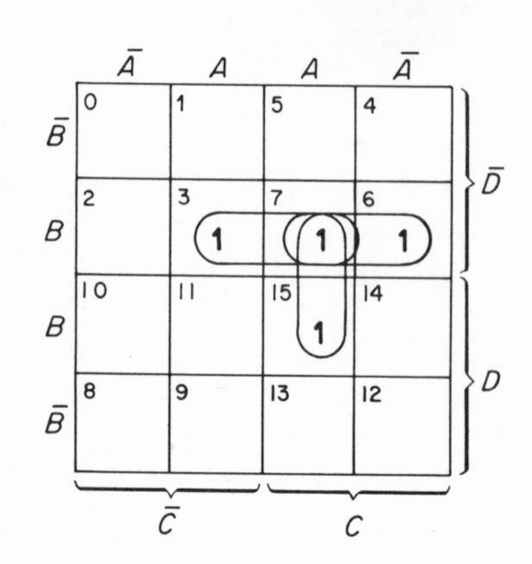

FIGURE 1-26. Map of the Function $F = \sum(3, 6, 7, 15)$.

corresponding pairs of terms can be combined, each with the elimination of one variable. This expression reduces to

$$F = \bar{A}\bar{B}\bar{D} + ABD$$

The next example illustrates the use of a single cell in forming more than one couple. Consider the expression

$$F = \sum(3, 6, 7, 15)$$

which is plotted in Fig. 1-26. Note that the cell for $ABC\bar{D}$ is shared in forming three couples. This expression reduces to

$$F = AB\bar{D} + BC\bar{D} + ABC$$

Adjacent Couples. Just as adjacent cells form a couple with the resultant elimination of one variable, adjacent couples form a pattern of four cells, or *quadruplet*, that result in the elimination of two variables. This is illustrated in Fig. 1-27 for the following function;

$$F = \sum(3, 7, 11, 15)$$

It is obvious from Fig. 1-27 that this expression can be reduced to

$$F = AB$$

The adjacent couples that make up the quadruplet can be formed in two ways, either $\sum(3, 7) + \sum(11, 15)$ or $\sum(3, 11) + \sum(7, 15)$. Either of these expressions can be reduced to AB. The important result is that a quadruplet is comprised of adjacent couples.

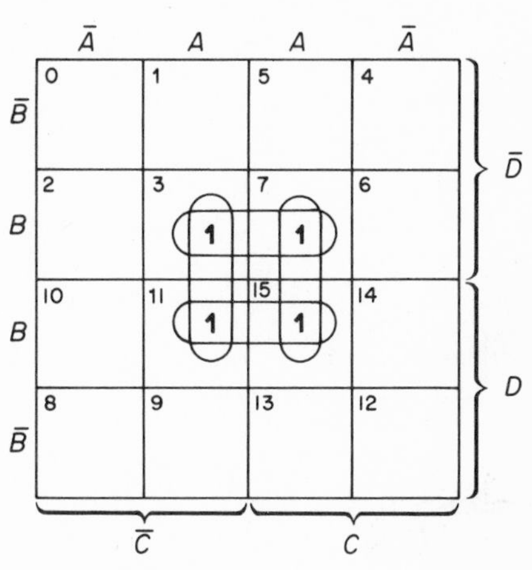

FIGURE 1-27. Map of Function $F = \sum(3, 7, 11, 15)$.

Some other quadruplets that can be represented with only two variables are shown in Fig. 1-28. The reader might wonder how many possible quadruplets there are in a four-variable map. This is left as an exercise.

An example will illustrate the use of the map in logic circuit design. It

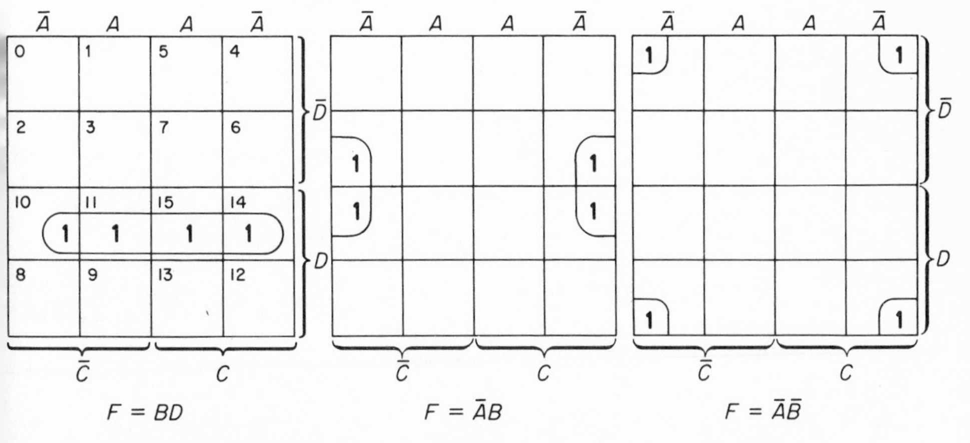

FIGURE 1-28. Typical Quadruplets Which Correspond to Two-Variable Terms.

is desired to use logic circuits to implement the switching function, Eq. (1-5).

$$F = \bar{A}\bar{B}\bar{C} + A\bar{B}C + AB\bar{D} + ACD + B\bar{C}D + \bar{A}B\bar{C} \tag{1-5}$$

The first task is to map this function and, if possible, reduce it to a smaller number of terms. The function is mapped in Fig. 1-29, and we see that three quadruplets can be formed. There is a choice in forming a quadruplet with the couple $\sum(3, 11)$, i.e., either the quadruplet $\sum(2, 3, 10, 11)$, as shown in Fig. 1-29, or the quadruplet $\sum(3, 7, 11, 15)$ can be used. This function, Eq. (1-5), is reduced to

$$F = \bar{A}\bar{C} + B\bar{C} + AC \tag{1-6}$$

$$= \bar{C}(\bar{A} + B) + AC \tag{1-7}$$

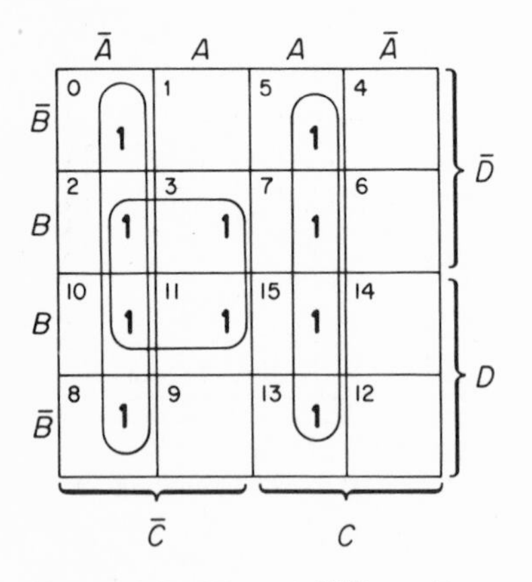

FIGURE 1-29. Map of $\bar{A}\bar{B}\bar{C} + A\bar{B}C + AB\bar{D} + ACD + B\bar{C}D + \bar{A}B\bar{C}$.

It also is worth noting that Eq. (1-5) is represented on the map (Fig. 1-29) by the cells that do *not* have a **1** marked in them. This is illustrated by using the NOT function, Eq. (1-8).

$$F = \overline{\bar{A}C + A\bar{B}\bar{C}} \tag{1-8}$$

These functions, Eqs. (1-6)–(1-8), are all equivalent, and it is the designer's choice as to which one is to be used. This usually depends upon what subfunctions are available or already have been generated. The NOT implementation is shown in Fig. 1-30, using standard NOR gates.

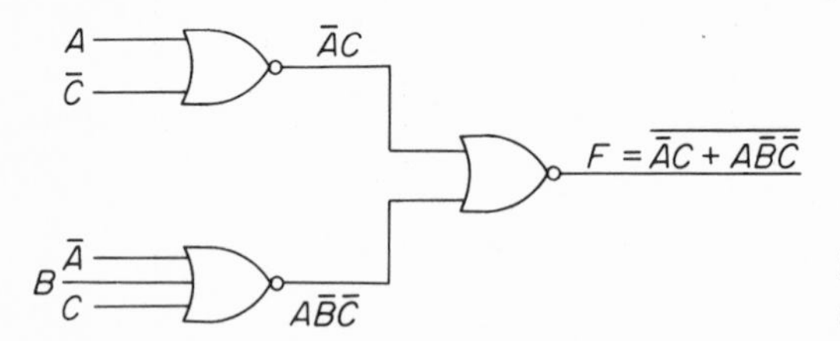

FIGURE 1-30. Circuit for Function $\overline{\bar{A}C + A\bar{B}\bar{C}}$.

The above example illustrates that the mapping technique is a very powerful tool for reducing Boolean expressions. It also gives considerable insight in the selection of couples, quadruplets, etc. in reducing the expression for practical circuit implementation. There is not necessarily an optimum or absolute minimization of the expression. The choice in forming couples, quadruplets, etc. gives many reduced expressions, and there is no "best" one per se. The designer's choice usually depends upon the type of digital building blocks

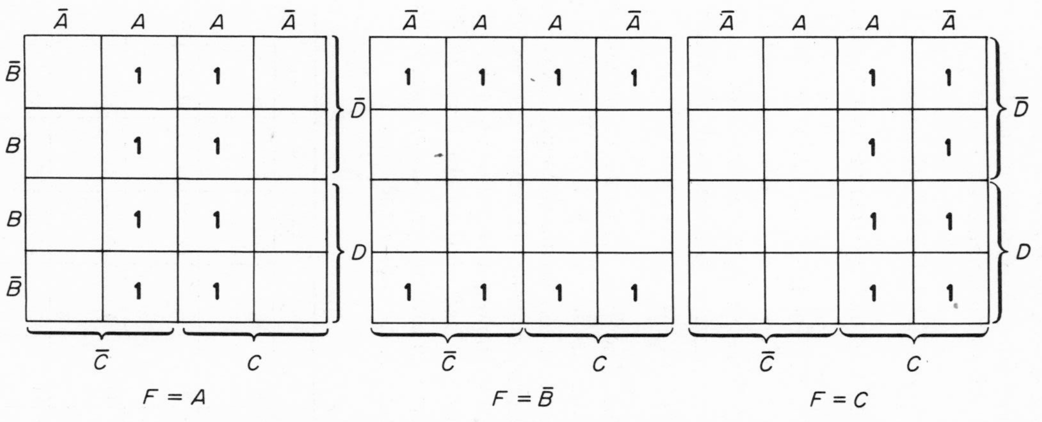

FIGURE 1-31. Typical Octuplets that Correspond to a Single-Variable Term.

that are being used in the system design, i.e., OR gates, NOR gates, etc. The above example was implemented with Eq. (1-8) using NOR gates. In Fig. 1-29 the map representing the function contains more cells with a **1** than with a **0** (shown as unmarked). Under these conditions it is desirable to consider the NOT function because it may result in fewer terms.

The NOT function, Eq. (1-8), can be operated upon using DeMorgan's theorem, giving the product-of-sum terms, Eq. (1-9).

$$F = (A + \bar{C})(\bar{A} + B + C) \tag{1-9}$$

This function also can be read directly from the map of Fig. 1-29.

Pattern of Eight Cells. The largest reduction can be made in a function of four variables when a pattern of eight cells, an *octuplet*, exists that represents a single variable. There are eight possible octuplets in a four-variable map, representing the four variables and their four complements. Three octuplets are shown in the maps of Fig. 1-31.

Five-Variable Map. The five-variable map is obtained by a continuation of the folding process and is shown in Fig. 1-32. With five variables the map has grown to such a size that couples are no longer always geometrically adjacent on a two-dimensional drawing. However, couples still differ in their numerical cell location by a power of two. For example, 5 and 21 are couples (differ by 2^4) and, similarly, 1 and 17, 0 and 16, 2 and 18, 11 and 27, etc. are couples. The couples are in mirror image locations about the axis, which includes, of course, coupling at the ends of the rows.

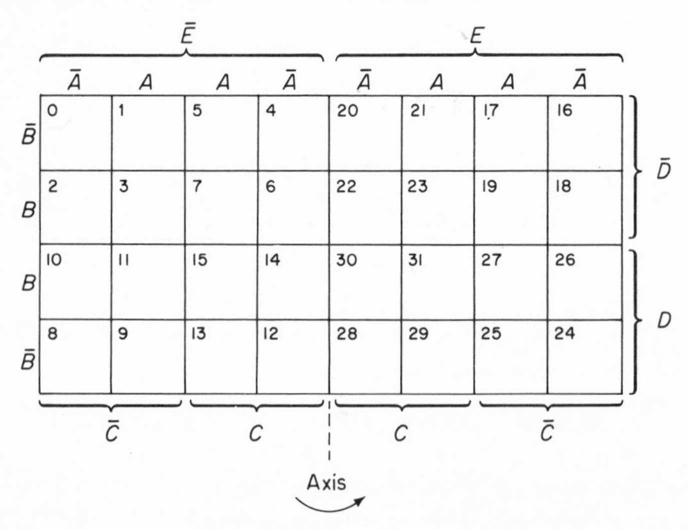

FIGURE 1-32. Five-Variable Map.

The use of the five-variable map is illustrated with the example shown in Eq. (1-10).

$$f(A, B, C, D, E) = \sum (0, 1, 3, 8, 9, 12, 13, 15, 16, 17, 19, 24, 25, 27, 31) \quad (1\text{-}10)$$

This function is mapped in Fig. 1-33. The cells with the **1**'s in Fig. 1-33 must

	$\bar{E}$				E				
	$\bar{A}$	A	A	$\bar{A}$	$\bar{A}$	A	A	$\bar{A}$	
$\bar{B}$	0 **1**	1 **1**	5	4	20	21	17 **1**	16 **1**	$\bar{D}$
B	2	3 **1**	7	6	22	23	19 **1**	18	$\bar{D}$
B	10	11	15 **1**	14	30	31 **1**	27 **1**	26	D
$\bar{B}$	8 **1**	9 **1**	13 **1**	12 **1**	28	29	25 **1**	24 **1**	D
	$\bar{C}$		C		C		$\bar{C}$		

FIGURE 1-33. Map of Eq. (1-10).

be grouped to reduce the function. Initially it is noted that there is one octuplet, Eq. (1-11).

$$\sum (0, 1, 8, 9, 16, 17, 24, 25) = \bar{B}\bar{C} \quad (1\text{-}11)$$

Other cells group into the following quadruplets:

$$\sum (1, 3, 17, 19) = A\bar{C}\bar{D} \quad (1\text{-}12)$$

$$\sum (8, 9, 12, 13) = \bar{B}D\bar{E} \quad (1\text{-}13)$$

$$\sum (17, 19, 25, 27) = A\bar{C}E \quad (1\text{-}14)$$

Finally, the remaining unused cells form a couple, Eq. (1-15).

$$\sum (15, 31) = ABCD \quad (1\text{-}15)$$

It is a good practice to check off each cell that is used in a group so that none is forgotten. The expression reduces to Eq. (1-16).

$$f(A, B, C, D, E) = \bar{B}\bar{C} + A\bar{C}\bar{D} + \bar{B}D\bar{E} + A\bar{C}E + ABCD \quad (1\text{-}16)$$

The above mapping technique may be extended to reduce Boolean expressions that contain six or more variables. The maps become rather large and care must be taken in selecting groups of cells, especially since couples are not always adjacent in the large-variable maps. However, the folding technique does place couples in mirror image locations that facilitate their location. A convenient check on the couples is noting that the numerical cell designations differ by a power of two.

1-5 MINIMIZATION BY TABULATION AND CHARTING

The basic principle used in the simplification of Boolean expressions with the mapping method is the formation of couples, quadruplets, octuplets, etc. of the terms in the function. This grouping of terms also can be accomplished using a tabulation method developed by W. V. Quine[6] and E. J. McClusky, Jr.[7] Mapping is a very useful tool for reducing Boolean expressions, but drawing maps for seven or more variables can become rather involved. The tabulation method also can become tedious for a large number of variables.

The tabulation method lists all the terms of the Boolean expression. If the expression is not in the standard sum-of-products form, it first must be expanded to that form. All possible couples are determined by comparing terms that differ by just one variable. In a similar manner, couples are compared to form quadruplets etc. It is convenient to list the binary number designation of the product terms, although the terms with the letter variables also can be used in the listing. The terms that do not form couples and couples that cannot form subsequent couples are entered in a chart along with the original terms to determine the reduced expression. We shall illustrate the tabulation method with the following example: Simplify the Boolean function in Eq. (1-17) using the tabulation method.

$$f(A, B, C, D, E) = \sum (0, 1, 3, 8, 9, 12, 13, 15, 16, 17, 19, 24, 25, 27, 31) \tag{1-17}$$

1. First list the product terms according to their binary designations. They are

   00000 00001 00011 01000 01001
   01100 01101 01111 10000 10001
   10011 11000 11001 11011 11111

2. Next group these binary number representations according to the number of 1's contained, as shown in the first column of Fig. 1-34. This is a "bookkeeping" technique that facilitates the location of couples by just comparing terms in successive adjacent groups.

Single Terms			Couples			Quadruplets			Octuplets	
0	00000	✓	0,1	0000-	✓	0,1,8,9	0-00-	✓	(0,1,8,9 16,17,24,25)	--00-*IX*
1	00001	✓	0,8	0-000	✓	0,1,16,17	-000-	✓		
8	01000	✓	0,16	-0000	✓	0,8,16,25	--000	✓		
16	10000	✓	1,3	000-1	***I***	1,9,17,25	--001	✓		
3	00011	✓	1,9	0-001	✓	1,3,17,19	-00-1	***VI***		
9	01001	✓	1,17	-0001	✓	8,9,12,13	01-0-	***VII***		
12	01100	✓	8,9	0100-	✓	8,9,24,25	-100-	✓		
17	10001	✓	8,12	01-00	✓	16,17,24,25	1-00-	✓		
24	11000	✓	8,24	-1000	✓	17,19,25,27	1-0-1	***VIII***		
13	01101	✓	16,17	1000-	✓					
19	10011	✓	16,24	1-000	✓					
25	11001	✓	3,19	-0011	✓					
15	01111	✓	9,13	01-01	✓					
27	11011	✓	9,25	-1001	✓					
31	11111	✓	12,13	0110-	✓					
			17,25	1-001	✓					
			24,25	1100-	✓					
			13,15	011-1	***II***					
			19,27	1-011	✓					
			25,27	110-1	***III***					
			15,31	-1111	***IV***					
			27,31	11-11	***V***					

FIGURE 1-34. Tabulation of Eq. (1-17).

3. All couples are now listed in the second column along with the decimal designations of the terms that form the couple. Recall that two terms form a couple when all the variables but one are the same; then that one variable can be eliminated in forming the new term representing the couple.

The term 00000 stands for $\bar{E}\bar{D}\bar{C}\bar{B}\bar{A}$, where the A position represents the least significant digit and 00001 stands for $\bar{E}\bar{D}\bar{C}\bar{B}A$. These terms combine to form the couple $\bar{E}\bar{D}\bar{C}\bar{B}$, which is represented as 0000-. The hyphen (-) is used to indicate the position of the variable that is eliminated. All possible couples must be formed by comparing each term in a group with every term in the adjacent group. As terms are used to form couples they are checked off as shown in Fig. 1-34. It is a worthwhile check on the couples to see that the decimal designations differ only by a power of two.

In a similar manner the groups of couples are compared, each term with all terms in adjacent groups, to form quadruplets, and the decimal designations of the single terms also are listed. The quadruplets are listed in the third column. This process is continued with the octuplets listed in the next column, etc., until no further combinations can be made. During this sequential process all terms that are used in forming combinations are checked off. All terms that are *not* checked, including those in the last column, represent the limit in forming successive combinations; and when these terms are logically summed together the resultant expression is equivalent to the original function, but not necessarily in the minimum form. The redundancy remaining in these unchecked terms is easily seen by listing them in chart form and looking for duplicate or multiple terms. This is shown in Fig. 1-35. First all unchecked

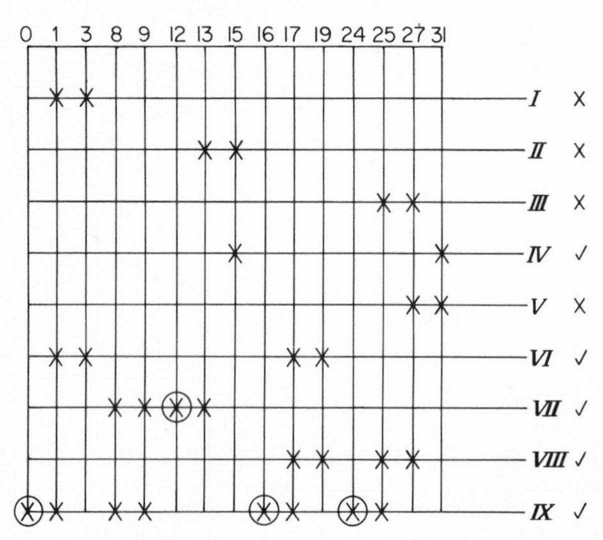

FIGURE 1-35. Chart for Tabulation of Fig. 1-34.

terms are identified with a Roman numeral designation. They are entered into Fig. 1-35 along with the original terms from which they were formed.

The original terms are entered at the top of Fig. 1-35, and the Roman numerals at the side. Crosses (x's) are placed at the intersections where the terms are contained in the Roman numeral expression; e.g., on line I, which represents a couple formed by terms 1 and 3, x's are placed at the 1 and 3 intersections. If a term contains only one x in a vertical line a circle is placed around that x. Roman numeral terms that contain circled x's must appear in the final expression. Therefore, terms VII and IX are required. There is a choice in selecting the remaining Roman numeral terms such that at least one x is contained in each vertical line; e.g., if VI is selected, since it contains 1 and 3, I is redundant and therefore not required. It is easily seen that there is not just one minimum expression, just as in the mapping technique. We place a checkmark beside the Roman numeral terms that are selected and a cross beside the others. The final chosen result is

$$f(A, B, C, D, E) = \text{IV} + \text{VI} + \text{VII} + \text{VIII} + \text{IX} \tag{1-18}$$

$$= (\text{-}1111) + (\text{-}00\text{-}1) + (01\text{-}0\text{-}) + (1\text{-}0\text{-}1) + (\text{-}\text{-}00\text{-})$$

$$= ABCD + A\bar{C}\bar{D} + \bar{B}D\bar{E} + A\bar{C}E + \bar{B}\bar{C} \tag{1-19}$$

The above example is the same problem that was done using mapping techniques, Eq. (1-10), and the same results are obtained, Eqs. (1-16) and (1-19). However, as pointed out previously, this is not the only solution.

The tabulation method really performs the same operations that are done using the mapping technique, only in a different way. The tabulation method is also tedious for a large number of variables, however, its systematic procedure is a form that can be programmed for solution on a digital computer. This is advantageous since computers are accessible to many engineers today.

REFERENCES

1. Boole, G., *Investigation of the Laws of Thought*, 1854, Dover, New York, 1951, (reprint).
2. Shannon, C. E., "A Symbolic Analysis of Relay and Switching Circuits," *Elec. Eng., Trans. Suppl.* 57; 713-723, 1938.
3. Karnaugh, M., "The Map Method for Synthesis of Combinational Logic Circuits," *Commun. Electron.*, pp. 593-599, November 1953.
4. Veitch, E. W., "A Chart Method for Simplifying Truth Functions," *Proc. Assoc. Comput. Mach. Conf.*, pp. 127-133, May 2-3, 1952.

5. Mahoney, M. V., Peticolas, A. B., and Laguzzi, M. C., *Logic Design*, 4th ed., RCA Institutes, 1966, 1967.
6. Quine, W. V., "A Way to Simplify Truth Functions," *Amer. Math. Mon.*, 62: 627-631, 1955.
7. McCluskey, E. J., Jr., "Minimization of Boolean Functions," *Bell. Syst. Tech. Jo.*, 35: 1417-1444, 1956.
8. Caldwell, S. H., *Switching Circuits and Logical Design*, John Wiley & Sons, New York, 1958.
9. Phister, M., Jr., *Logical Design of Digital Computers*, Chaps. 3 and 4., John Wiley & Sons, New York, 1958,
10. Marcus, M. P., *Switching Circuits for Engineers*, 2nd ed., Prentice-Hall, Englewood Cliffs, New Jersey, 1967.

PROBLEMS

1. Draw the switch contact representations to show the equivalency of the following expressions:
   (a) $A(B + C) = AB + AC$
   (b) $A + AB = A$
   (c) $A(A + B) = A$
   (d) $A + \bar{A} = 1$ (Note: When switch A is closed, switch $\bar{A}$ is open, and vice versa.)
   (e) $A\bar{A} = 0$
   (f) $A + \bar{A}B = A + B$

2. Simplify the following expressions:
   (a) $\bar{A}\bar{B} + A\bar{B} + \bar{A}B + AB$
   (b) $A + B + \bar{A}\bar{B} + \bar{A}B(A + \bar{B})$
   (c) $(A + AB)(\bar{A} + \bar{A}B)$
   (d) $(AB + \bar{A}\bar{B})(\bar{A} + B)A\bar{B}$
   (e) $AB + \bar{A}B\bar{C} + \bar{A}BC$
   (f) $(A + \bar{B} + \bar{A}B)\bar{C}$
   (g) $(A + \bar{A}B)(A + AB + C)$
   (h) $AC\bar{D} + \bar{A}C$

3. Complement and simplify the following:
   (a) $\bar{A}\bar{B} + A\bar{B} + \bar{A}B$
   (b) $\bar{A}B(A + \bar{B})$
   (c) $\bar{A} + \bar{B} + AB$
   (d) $A\bar{B}(A + B) + B + \bar{A}C$

4. Write the complements of
   (a) $[A + \bar{B}\bar{C}][\bar{A} + (B\bar{C} + D)]$
   (b) $A\bar{B}C + (\bar{A} + B + D)(AB\bar{D} + AC)$
   (c) $A + [(B + \bar{C})D + \bar{E}]F$

5. Show that the circuit symbols in Fig. P1-5 are equivalent (note that the small circle denotes complementation or inversion):

(a) =

(b) =

FIGURE P1-5

6. Construct truth tables and show by induction that the following expressions are true:
   (a) $\overline{AB} = \bar{A} + \bar{B}$
   (b) $\overline{A + B} = \bar{A}\bar{B}$
   (c) $A\bar{B} + AC + BC = A\bar{B} + BC$

7. Show by algebraic manipulation (using theorems) that the following equation is valid:

$$\overline{\bar{A}B + A\bar{B}} = \overline{A \oplus B}$$

8. Synthesize the Boolean functions that satisfy the truth tables in Fig. P1-8.

B	A	f (A,B)
0	0	0
0	1	1
1	0	1
1	1	1

(a)

C	B	A	f (A, B, C)
0	0	0	1
0	0	1	1
0	1	0	0
0	1	1	1
1	0	0	0
1	0	1	0
1	1	0	0
1	1	1	1

(b)

FIGURE P1-8

9. Using Venn diagrams, simplify
   (a) $A\bar{B} + AB$
   (b) $A\bar{B} + \bar{A}B + AB$
   (c) $\bar{A}\bar{B} + A\bar{B} + \bar{A}B$
   (d) $A\bar{B} + A\bar{C} + B\bar{C}$

10. (a) Write the Boolean expression for the output of the following circuit:
    (b) Express the function F in a product-of-sums form. Draw a logic diagram using only NAND gates that will implement the function F.

FIGURE P1-10

11. Using mapping techniques, simplify the following functions:
    (a) $A\bar{B} + B\bar{A} + AB$
    (b) $A\bar{B}\bar{C} + AB\bar{C} + ABC$
    (c) $A\bar{B}\bar{C} + A\bar{B}C + ABC + \bar{A}\bar{B}C$
    (d) $\bar{A}\bar{B}\bar{C} + \bar{A}B\bar{C} + \bar{A}C$
    (e) $\bar{A}B + AB\bar{C} + AC + \bar{A}\bar{B}C$
    (f) $A\bar{B}\bar{C}\bar{D} + AB\bar{C}D + AB\bar{C}\bar{D} + A\bar{B}\bar{C}D$
    (g) $\bar{A}\bar{B}\bar{C}\bar{D} + \bar{B}\bar{C}\bar{D} + \bar{A}\bar{B}CD + A\bar{C}\bar{D} + \bar{A}B\bar{C} + \bar{A}\bar{B}C\bar{D} + \bar{A}\bar{C}D$
    (h) $\bar{A}\bar{B}(\bar{C}\bar{D} + CD) + (A + \bar{B})\bar{C}\bar{D} + (\bar{A}C + A\bar{C})D$

12. (a) Show that the negative of a switching function can be obtained by joining the standard products not contained in the original function by OR operators.
    (b) Show that the negative of a function can be obtained by joining the standard sums not contained in the original function by AND operators.

13. A standard sum of products is represented by $F_1 = \sum m_i$; e.g., $F_1 = 0000 + 0100 + 0101 + 1001$ or $F_1 = \sum (0, 4, 5, 9)$. Similarly, a standard product of sums is represented by $F_2 = \prod M_k$; e.g., $F_2 = \prod (1, 7, 9)$, where M_1 denotes $(A + \bar{B} + \bar{C} + \bar{D})$. Find the relationship between m and M. (One solution is to take a sum-of-products example, with at least three variables, and convert it to a product of sums.)

14. Simplify the following functions by the map method:
    (a) $F = \sum (0, 1, 2, 5, 6, 7)$
    (b) $F = \sum (3, 5, 7, 11, 15)$
    (c) $F = \sum (1, 2, 3, 4, 5, 7, 9, 15)$
    (d) $F = \sum (1, 3, 9, 10, 11, 14, 15)$
    (e) $F = \sum (5, 7, 10, 11, 24, 25, 28, 29)$
    (f) $F = \prod (1, 3, 7, 9, 11, 15)$
    (g) $F = \prod (0, 2, 4, 8, 9, 10, 12, 13)$

15. Use the map method to simplify the following product-of-sums functions:
    (a) $F = \prod (1, 3, 7, 9, 11, 15)$
    (b) $F = \prod (0, 2, 4, 8, 9, 10, 12, 13)$
    (c) $F = \prod (4, 6, 7, 12, 14, 15)$

16. Write the transmission function in simple form for the network in Fig. P1-16.

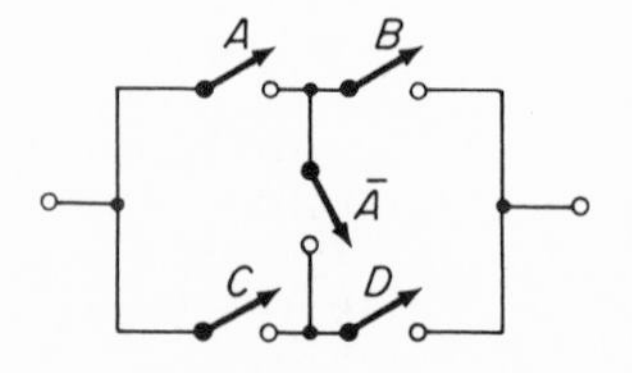

FIGURE P1-16

17. Use the map method to simplify the following functions:
    (a) $F = \sum (0, 1, 4, 5, 10, 11, 14, 15, 18, 19, 22, 23, 24, 25, 28, 29)$
    (b) $F = \sum (1, 5, 9, 13, 17, 21, 25, 29)$
    (c) $F = \sum (3, 5, 7, 12, 19, 21, 23, 28, 35, 39, 44, 51, 55, 60)$

18. Simplify the following switching functions using the tabulation method:
    (a) Prob. 14(c)
    (b) Prob. 17(b)
    (c) Prob. 17(c)
    (d) $F = \sum (5, 10, 13, 14, 21, 22, 29, 30, 34, 37, 42, 45, 50, 53, 54, 61)$

CHAPTER 2

DIGITAL ELECTRONIC CIRCUITS

Basic to the design and implementation of a digital system is knowledge of the individual circuits that make up that system. In the previous chapter relay and switch contacts were used to illustrate the switching functions. In many applications relays are not suitable because of their size, weight, and relatively

FIGURE 2-1. Integrated Circuit Wafer. Courtesy Raytheon Company

slow speed of operation. Most modern digital data processing equipment uses semiconductor circuits fabricated on a single silicon chip or wafer. The individual circuit components are integrated on a monolithic slab of silicon. A typical digital circuit containing approximately 8 transistors, 10 resistors, and 10 diodes is fabricated on a chip of 40 by 40 mils. A photograph of a 1-in. silicon wafer containing 300 such circuits is shown in Fig. 2-1.

Logic functions also can be implemented in small mechanical structures with pneumatic control known as *fluidics*. Fluidic devices contain no moving mechanical parts, and the states of these digital elements are determined by airflow. The operational speed of fluidic devices is not as fast as semiconductor integrated circuits, but the devices are not susceptible to errors due to electrical noise, radiation, and mechanical vibration. Magnetic circuits also can be used to implement logic functions.

This chapter is concerned primarily with semiconductor logic circuits, especially the types that are fabricated using integrated circuit technology. Although many devices can be used to implement logic functions, semiconductor circuits are the most versatile and are in wide use. The technology of semiconductor circuits has reached the point where component design and circuit design have merged into a single discipline. The engineer performing systems design will use integrated circuits and complete functions as building blocks in constructing a digital system. However, regardless of whether the engineer designs his own circuits or uses blocks designed by someone else, he should have a thorough knowledge of the circuit design principles, including input, output, and waveform requirements, to facilitate proper interconnection of the circuit modules. To provide this understanding we shall investigate the individual circuit functions. Logic circuits can be generally categorized as *passive* and *active* circuits. The passive logic circuits contain only passive elements, i.e., resistors, diodes, and capacitors. The active logic circuits contain transistors as well as passive components. Some of the more common configurations are presented below.

2-1 DIODE-RESISTOR LOGIC CIRCUITS

Diodes, used in conjunction with resistors, can be used to implement the basic OR and AND Boolean functions. However, since diodes cannot be used to invert electrical waveforms, they cannot be used to perform the NOT function.

The characteristics and symbolic representation of a typical low-level diode are shown in Fig. 2-2. The diode does not exhibit ideal characteristics, as indicated by the finite slope of the characteristics in the first quadrant. This slope is due to the forward resistance of the diode. There is also a potential

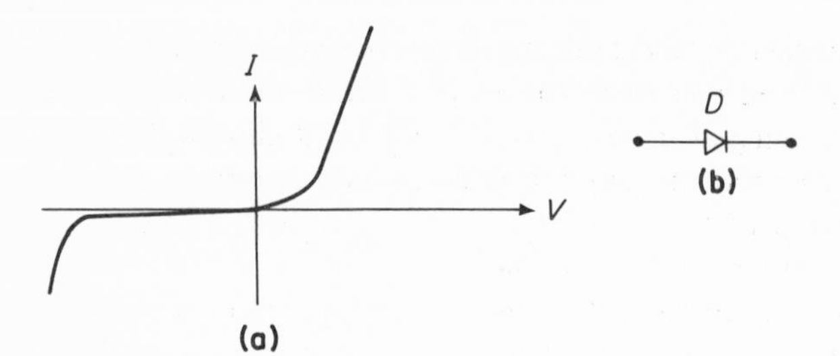

FIGURE 2-2. (a) Characteristics of a Typical Diode. (b) Diode Symbol.

drop of V_γ volts across the semiconductor junction before significant forward current flows. A piecewise linear representation of a semiconductor diode provides a very useful model for analysis. The equivalent circuit and characteristics are shown in Fig. 2-3.

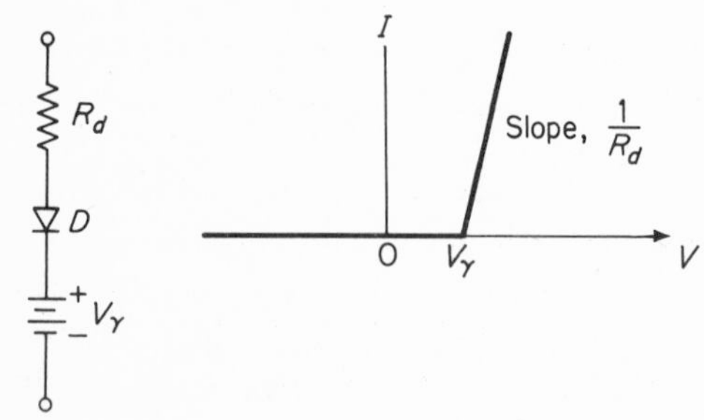

FIGURE 2-3. (a) Equivalent Circuit of Semiconductor Diode. (b) Piecewise Linear Diode Characteristics.

The numerical values of V_γ and R_d depend upon the type of diode used and also upon the operating point. For small-signal diodes V_γ is approximately 0.2 V for germanium and 0.6 V for silicon; R_d is in the order of 20 Ω. At high values of current the junction voltage drop V_γ increases slightly by approximately 0.1 V and R_d decreases. In some digital circuits V_γ is small compared to the supply voltage and can be neglected. Also, frequently R_d is in series with a larger resistance and it can be neglected. However, the equivalent circuit for each condition should be analyzed.

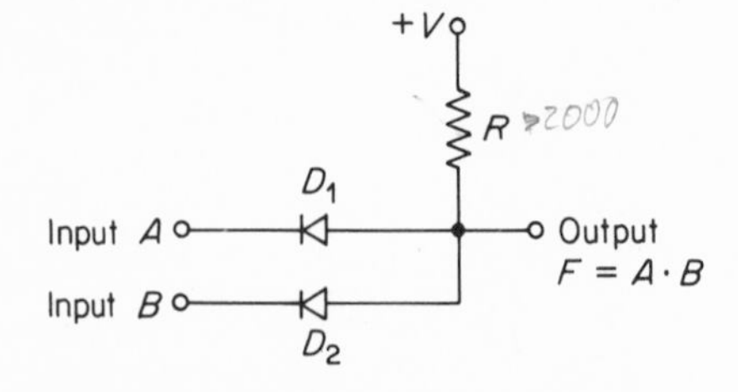

FIGURE 2-4. Diode AND Gate (Positive Logic).

A	B	F
−V	−V	−V
−V	+V	−V
+V	−V	−V
+V	+V	+V

FIGURE 2-5. Voltage Truth Table for AND Gate of Fig. 2-4.

A diode AND gate is shown in Fig. 2-4. The inputs A and B represent the switching variables, and each is assumed to take on either $+V$ or $-V$ Volts. Neglecting the voltage drop across the diodes, the output F takes on the values indicated in the truth table in Fig. 2-5. Under the conventions of positive logic, the more positive voltage $+V$ represents the value **1** and the lower voltage $-V$ the value **0**. The truth table is rewritten in Fig. 2-6 with the positive logic representation. The table clearly indicates that the AND function $F = AB$ is realized.

A	B	F
0	0	0
0	1	0
1	0	0
1	1	1

FIGURE 2-6. AND Gate Truth Table for Positive Logic.

A	B	F
1	1	1
1	0	1
0	1	1
0	0	0

FIGURE 2-7. OR Gate Truth Table for Negative Logic.

It should be noted that if a negative logic convention were employed (i.e., in the table of Fig. 2-5 if the $-V$ represents a **1** and the $+V$ a **0**), then the truth table represents the OR function, $F = A + B$, and is shown in Fig. 2-7.

Although the same circuit can be used to implement the AND and OR gates with a change in logic mode, it is much simpler to maintain a consistent set of logic definitions and use a different circuit for the OR gate. Such a circuit, a diode-resistor OR gate for positive logic, is shown in Fig. 2-8. The corresponding truth tables, using the different permutations of voltage inputs, are shown in Fig. 2-9. The tables represent the function $F = A + B$. If

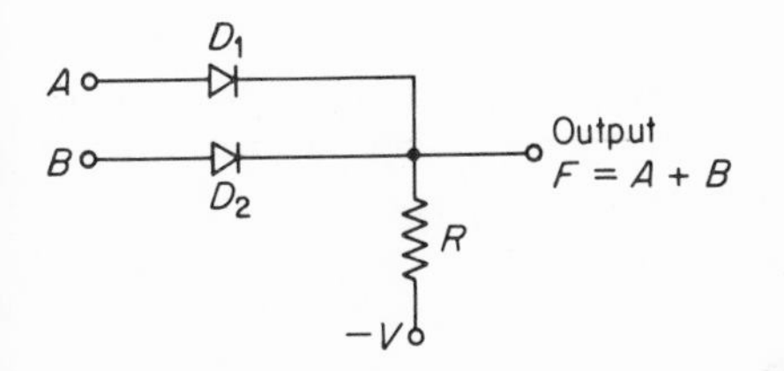

FIGURE 2-8. Diode OR Gate (Positive Logic).

(a)

A	B	F
−V	−V	−V
−V	+V	+V
+V	−V	+V
+V	+V	+V

(b)

A	B	F
0	0	0
0	1	1
1	0	1
1	1	1

FIGURE 2-9. (a) Voltage Truth Table for Diode OR Gate of Fig. 2-8. (b) OR Gate Truth Table for Positive Logic.

negative logic were employed, similarly, this gate would implement the AND function. This duality of functions is, of course, consistent with De Morgan's theorem. A positive logic convention will be assumed in this book unless specified otherwise.

Analysis of the Diode Gate

We shall analyze the diode AND gate using the conditions indicated for the circuit shown in Fig. 2-10. The forward resistance of the diode R_d is included in the equivalent circuit. With the conditions shown, diode D_1 is conducting and D_2 is cut off. Therefore, the output voltage V_f is

$$V_f = -V + V_\gamma + IR_d \tag{2-1a}$$

$$= -V + V_\gamma + \left[\frac{V_{CC} - (-V) - V_\gamma}{R + R_d}\right] R_d \tag{2-1b}$$

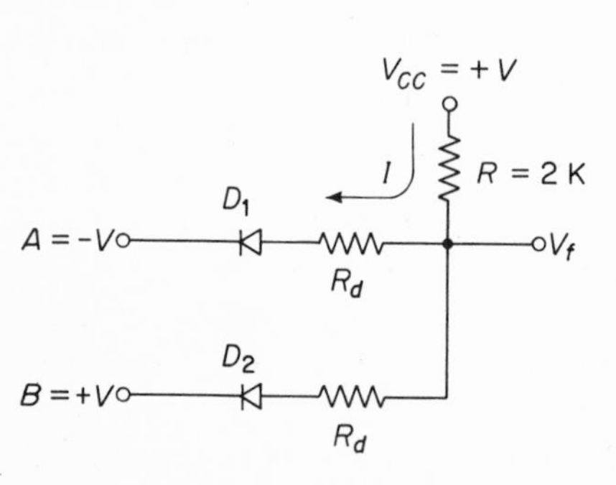

FIGURE 2-10. AND Gate Including Diode Forward Resistances.

where V_γ is the voltage drop across the diode semiconductor junction. The typical forward resistance of a high-conductance diode is 20 Ω; with $R = 2\text{K}$, Eq. (2-1b) becomes

$$V_f = -V + V_\gamma + [V_{CC} - (-V) - V_\gamma]0.01 \tag{2-2}$$

Now, for logic level **0** let $-V = 0$ Volts; and if $V_{CC} = +V$ Volts, then

$$V_f = V_\gamma + (V - V_\gamma)0.01 \tag{2-3a}$$

$$= V_\gamma + 0.01V \tag{2-3b}$$

for $V > V_\gamma$. If $V = 5$ V, then $V_f = (V_\gamma + 0.05)$ Volts, which is sufficiently close to 0 V for a logical **0** output. Obviously, if the diode resistance were larger and the load resistor R were smaller, then the output voltage would be a larger percentage of V_{CC}. If this **0** output were too large, then an ambiguity would result. It is good design practice to limit the **0** output level to approximately 25% of the **1** output level to maintain a good margin or separation between the **0** and **1** states. Therefore,

$$V_f(\mathbf{0}) \leq 0.25V_f(\mathbf{1}) \tag{2-4}$$

When V_f is a logical **1**, assume the output is $+V$ Volts, although we will see that this actually depends upon the loading of the output circuit. The **0** level of V_f is obtained from Eq. (2-1b). Assuming the same conditions as for the derivation of Eq. (2-3b) and substituting Eq. (2-16) into Eq. (2-4) gives Eq. (2-5a).

$$V_\gamma + \frac{R_d V}{R + R_d} \leq 0.25V \tag{2-5a}$$

If the voltage drop across the diode junction is small compared to the voltage level for a logical **1**, i.e., if $V \ll 0.25V$, then Eq. (2-5a) reduces to

$$\frac{R_d}{R + R_d} \leq 0.25 \tag{2-5b}$$

Although Eq. (2-5a) indicates that R should be very large to satisfy the inequality, practically, very large values of R will result in poor response times for the circuit. Note that when the AND gate output goes to a **1** the diodes are back biased and the junction and stray capacitances must charge through the resistance R. Therefore, Eq. (2-5a) can be used to determine a lower bound on the value of R. An output load on the gate, of course, must be taken into account in the equivalent circuit for determining the value of R to be used.

In the above discussion ideal voltage sources were assumed for the logic input voltages in Fig. 2-10. In practice the source resistance is nonzero, and this value must be included in series with the diode resistance in the analysis. This situation is compounded through several stages of passive logic, ultimately leading to the point where the voltage level separation is too small to produce two distinct logic states. Obviously, before this condition is reached signal reshaping must be provided.

Improper Connection of Diode Gates. Extreme care should be taken in connecting passive logic gates together because of the various impedance levels encountered. The circuits should render the proper logic functions for all possible combinations of input states. An improper connection of passive logic gates is shown in Fig. 2-11. This circuit is an AND gate followed by an OR gate. The input conditions shown in Fig. 2-11 indicate that the output of the

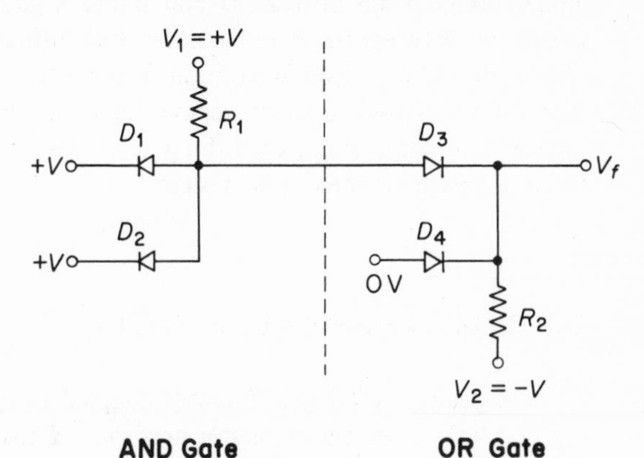

FIGURE 2-11. Example of Improper Connection of Diode Gates.

AND gate should be a logical **1** $(+V)$, with both inputs **1**'s. Since the AND gate output is connected to the input of an OR gate, the OR gate output voltage V_f also should be a **1**. We will now examine the circuit to see if this result is actually achieved.

With the voltage conditions shown in Fig. 2-11, diodes D_1 and D_2 are nonconducting and therefore exhibit very large resistances. Diode D_4 also is nonconducting if V_f is not a negative voltage. We will make this assumption and then check the resultant value of V_f. The circuit of Fig. 2-11 is redrawn in Fig. 2-12 with the nonconducting diodes removed. We also will assume that the value of the forward resistance of D_3 is negligible with respect to the resistors R_1 and R_2, which typically are in the order of 2 K. If $R_1 = R_2$, then $V_f \approx 0$V which clearly is not a **1** output. The circuit does not perform the required operation. R_1 should be small compared to R_2 for the output voltage to be within the region for a positive **1**. R_1 cannot be made too small or improper AND gate operation and excessive power dissipation will result. Also if R_2 is made very large, the circuit will have a slow response time due to the charging of the junction and stray capacitances through this resistance. The equivalent value of R_2 is also restricted because it includes the loading of the input impedance of the next stage. This example indicates the limitations in interconnecting passive gates in this manner.

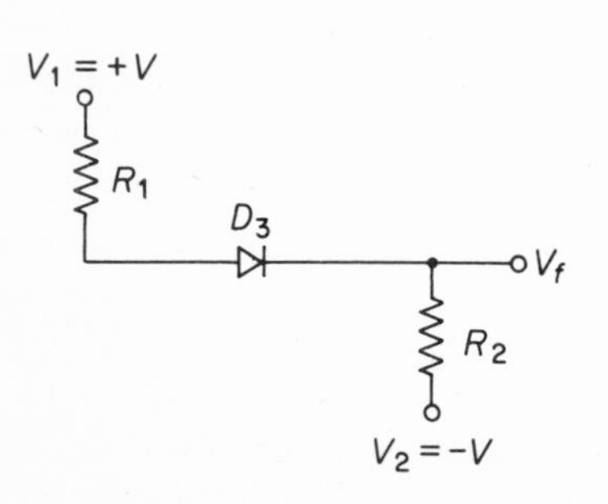

FIGURE 2-12. Equivalent Circuit for Condition Shown in Fig. 2-11.

In addition to the above restriction, proper successive connections of passive logic circuits result in voltage level deterioration because of the successive diode voltage drops through several stages of gates. This decreases the voltage separation between the **0** and **1** states and ultimately could lead to the point where the voltage level separation is too small to produce two distinct logic states. The disadvantages of passive logic can be overcome by the use of a transistor in conjunction with the gate to ensure a constant voltage level separation throughout the digital system.

2-2 DIODE-TRANSISTOR LOGIC (DTL)

A passive AND gate directly coupled to an npn transistor is shown in Fig. 2-13(a). Since the transistor performs an inverting, or NOT function, this complete circuit is a NAND gate. The output of the diode AND gate is applied to the base of the npn transistor, and the output voltage at F is inverted and isolated from the diode gate. This circuit maintains a voltage level separation by virtue of the transistor's bias supplies.

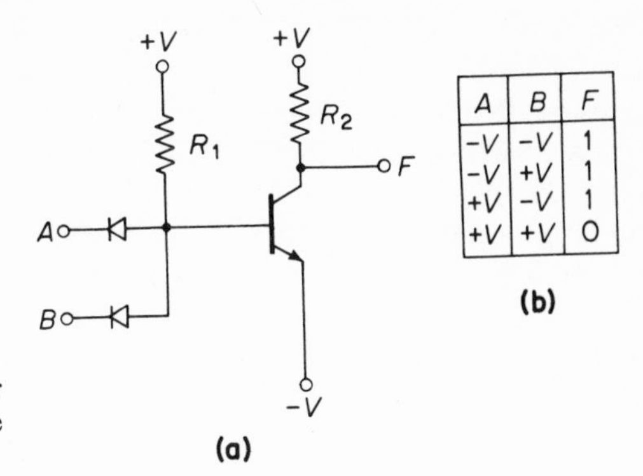

A	B	F
-V	-V	1
-V	+V	1
+V	-V	1
+V	+V	0

IGURE 2-13. (a) Diode-Transistor ogic NAND Gate. (b) Voltage 'ruth Table for NAND Gate.

Transistor Digital Circuits

In digital electronic circuits transistors are usually operated in either the :utoff or staturation states. These nonlinear regions of operation provide reatively stable values of output voltage over a wide range of operating condiions, including changes in loading and temperature. The transistor circuit :onfiguration that is most frequently used in digital circuits is the commonmitter stage, which provides signal inversion.

Typical collector characteristics for a switching-type transistor are shown in Fig. 2-14, with two load lines drawn for load resistances R_L and R'_L. Initially

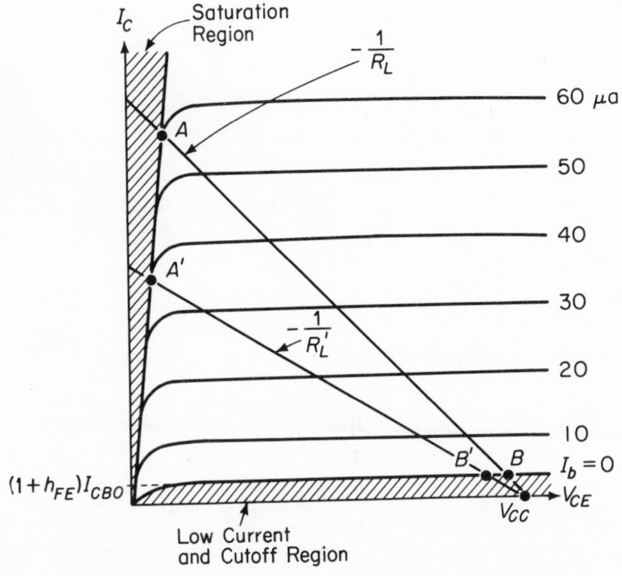

FIGURE 2-14. Transistor Collector Characteristics.

let us consider the load line of slope $-1/R_L$ drawn through points A and
Points A and B correspond to the saturation and cutoff operating points of th
transistor. It is seen that the output voltage V_{CE} is very small when the tran
sistor is in saturation, approximately 0.1 V for germanium transistors, and 0.3
for silicon. When the transistor is in the cutoff region the output voltage
very large; at point B it is approximately equal to the effective supply voltag
V_{CC}. As indicated in Fig. 2-14, when the base current is zero, $I_B = 0$, th
transistor is not cut off, and the collector current could be as high a
$(1 + h_{FE})I_{CBO}$, where I_{CBO} is the reverse saturation current across the collecto
to-base junction with the emitter open. To assure operation in the cutoff regio
it is good design practice to reverse bias the base-to-emitter junction. This ca
be accomplished with a negative bias supply, such as $-V_{BB}$ shown in Fig. 2
15(a). When the base-to-emitter junction is reverse biased, the transistor is cu
off and the collector cutoff current I_{CBO} flows in the base circuit to the bia
supply. At room temperature I_{CBO} is typically 1 μA for germaniu
transistors and 5 nA for silicon. This reverse saturation current is temperatur
sensitive and approximately doubles for every 10°C rise in temperature. Thi
causes design difficulties in using germanium transistors at high temperature
Silicon transistors are preferred for high temperature operation, and digita
integrated circuits are fabricated on silicon wafers.

If the transistor load resistance should change, as shown by the load lin
through points A' and B' in Fig. 2-14, it is seen that the collector-emitte
voltage does not change appreciably. This relatively stable output voltage wit
varying loads makes the common-emitter transistor operating in the saturatio
and cutoff regions attractive for digital circuit applications.

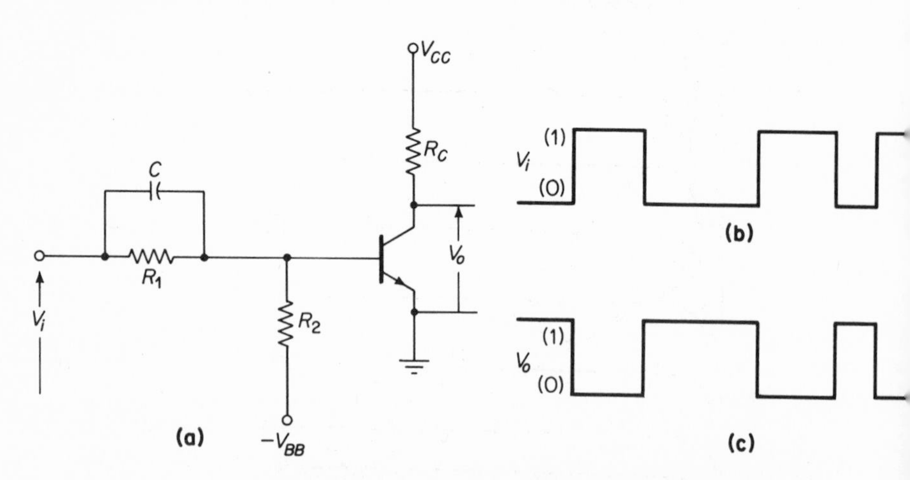

FIGURE 2-15. (a) A Transistor NOT Circuit. (b) Input Waveform. (c) Output Waveform.

The NOT or Inverter Circuit. Because the common-emitter amplifier provides signal inversion, it is a NOT circuit for binary input signals; i.e., the output is in the **1** state if the input signal representation is a **0**, and vice versa. Since the transistor operates in saturation and cutoff, it reshapes the binary signals, i.e., provides signal restoration. A transistor NOT circuit is shown in Fig. 2-15(a). Typical input and output waveforms, illustrating the signal inversion, are shown in Fig. 2-15(b) and (c).

The calculations required to determine the quiescent conditions are illustrated by the following example. A transistor inverter circuit is shown in Fig. 2-16, with the input voltages for levels **0** and **1** given as 0 and 12 V, respectively. We shall determine the output voltages for these inputs. Suppose

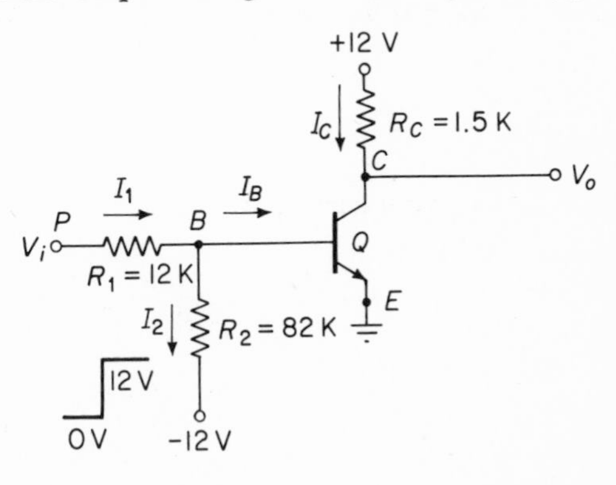

FIGURE 2-16. Inverter Circuit Example.

we are given a silicon transistor with a minimum h_{FE} of 30, $V_{CE}(\text{sat}) = 0.4$ V and $V_{BE}(\text{sat}) = 0.7$ V. We shall employ the method of assumed conditions and then perform calculations to verify these initial assumptions.

1. With V_i equal to 0 V, assume that the transistor Q is cut off. Therefore, $I_B = 0$, neglecting the very low collector cutoff current I_{CBO}. At high temperatures and with germanium transistors the collector cutoff current will have to be included in the calculations for worst-case analysis. I_{CBO} can be obtained from the manufacturer's data sheet. However, with zero base current, the base voltage V_B is simply obtained by the voltage division, Eq. (2-6). This calculation assumes that

$$V_B = -12\left(\frac{12}{12 + 82}\right) = -1.5 \text{ V} \tag{2-6}$$

the source impedance for V_i is very small compared with the 12-K resistor. This is a reasonable assumption since the input, $V_i = 0$ V, would be obtained from a saturated transistor that has a typical source resistance of 50 Ω [calculated by Eq. (2-13)]. Since Eq. (2-6) indicates that the base-to-emitter voltage is -1.5 V, the transistor is cut off and the output voltage V_0 is 12 V with a source resistance of 1.5 K.

2. The next step is to assume that the transistor is in saturation when $V_i = 12$ V and to make calculations to verify this condition. We also will assume that the driving source V_i is a cutoff transistor with a 1.5-K load resistor. The equivalent circuit showing the source generator and the base input circuit is given in Fig. 2-17.

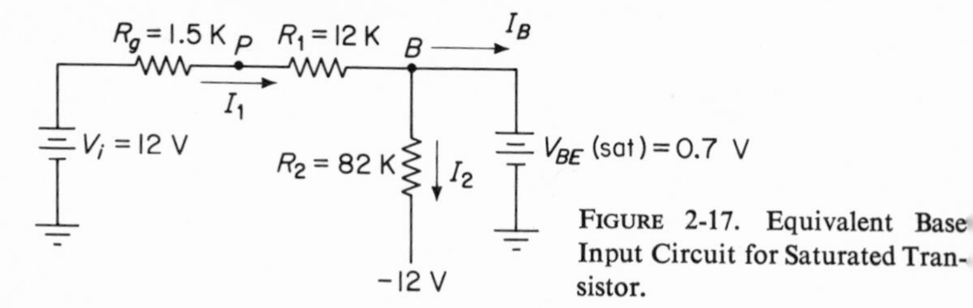

FIGURE 2-17. Equivalent Base Input Circuit for Saturated Transistor.

The input current is

$$I_1 = \frac{V_i - V_{BE}(\text{sat})}{R_g + R_1} = \frac{11.3}{13.5} = 0.84 \text{ mA} \tag{2-7}$$

$$I_2 = \frac{V_{BE}(\text{sat}) - (-12)}{R_2} = \frac{12.7}{82} = 0.15 \text{ mA} \tag{2-8}$$

$$I_B = I_1 - I_2 = 0.84 - 0.15 = 0.69 \text{ mA} \tag{2-9}$$

It is necessary to check if a base current of 0.69 mA is sufficient to have the transistor in saturation. The minimum base current required for saturation is

$$(I_B)_{\min} = \frac{I_C}{h_{FE}} \tag{2-10}$$

where the collector current is

$$I_C = \frac{V_{CC} - V_{CE}(\text{sat})}{R_C} = \frac{11.6}{1.5} = 7.7 \text{ mA} \tag{2-11}$$

We have

$$(I_B)_{\min} = \frac{7.7}{30} = 0.26 \text{ mA} \tag{2-12}$$

Since the actual base current of 0.69 mA is more than $2\frac{1}{2}$ times the minimum base current required for saturation, the transistor is indeed in saturation. The output voltage of the inverter, V_o, therefore is 0.4V, with a source resistance that is equal to the transistor saturation resistance R_{CE} (sat), where

$$R_{CE}(\text{sat}) = \frac{V_{CE}(\text{sat})}{I_C} = \frac{0.4}{7.7 \times 10^{-3}} = 52\,\Omega \qquad (2\text{-}13)$$

The capacitor C across R_1 in Fig. 2-15(a) is called a *speed-up* capacitor and is used to improve the transient response of the inverter. This capacitor aids in removing the minority carrier charge stored in the base when the signal changes abruptly to cut the transistor off. When the transistor is being turned on, its input capacitance shunts the total input resistance and the capacitor C provides an ac voltage divider to speed up the turn-on time. This capacitor is on the order of 100 pF; however, its exact value depends upon the transistor used.

The total charge stored in the base region of a transistor for given operating conditions is usually listed on the manufacturer's data sheets for switching transistors and is designated Q_T. When the transistor is being cut off, all the stored base charge must be withdrawn and transferred to the capacitor C. If C is just large enough to transfer the charge Q_T, then its value is

$$C = \frac{Q_T}{\Delta V} \qquad (2\text{-}14)$$

where ΔV is the change in input voltage levels. As an example, the maximum storage charge for a typical medium-speed switching transistor in saturation is 1,000 pC. For an input voltage swing of 10 V, the required speed-up capacitor is $C = 1{,}000/10 = 100$ pF.

Modern high-speed switching transistors are constructed with very thin base regions, and there is very little stored base charge to cause a long storage delay time. This is especially true in the design and fabrication of integrated circuits.

The Diode-Transistor Logic NAND Gate (DTL)

A diode-resistor AND gate followed by a transistor inverter circuit to perform the NOT-AND or the NAND function is shown in Fig. 2-18. The NAND gate has three input leads, A, B, and C, with inputs of either 0.3 or 12 V. We will assume that the inputs also are obtained from similar transistor inverters and, therefore, the source resistances are 50 Ω and 1.5 K for the transistors in saturation and cutoff, **0** and **1**, respectively. The truth table for this NAND gate is shown in Fig. 2-19.

We will verify that the operation of this circuit, i.e., the transistor Q

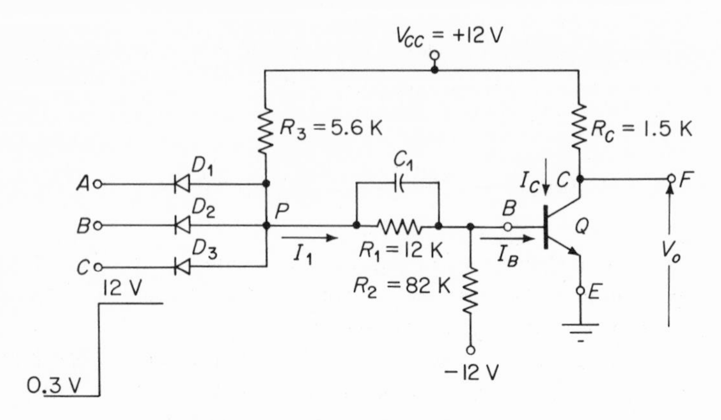

FIGURE 2-18. A Three-Input NAND Gate.

A	B	C	F
0	0	0	1
0	0	1	1
0	1	0	1
0	1	1	1
1	0	0	1
1	0	1	1
1	1	0	1
1	1	1	0

FIGURE 2-19. Truth Table for the Three-Input NAND Gate.

is in saturation, $V_0 = 0.3$ V, when all inputs are **1**'s; otherwise the transistor is cut off.

1. With all **1**'s at A, B, and C, $+12$ V, we will assume the condition in which the diodes D_1, D_2, and D_3 are reverse biased and Q is in saturation and then will check to verify that these conditions are true. If Q is in saturation, then the base voltage is

$$V_B = V_{BE}(\text{sat}) = 0.7 \text{ V} \tag{2-15}$$

and the voltage at point P is

$$V_P = V_{CC} - I_1 R_3 \tag{2-16}$$

I_1 also flows through R_3 when the diodes are reverse biased.

$$I_1 = \frac{V_{CC} - V_B}{R_1 + R_3} = \frac{12 - 0.7}{17.6} = 0.64 \text{ mA} \tag{2-17}$$

$$V_P = 12 - (0.64)(5.6) = 8.4 \text{ V} \tag{2-18}$$

With points A, B, and C at 12 V and point P at 8.4 V the diodes are indeed reverse biased. The transistor base current I_B is

$$I_B = I_1 - \frac{12.7}{82} = 0.64 - 0.15 = 0.49 \text{ mA} \tag{2-19}$$

The collector current I_C is

$$I_C = \frac{V_{CC} - V_{CE}(\text{sat})}{R_C} = \frac{11.7}{1.5} = 7.8\ \text{mA} \tag{2-20}$$

Therefore,

$$(h_{FE})_{\min} = \frac{7.8}{0.49} = 16 \tag{2-21}$$

That is, the transistor will be in saturation if $h_{FE} \geq 16$. To apply a reasonable safety factor, a transistor with an h_{FE} of 25 or greater should be used.

2. If any input is a **0**, i.e., at 0.3 V, then a diode conducts and the voltage at point P is 1 V, which is the input voltage of 0.3 V plus the 0.7 V drop across the diode.

   Assuming the transistor is cut off, and neglecting leakage current, the transistor base voltage is

$$V_B = V_P - I_1' R_1 \tag{2-22}$$

where I_1' is the current through R_1 when Q is cut off.

$$V_B = 1 - \left(\frac{13}{12 + 82}\right)(12) = 1 - 1.66 \doteq -0.66\ \text{V} \tag{2-23}$$

With a base-to-emitter voltage of -0.66 V the transistor is indeed cut off.

Fan-In and Fan-Out in Logic Circuits

Logic circuits have to be examined relative to the maximum number of inputs and outputs that can be accommodated. The maximum number of input leads that can be used with a logic circuit is called the maximum fan-in. If a certain logic gate is limited to six inputs, it is said that the gate has a fan-in of six.

The maximum number of logic circuits that can be driven by another circuit is called the maximum fan-out. This limitation often is determined by the maximum output current capability of a transistor inverter. Fan-out has to be considered as a maximum load in the basic circuit design of the inverter circuit.

In designing a digital system it is desirable to have a standard set of basic logic circuits called digital building blocks. Loading is considered in the circuit design, and maximum fan-in and fan-out rules are established for use by the logic designer. Digital systems are implemented by the interconnection of these building blocks.

The fan-in and fan-out limitations of each circuit type have to be considered in detail during the circuit design. There are no critical limitations to the number of diode inputs, fan-in, in the basic diode AND and OR gates shown in Figs. 2-4 and 2-8, respectively. This also is true when these circuits are used in the input stages of DTL NAND and NOR gates. However, the fan-out or output loading is severely restricted, as was previously discussed for the diode gates.

We shall now investigate the fan-out limitations of the transistor inverter circuit. In doing so we shall refer to the analysis of the NAND gate of Fig. 2-18. The minimum h_{FE} for the transistor is 16, as established by Eq. 2-21, in order to maintain saturation with a collector load resistance, $R_C = 1.5$ K, and no other load. Now we will assume that the transistor output, point F, is connected to the input of an identical NAND gate to the one shown in Fig. 2-18. When the transistor is conducting, the effective load is now 1.5 K in parallel with 5.6 K, giving a total resistance of 1.18 K. The loading resulting from resistors R_1 and R_2, and the effect of the diode forward conductance are considered negligible. The collector current, transistor in saturation, is

$$I'_C = \frac{11.7}{1.18} = 9.9 \text{ mA} \tag{2-24}$$

Therefore, with a transistor base current of 0.49 mA (see Eq. 2-19) the minimum h_{FE} required is

$$(h_{FE})_{\min} = \frac{9.9}{0.49} = 20 \tag{2-25}$$

Assuming a rated $(h_{FE})_{\min}$ of 30 for this transistor, we shall determine the maximum fan-out into similar NAND gate loads. The maximum allowable collector current is

$$(I_C)_{\max} = (I_B)_{\max}(h_{FE})_{\min} = (0.49)(30) = 22 \text{ mA} \tag{2-26}$$

Also,

$$(I_C)_{\max} = \frac{V_{CC} - V_{CE}(\text{sat})}{R'_L} = \frac{11.7}{R'_L} \tag{2-27}$$

where R'_L is the equivalent transistor collector load resistance.

$$\frac{1}{R'_L} = \frac{1}{R_C} + n\left(\frac{1}{R_3}\right) = \frac{1}{1.5} + \frac{n}{5.6} \tag{2-28}$$

Upon substituting Eq. 2-28 into Eq. 2-27,

$$22 = 11.7\left(\frac{1}{1.5} + \frac{n}{5.6}\right) \tag{2-29}$$

Solving for n gives $n = 6.8$. Therefore, the fan-out limitation, to maintain the transistor in saturation, is a maximum of six. To provide a safety factor, the maximum fan-out should not exceed five of these loads. A worst-case analysis should include the effects of high temperature operation.

Sink and Source Loads

Loads are classified as *sink loads* if current flows out of the inputs and as *source loads* if current flows into the inputs. The NAND gate of Fig. 2-18 is a sink load. The current flowing out of the input leads was considered in calculating the maximum fan-out in the above example.

The diode OR gate, shown in Fig. 2-8, is a source load. The driving circuit acts as a power source, driving current into the diode OR gate. The fan-out capabilities of a circuit also should be determined for standard source loads.

The DTL NOR Gate

The DTL NOR gate is implemented by connecting the output of the diode OR gate of Fig. 2-8 to the input of the NOT circuit of Fig. 2-15. This combined circuit is shown in Fig. 2-20(a). The resistor R used in the basic OR

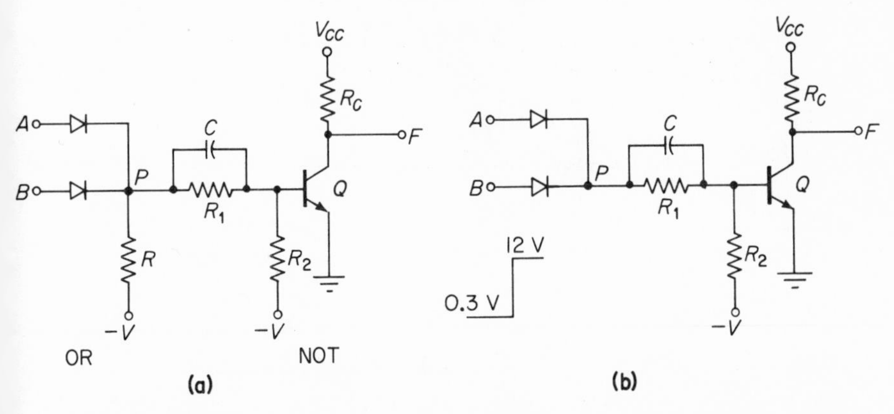

FIGURE 2-20. (a) NOR Gate Using Basic Circuits. (b) A More Practical NOR Gate.

gate provides a return path for the input signals A and B. We see in Fig. 2-20(a) that a return path also is provided through resistors R_1 and R_2; therefore, R is not required. A simplified circuit is shown in Fig. 2-20(b), with the input levels of 0.3 V for a **0** and 12 V for a **1**.

Although only two inputs are shown in Fig. 2-20, additional diodes can be connected to point P to form a multiple-input NOR gate. The truth table for a three-input NOR gate is given in Fig. 2-21. If all inputs are **0**'s, then all the diodes conduct and the voltage at point P is approximately 0 V, thereby holding the transistor in cutoff. This is shown in the inverter calculations by Eq. (2-6). Similarly, if the input to any diode is high, i.e., a **1**, point P is also high. Diodes with low or **0** inputs are reverse biased. The calculations to show that the transistor is in saturation, and therefore that the output is a **0**, are similar to the inverter circuit calculations. Reference is made to Fig. 2-17 and Eqs. (2-7), (2-8), and (2-9). If more than one diode has a **1** input, then the equivalent generator resistance, R_g in Fig. 2-17, will be

A	B	C	F
0	0	0	1
0	0	1	0
0	1	0	0
0	1	1	0
1	0	0	0
1	0	1	0
1	1	0	0
1	1	1	0

FIGURE 2-21. Truth Table for a Three-Input NOR Gate.

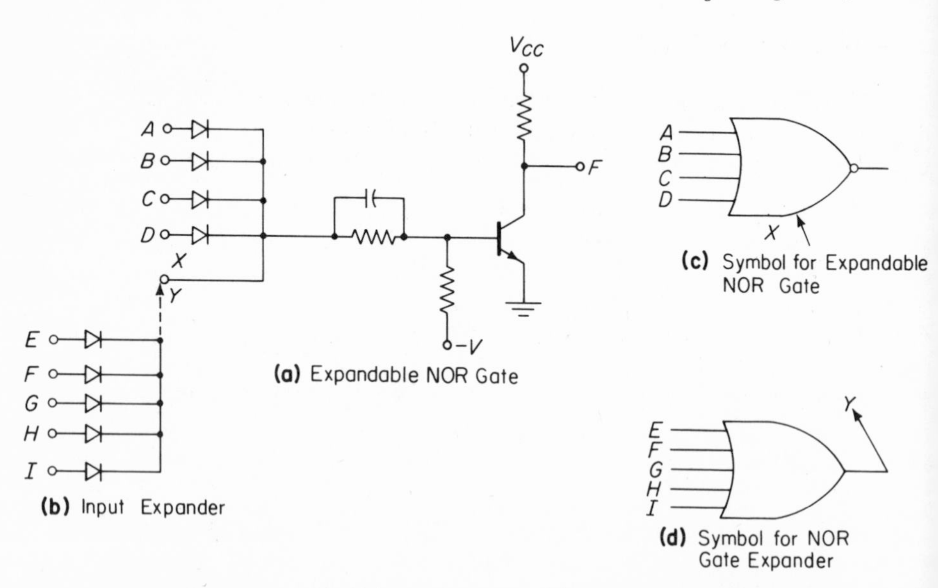

FIGURE 2-22. Expandable NOR Gate. (a) Expandable NOR Gate. (b) Input Expander. (c) Symbol for Expandable NOR Gate. (d) Symbol for NOR Gate Expander.

educed by the parallel combinations. This would provide more base current or increased transistor saturation. Therefore, additional diodes can be added o point P to accommodate a very large fan-in. An exact analysis would nclude the diode voltage drop in the equivalent circuit. However, because of he similarity to the example analyzing the inverter circuit, it is not necessary o present any further calculations.

Expandable Gates. We have shown how the fan-in of NOR and NAND gates (also OR and AND gates) can be increased by the addition of more diodes at the inputs. The commerically available digital building blocks, especially integrated circuits, have expandable gates and input expander modules o accommodate large numbers of inputs. The use of expandable gates and the ssociated symbols is illustrated using NOR logic in Fig. 2-22.

2-3 NOISE-MARGINS

Noise may be induced in digital circuits in many different ways. Outside sources of noise are (1) electromagnetic radiation (EMI, electromagnetic interference) that is picked up by the circuit conductors and (2) conducted interference, noise from the power line which is due to other operating equipment such as rotating machinery and equipment with electrical contactors. These outside noise sources can be attenuated by the use of shields and filters.

Noise also may be generated within the digital equipment itself due to electrostatic and magnetic coupling. High frequency, large voltage pulses can be capacitively coupled along long adjacent conductors. The effect is more severe when the voltage change is coupled into a high-impedance line. Therefore, long leads should be avoided for high frequencies and for fast rise times. Also, the source and load impedances should be low, which results in a high power consumption. Current loops also should be avoided to prevent magnetic coupling of pulses into other lines. These problems can be reduced appreciably by using twisted pairs of leads for long signal runs.

Noise also is generated by fast switching of high currents on ground return busses. The resultant voltage drops developed across lead inductances can be amplified and cause false signals in the logic circuits. These noise effects can be reduced by the use of a ground plane to which the logic circuits are connected with short leads. The plane should have sufficient thickness to provide a low resistance for the common return path for the circuits.

The noise immunity of a logic gate is specified in terms of dc margins determined under worst-case conditions for both the **0** and **1** level. Assuming positive logic, the margin for a **1** input applies to negative-going noise on the high logic level or the power supply line. The margin for a **0** input applies to positive-going noise on the low logic level or the ground line.

The dc margin is the difference between the worst-case output level an the worst-case input threshold. We will determine these regions for a NO gate from the transfer characteristic shown in Fig. 2-23. The characteristic are plotted for both high and low va ues of temperature. We will assum that the circuit specifications requir an operation over this temperatur range. From Fig. 2-23 we will deter mine the threshold values. Actually, fc worst-case considerations, the $-15°$ curve is used for the **1**-level threshol and the 90°C curve for the **0**-leve threshold. These values are 2.5 V fc the **1**-level input threshold and 1.5 for the **0**-level input. If the fan-ou or circuit loading, is restricted so tha the output level is 3.3 V or greater fo a **1** and 0.6 V or less for a 0, then the minimum dc margins can be determine These minimum noise-margins are 0.8 V for a **1** and 0.9 V for a **0**. The variou levels are shown in Fig. 2-24.

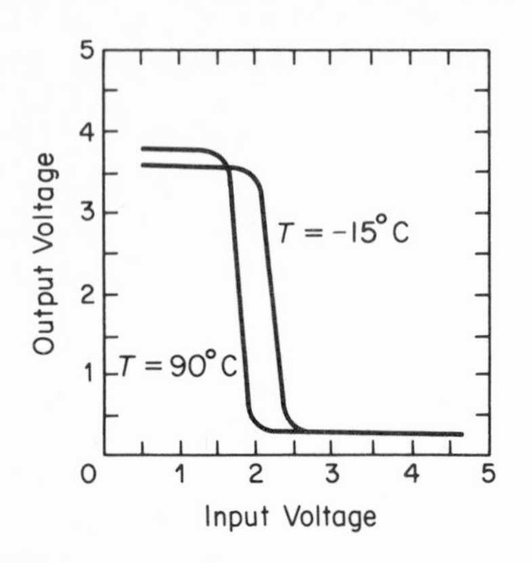

FIGURE 2-23. NOR Gate Transfer Characteristics.

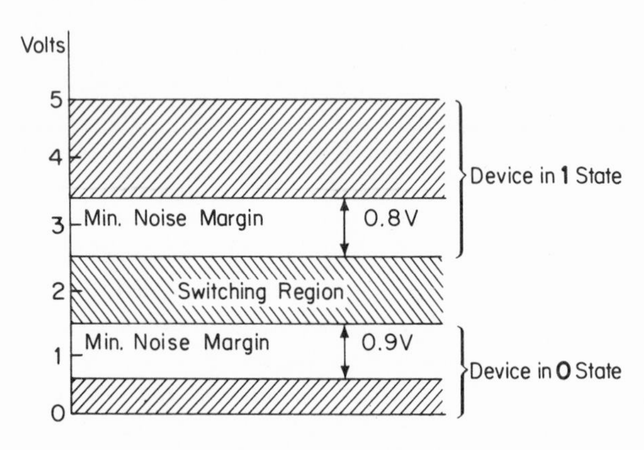

FIGURE 2-24. Logic Levels and Noise-Margins.

The dc noise-margin can be increased in DTL circuits by the addition o silicon diodes in series with the base of the transistor, as shown in the NOF gate of Fig. 2-25. Any input voltage has to be of sufficient magnitude to over come the forward voltage drop of the diode D, in addition to the transisto base-emitter junction voltage drop, prior to transistor conduction. This in creases the switching threshold, and therefore the noise immunity, by 0.6 V The addition of diodes in integrated circuits is not considered significant becaus there is no more effort in fabricating a diode than there is a resistor using this technology.

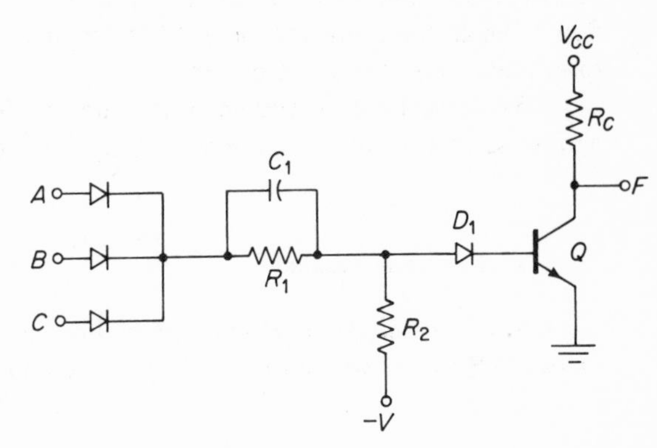

FIGURE 2-25. DTL NOR Gate with Base Diode for Improved Noise Immunity.

2-4 DIRECT-COUPLED TRANSISTOR LOGIC (DCTL)

Transistors in Series

Transistors can be considered analogous to switch contacts, with the restriction that the current flow is unilateral. Transistors in saturation are closed switches, and transistors in cutoff are open. Since switch contacts in series perform the AND operation, transistors in series function similarly. However, since common-emitter transistor circuits invert or negate, transistors

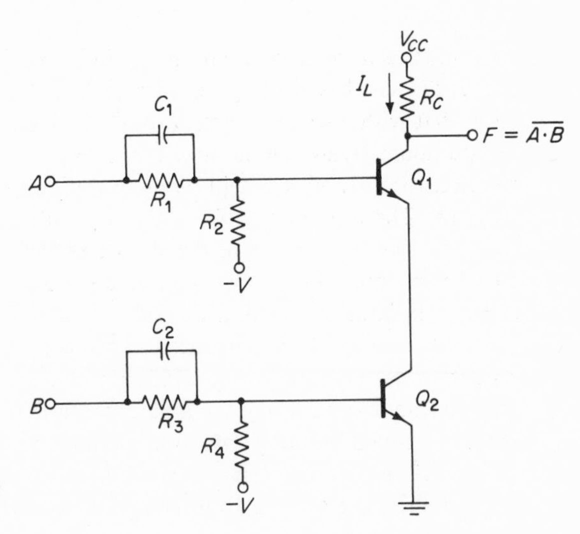

FIGURE 2-26. Transistor NAND Gate for Positive Logic.

in series perform the NAND operation. A series transistor NAND gate for positive logic is shown in Fig. 2-26. With **0** inputs both transistors are biased in the cutoff region and the output voltage is $+V$, a logical **1**. Current will flow through the transistors only if both inputs are **1**'s, causing both transistors to saturate, and the output becomes a **0**.

The actual output voltage is the sum of the two transistor collector-to-emitter saturation voltages, approximately 0.3 V each for silicon transistors.

Transistors in Parallel

A positive-logic NOR gate can be implemented by connecting transistors in parallel, as shown in Fig. 2-27. For a multiple-input NOR gate many

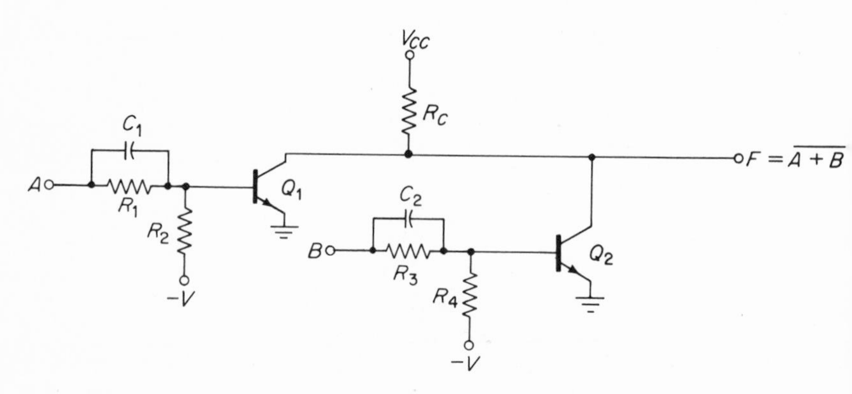

FIGURE 2-27. Transistor NOR Gate.

transistors can be connected in parallel, although only two are shown in Fig. 2-27. The base biasing network and the load resistor R_C should be selected so that with a single **1** input (all other inputs **0**) the associated transistor is in saturation, giving a **0** output. Then, with two or more inputs being **1**'s, the associated transistors also will be in saturation because of the reduced collector currents. The load current $I_L \approx V_{CC}/R_C$ will be shared by all the ON transistors. The sharing may not be equal because of slight differences in the transistor parameters, but this is not significant since each transistor alone is capable of being in saturation with a **1** input and therefore providing a **0** output.

The important consideration in the parallel operation of transistors, with a single collector load, is that each transistor alone must be capable of going into saturation with a single **1** input. Because of this we cannot arbitrarily expand integrated circuit NOR gate modules by tying their outputs together. This, of course, parallels the collector resistors, and the increased loading may prevent a single transistor from going into saturation and thereby giving incorrect output voltage levels.

2-5 CURRENT MODE LOGIC (CML)

Transistors connected as shown in Fig. 2-28 make use of a constant current that is switched from one transistor to another and are known as current mode logic. The transistors in CML circuits may be operated in either the saturation or linear regions. When operation is performed in the linear or active region, switching speeds in the order of 1 nsec can be achieved.

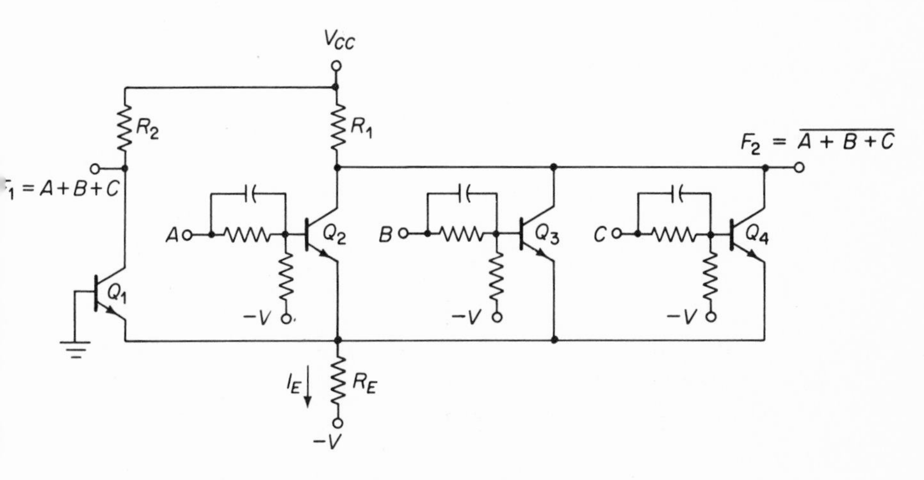

FIGURE 2-28. NOR Gate Using Current Mode Logic.

The operation of the CML circuit is similar to that of a differential amplifier, with a large signal input to one side and the other side connected to ground potential. Using positive logic, with a **1** input to either A, B, or C, there is conduction on the right-hand side of the circuit and Q_1 is cut off. This makes the F_1 output a **1** and the F_2 output a **0**. If all inputs are **0**'s, transistor Q_1 will conduct and transistors Q_2 through Q_4 will be cut off because the base of Q_1 is grounded. This causes the approximate constant current I_E to switch from one side of the circuit to the other. As with differential amplifiers, it is common to replace R_E with a fixed-biased transistor.

CML circuits have the disadvantages of a large power requirement and variations in the output logic levels if the transistors are not operated in saturation. The circuits are usually used where very high speed is required.

2-6 TRANSISTOR-TRANSISTOR LOGIC (TTL)

A TTL NAND gate is shown in Fig. 2-29. The TTL designation is given to the multitransistor AND gate, transistors Q_1 through Q_3, connected in the common-base configuration. The output of this gate drives the output

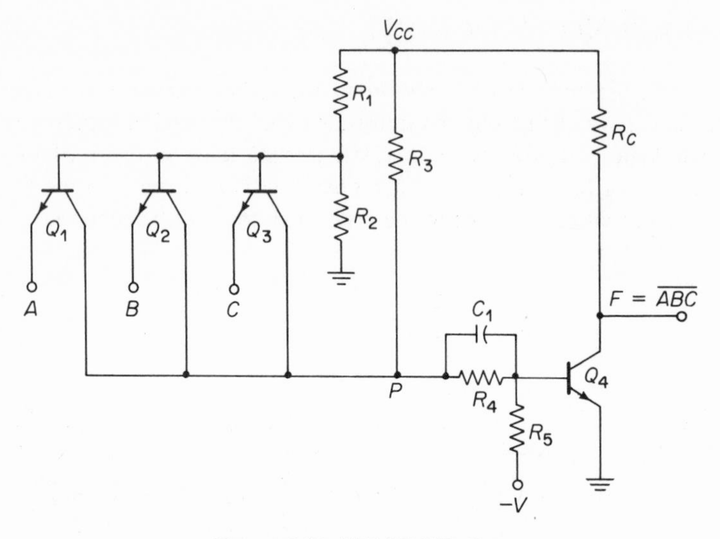

FIGURE 2-29. TTL NAND Gate.

transistor Q_4, which is either in cutoff or saturation and provides inversion. The inputs to the TTL gate are the emitters of the transistors. If any input is a **0**, an input transistor conducts and the voltage at point P is low, causing Q_4 to be cut off. Point P will rise in voltage only when all inputs are high, **1**'s, therefore generating the AND function. The **1** input at point P causes Q_4 to go into saturation and provides inversion. Transistor Q_4 is part of a conventional inverter circuit, as illustrated in Fig. 2-15.

We observe that the input transistors in Fig. 2-29 have their bases and collectors connected together. This same function can be performed in a specially constructed transistor that has multiple emitters. Multiemitter transistors are frequently used in integrated circuit fabrication.

Integrated-Circuit Gates

An integrated TTL gate is shown in Fig. 2-30. The multiemitter transistor is designated as Q_1. Transistor Q_2 is used as a driver stage to drive a power output stage Q_3 and Q_4. Transistors Q_3 and Q_4 are connected to deliver large currents in either direction to the load. This allows all external capacitances to be charged rapidly, and therefore both rise and fall times are quite small.

When one or more of the inputs to Q_1 are low, **0**'s, transistor Q_1 is in saturation. This causes Q_2 and Q_3 to be cut off. The output of the gate is in the

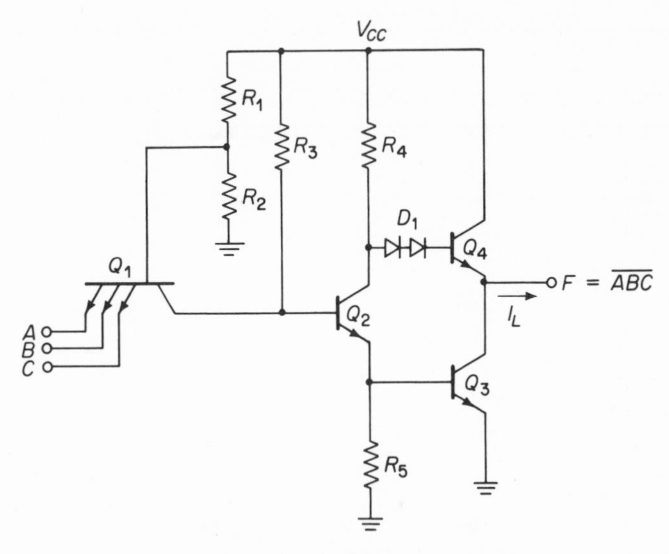

FIGURE 2-30. Integrated Circut TTL NAND Gate.

1 state, and current is delivered to the load via Q_4. Transistor Q_4 receives its base current drive through resistor R_4 and the double diode D. In this condition the actual output voltage is a function of the load current I_L, the current amplification factor of Q_4 (to determine the base current), the voltage drop across the double diode, and the value of resistor R_4.

When all inputs to the gate are high, **1**'s, transistor Q_1 is cut off, causing Q_2 and Q_3 to go into saturation. The output voltage is then the saturation voltage of Q_3. In this condition the current direction is from the load. The double diode D is provided to bias off the current in Q_4 in this mode. This is easily seen by summing the voltage drops around transistors Q_2, Q_3, and Q_4. If D were not present, Q_4 would be forward biased and conducting heavily.

In integrated circuit fabrication it is relatively easy to produce diodes and transistors, therefore, designs exist that contain many more transistors and diodes than normally would be used in discrete component circuits. Also, integrated circuit fabrication techniques can produce circuits that do not have an exact discrete component equivalent, e.g., the multiemitter TTL gate.

An attempt has been made to categorize logic circuits into different classifications. Some of these types have been discussed in this chapter, i.e., DTL, DCTL, CML, and TTL. Different circuits in the different classifications can have advantages in certain applications, so modern integrated circuit modules contain a hybrid design using portions of the different classifications. Obviously, different types of circuits cannot be interconnected indiscriminately. The logic levels and loading requirements must be compatible.

2-7 THRESHOLD LOGIC

In addition to the AND, OR, NAND, and NOR gates described earlier, digital or Boolean functions can be implemented using gates containing thresh-

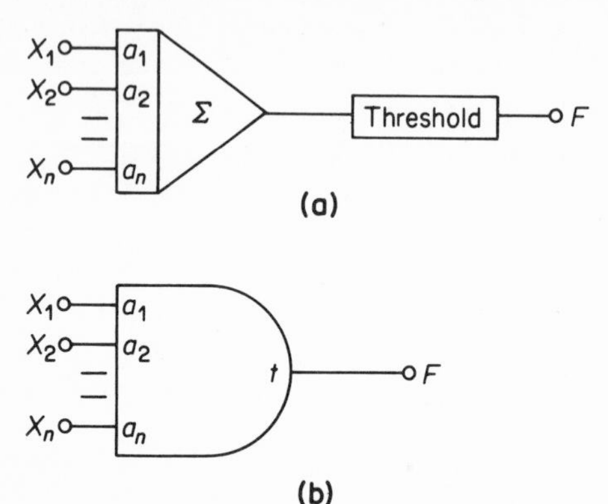

FIGURE 2-31. (a) Block Diagram of Threshold Gate. (b) A Logic Representation of the Threshold Gate.

old circuits, known as *threshold logic.* Threshold logic sums the weighted values of the input binary variables; if the sum is equal to or greater than a set threshold value, the output is a **1**; otherwise the output is a **0**. A block diagram of a basic threshold gate is shown in Fig. 2-31. The input binary variables, X_1, X_2. . . . , X_n, are multiplied by the weighting factors, a_1, a_2, . . . , a_n, and are summed together. The a's are real numbers, positive, negative, or zero. The weighted sum is compared to a threshold t to generate the output binary variable F. The operation of the gate is as follows:

$$F = \mathbf{1} \text{ if } \sum_{i=1}^{n} a_i X_i \geq t \tag{2-30}$$

and

$$F = \mathbf{0} \text{ if } \sum_{i=1}^{n} a_i X_i < t \tag{2-31}$$

It is also convenient to represent the gate output in the following manner:

$$F = \langle a_1 X_1 \boldsymbol{+} a_2 X_2 \boldsymbol{+} \cdots \boldsymbol{+} a_n X_n \rangle_t \tag{2-32}$$

It is necessary to use a boldface plus sign ($\boldsymbol{+}$) to denote arithmetic addition; the regular plus sign ($+$) will continue to represent the Boolean OR connective. (Perhaps the engineer should have retained the use of the logician's cup and cap (∪ and ∩) for the Boolean connectives to avoid the confusion with the plus sign.)

The OR and AND functions are actually special cases of the very general threshold function. For example, if all the a's are of weight 1 and the threshold t is 1, Eq. (2-32) represents an OR gate. Also, when all the a's are 1's and the threshold is n, the equation represents an AND gate. These relations are shown symbolically in Fig. 2-32. The complementary outputs of these threshold gates of course would produce the NOR and NAND relationships. When the weights and threshold are chosen differently, the gate can realize more complicated logical functions.

As an example, we will consider the following Boolean function:

$$F_1 = X_1 + X_2X_3 \tag{2-33}$$

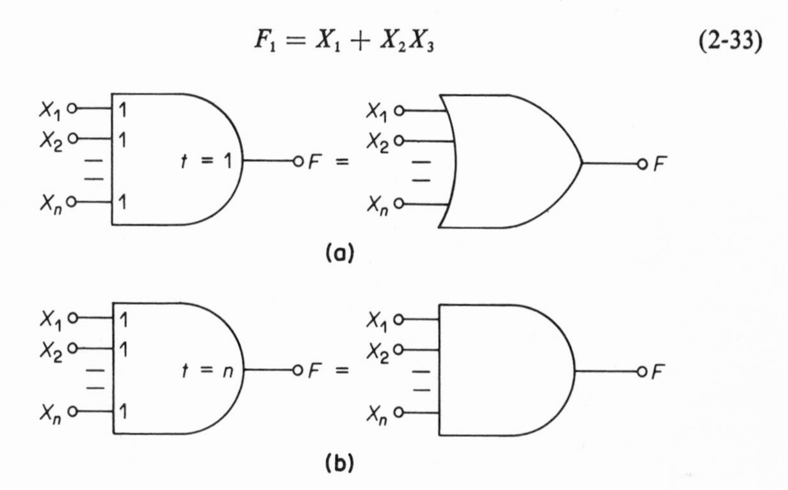

FIGURE 2-32. (a) Threshold Gate Equivalent to OR Gate. (b) Equivalent to AND Gate.

This function can be realized with a single threshold gate with $t = 2$ and the following weights:

$$F_1 = \langle 2X_1 + X_2 + X_3 \rangle_{t=2} \tag{2-34}$$

We can verify Eq. (2-34) by noting that when X_2 and X_3 are **1**'s the sum is 2, which gives a logical **1** output; or if X_1 alone is a **1**, the weighted value is 2, which also gives a logical **1** output. Two gates are required to realize this function with AND/OR logic circuits, whereas only one threshold gate is required.

As another example, we will consider the following function:

$$F_2 = X_1X_2 + X_3X_4 \tag{2-35}$$

which can be implemented with two AND gates and one OR gate. The two-input AND function, X_1X_2, can be realized using threshold logic as

$$Y_1 = X_1X_2 = \langle X_1 + X_2 \rangle_{t=2} \tag{2-36}$$

Substituting Y_1 for X_1X_2 in Eq. (2-35) gives

$$F_2 = Y_1 + X_3X_4 \tag{2-37}$$

Now, by comparison with Eq. (2-33), and using the results of Eq. (2-34), we obtain

$$F_2 = \langle 2Y_1 + X_3 + X_4 \rangle_{t=2} \tag{2-38}$$

Finally, substituting Eq. (2-36) into Eq. (2-38) gives

$$F_2 = \langle 2\langle X_1 + X_2 \rangle_{t=2} + X_3 + X_4 \rangle_{t=2} \tag{2-39}$$

A logic diagram of this function using both AND/OR gates and threshold gates is shown in Fig. 2-33.

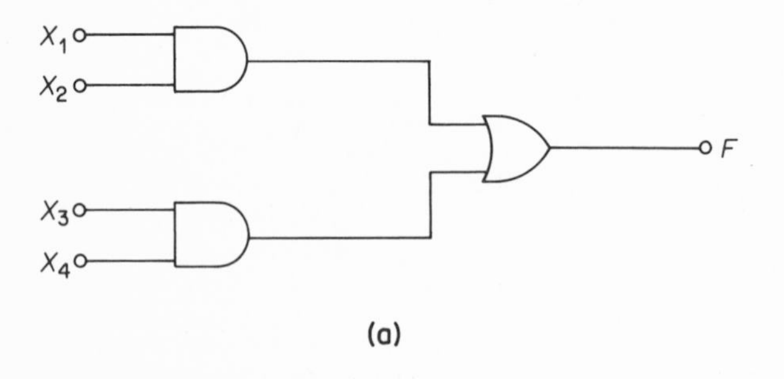

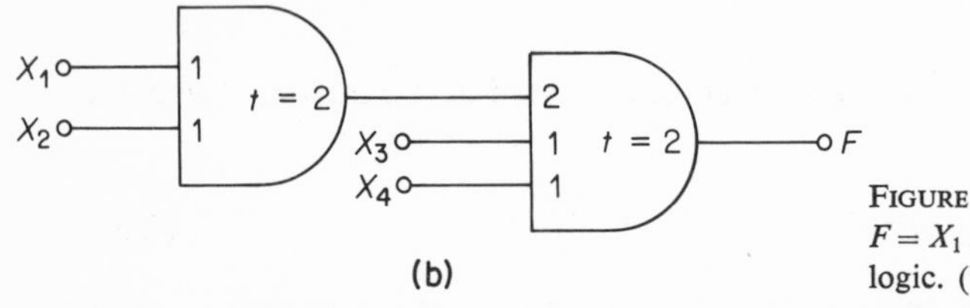

FIGURE 2-33. Implementation of $F = X_1 X_2 + X_3 X_4$. (a) AND/OR logic. (b) Threshold logic.

Threshold circuits can be constructed with resistor summing networks and semiconductor threshold circuits such as differential amplifiers. The gates also can be fabricated using magnetic circuits, where the weighting factors are represented by different numbers of turns on the magnetic core. Threshold gates require stringent component tolerances. The actual tolerances depend upon the resolution required in the circuit. For example, if the various input permutations to a threshold gate result with a sum of all integral values up to a maximum summation of 10 then the resolution required by the threshold circuit is 1 part in 10. This is not difficult to achieve using standard circuit techniques. Using a 10-V reference supply, this is a resolution of 1 in 10 V. However, if a threshold gate for a logic function requires a resolution of 1 part in 100, then the circuit requires precision and stable components to meet the requirements. Using a 10-V power supply, this is a resolution of 0.1 in 10 V and voltage drifting in the circuit becomes a severe problem. In addition to the precision circuit components, very stable reference voltages are required for the threshold reference and for limiting or reshaping the input signals.

It is not practical to construct threshold gates using discrete circuit components because of the circuit complexity and stability required. However, advances in integrated-circuit technology may mass produce inexpensively stable and reliable threshold gates. The use of threshold gates promises significant reductions in numbers of gates and interconnections required in logic systems.

REFERENCES

1. Millman, J., and Taub, H., *Pulse, Digital, and Switching Waveforms*, Chap. 9, McGraw-Hill Book Company, New York, 1965.
2. Harris, J. N., Gray, P. E., and Searle, C. L., *Digital Transistor Circuits*, Chaps. 5, 6, and 7, John Wiley & Sons, New York, 1966.
3. Strauss, L., *Wave Generation and Shaping*, McGraw-Hill Book Company, New York, 1960.
4. Littauer, R., *Pulse Electronics*, Chap. 9, McGraw-Hill Book Company, New York, 1965.
5. Lewis, P. M., and Coates, C. L., *Threshold Logic*, John Wiley & Sons, New York, 1967.

PROBLEMS

1. The circuit shown in Fig. P2-1 is a diode-resistor OR gate driving an AND gate. Assume that silicon diodes are used with the following conditions: zero forward resistance, infinite reverse resistance, and a forward voltage drop of 0.7 V. The following definitions apply for inputs at A, B, and C: **0**, 0.3 V at zero source resistance and **1**, 5 V at 1 K source resistance.
   (a) Will the circuit perform the required function for all possible input conditions for *A*, *B*, and *C*?
   (b) Discuss the operation of the circuit if $V_2 = 0$ V.
2. Note the inverter circuit shown in Fig. P2-2. At very high temperatures the effects of the reverse collector saturation current, I_{CBO}, cannot be neglected. Assume the transistor is still cut off when $V_{BE} = 0$ V. With the input at 0 V, at cutoff $I_1 = 0$ and I_{CBO} flows opposite I_B and in the same direction as I_2.

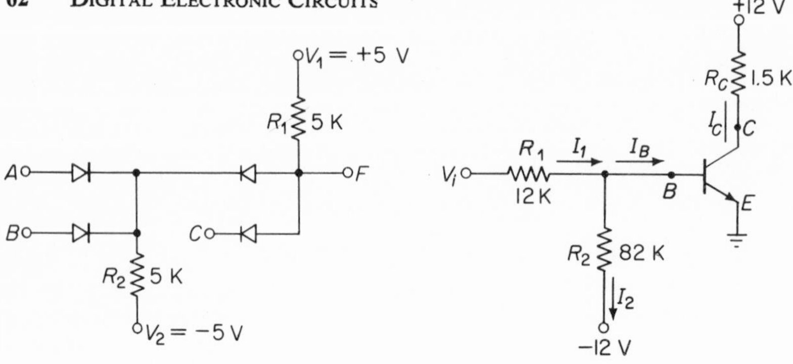

FIGURE P2-1

FIGURE P2-2

(a) Find the value of I_{CBO} that just brings the transistor to the point of cut-off.
(b) At 25°C the value of I_{CBO} is 10 nA for a given silicon transistor. Find the maximum operating temperature for this inverter. Assume that the value of I_{CBO} doubles for every 10°C rise in temperature.

3. The network in Fig. P2-3 is an ac voltage divider. Resistor R_2 might represent the base bias resistance of a transistor circuit and C_2 the input capacitance of the transistor. R_1 is a current-limiting input resistor, and

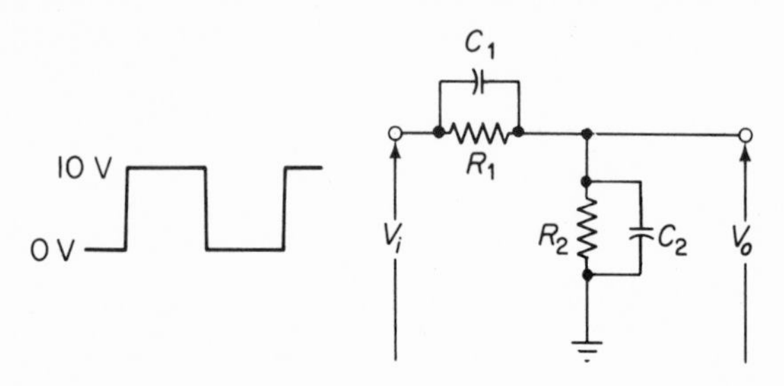

FIGURE P2-3

C_1 is a speedup capacitor. With the square wave input voltage waveform shown, assume a zero source resistance and sketch the output waveform for the following conditions:

$$R_1C_1 < R_2C_2 \qquad R_1C_1 > R_2C_2 \qquad R_1C_1 = R_2C_2$$

In all cases assume that the period of the waveform is much larger than the time constants.

4. Note the NOR circuit in Fig. P2-4. Assume A, B, and C are driven from zero source resistances and that the diodes are ideal. Q_1 is a silicon transistor. Define state **1** as +9 V to +12 V and state **0** as 0 V to +2 V. Making reasonable assumptions for V_{BE}(sat), etc.
(a) What is the minimum h_{fe} to assure that the transistor is in saturation in the **0** output state?
(b) How much I_{CBO} is permissible in the **1** output state?
(c) What is the minimum permissible value of R_L, i.e., maximum loading?

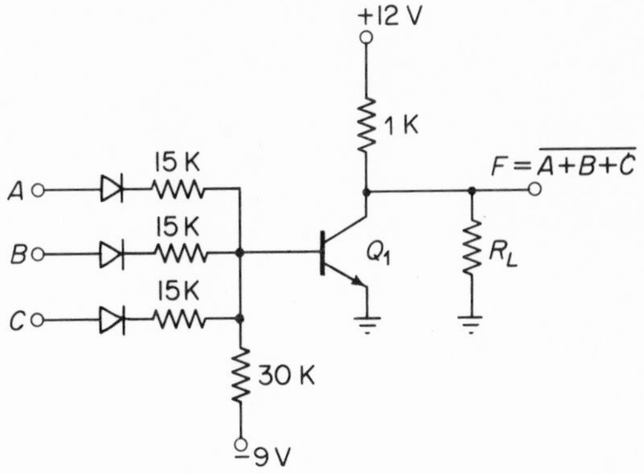

FIGURE P2-4

5. The circuit in Fig. P2-5 uses silicon transistors, V_{CE} (sat) = 0.25 V and V_{BE}(sat) = 0.7 V.
   (a) Show that this circuit perfoms the NAND operation.
   (b) Calculate the collector currents in each transistor when the inputs are high. Assume the inputs are taken from the output of a similar gate. What are the output logic levels?
   (c) Show how to modify this circuit so that it becomes a NOR gate for negative logic.

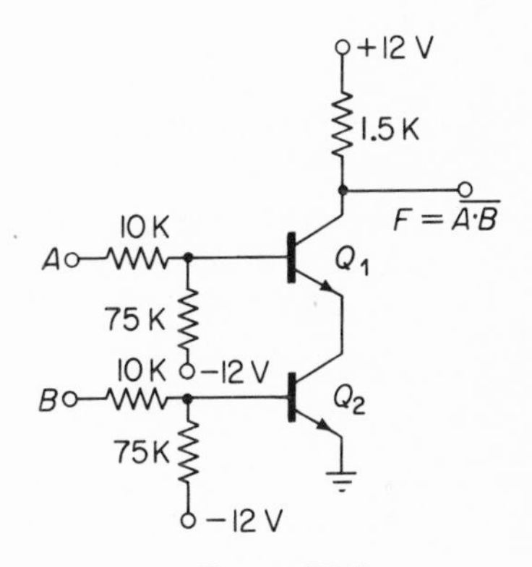

FIGURE P2-5

6. The circuit shown in Fig. P2-6 uses silicon diodes and a silicon transistor.

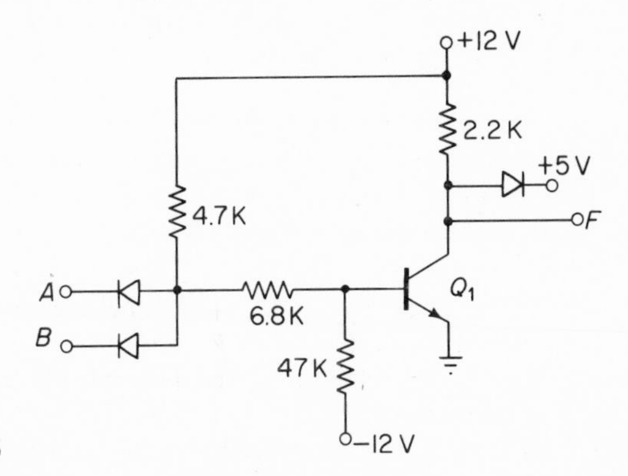

FIGURE P2-6

The output is clamped through a diode to 5 V. The inputs A and B are obtained from the F outputs of similar gates.

(a) What are the logic levels? Take junction voltages into account. Make reasonable assumptions and state them.

(b) Verify that the circuit satisfies the NAND operation. Assume $(h_{FE})_{min} = 15$.

(c) What is the maximum allowable value of I_{CBO}?

7. The outputs from two NAND gates, identical to Fig. P2-6, are to be combined by an OR operation. In lieu of using a separate OR gate to perform this operation, can the collectors of the two output transistors be wired together to perform this function? Analyze the operation of the circuit with the additional loading.

8. The circuit shown in Fig. P2-8 uses silicon transistors. The inputs A and B are obtained from the output F of similar gates. Neglect I_{CBO} and the forward-biased junction voltage.

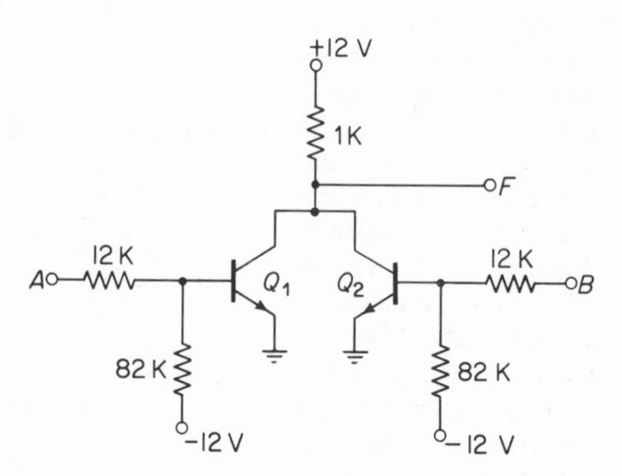

FIGURE P2-8

(a) What are the logic levels?

(b) Verify that the circuit satisfies the NOR operation for positive logic. What is $(h_{FE})_{min}$?

(c) Show how to modify this circuit so that it becomes a NAND gate for negative logic.

9. Resistor logic is often used to construct a majority gate for use in threshold logic, as shown in Fig. P2-9. The binary inputs, with zero source resistance, are summed in a resistor network and applied to a threshold circuit. The threshold circuit is an ideal Schmitt trigger or voltage comparator (Chap. 5). The voltage transfer characteristics for the threshold circuit are shown in Fig. P2-9. The output is 0 V until the threshold value is reached, a V, and then the output is +5 V, a **1**.

(a) For the given circuit, determine the value of the threshold voltage a for the output to be a **1** when two or more of the inputs are **1**'s.

(b) Give a Boolean function that can be implemented with this circuit.

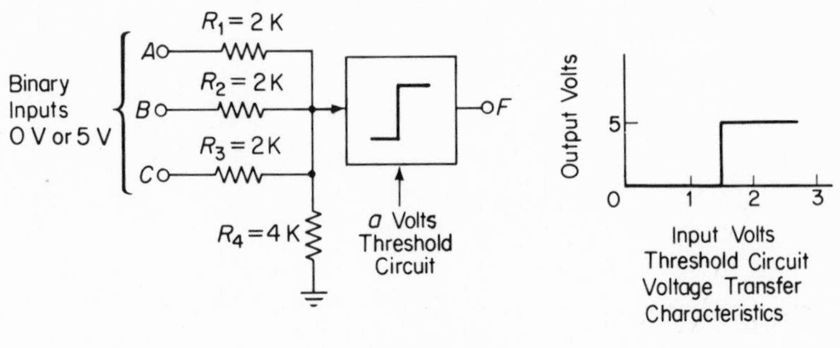

FIGURE P2-9

10. Design a circuit as shown in Fig. P2-9, i.e., determine values for R_1, R_2, R_3, R_4 and a to satisfy the following function:

$$F = A + BC$$

CHAPTER 3

INTRODUCTION TO SEQUENTIAL CIRCUITS: THE FLIP-FLOP

The basic logic functions used in the analysis and synthesis of combinational circuits are developed in the first two chapters. From the discussion it should be apparent that combinational logic is actually a decoding function. Boolean relationships may be used to describe the input logic that results in a prescribed output function. The output function is independent of all prior input or output conditions; i.e., it is uniquely determined by the current set of inputs. In terms of its logic, then, the combinational circuit is without memory and cannot be used to store digital data.

With the addition of appropriate feedback, a new class of logic functions, called *sequential circuits*, may be developed. The output of sequential circuits as the name implies, are determined by a sequence of input conditions. To digress for a moment, a somewhat humorous and excellent analogy to the relationship between sequential and combinational circuits may be made with the operation of combination and tumbler locks. Note that in reality the tumbler lock is a combinational lock, requiring a given input code (teeth on a key) to open, whereas a combinational lock is a sequential lock since it requires a specific sequence of input motions to operate. Also, the combination lock must possess the means to mechanically store several input motions prior to the last input. Likewise, a sequential circuit must store in its internal circuitry prior inputs (and outputs) that relate to its logical functioning. A means therefore must be provided to store or remember input sequences in the internal states of the circuit.

3-1 THE FLIP-FLOP AS A MEMORY UNIT—THE BIT

The flip-flop is a sequential circuit that functions as a basic logic memory element. It has two distinct states of equilibrium; therefore, the flip-flop may be used as a single binary-digit storage device. The two states of equilibrium are referred to as the *stable states* and are called by various names such as HIGH and LOW, SET and RESET, ONE and ZERO, UP and DOWN, or TRUE and FALSE. All these terms are used in the literature, have the same basic meaning, and may be used interchangeably; e.g., if one states that a flip-flop is SET or that the output is HIGH or TRUE or that the flip-flop is in the ONE state, the meanings are identical.

In binary digital systems the basic unit of memory is called the *bit*, derived from "binary digit." In terms of information content, the appearance of a single binary digit has associated with it one bit of information. (In a strict sense this definition is true only when the probability of a **1** or **0** being true is equal.) Therefore, the term bit is used synonymously with the storage capacity of any bistable element such as a flip-flop.

3-2 STATE DIAGRAMS

In order to give a more orderly description of input sequences, internal states, and output states of sequential circuits, we will now make use of a graphical tool known as the *state diagram.* State diagrams are useful in representing the behavior of sequential circuits. In a state diagram nodes are used to represent the different internal and external states of a circuit, and connections between nodes are used to show the transition between states with different input signals.

We will illustrate the use of state diagrams by first considering the operation of a binary element. The RS flip-flop is referred to as RESET when in the **0** state and SET when in the **1** state. Regardless of the current state that the flip-flop is in, it will be established in the **1** state by a logical **1** input to the set (S) input. We symbolize this input by the binary designation (10). Likewise, energizing the reset (R) input with a logical **1** always will establish the flip-flop in the **0** state. The reset input is designated (01). With logical **0** inputs to both S and R, the flip-flop *rests;* i.e., it retains its current state. The rest input designation is (00). By means of the simple diagram of Fig. 3-1, the operation of the RS flip-flop is indicated.

A more complex sequential function will now be discussed to illustrate further the use of state diagrams. Consider a coded sequential switch that is required to yield an output whenever the three inputs *A*, *B*, and *C* are received in sequence. In this system there is no time limitation imposed on the arrival or intervals between inputs. The only requirement is on the sequence of arrival. (It is possible to design a clocked circuit where the inputs are required to arrive at specific time intervals.) Any out-of-sequence input, such as *A-B-B-C* or *A-C-A-B-A-C*, will not produce an output.

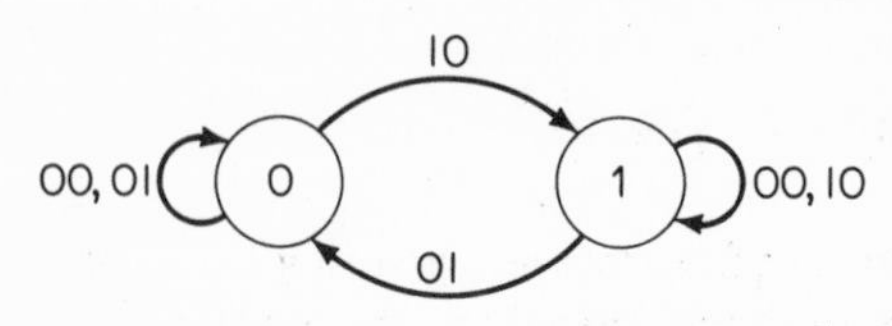

FIGURE 3-1. State Diagram for the *RS* Flip-Flop.

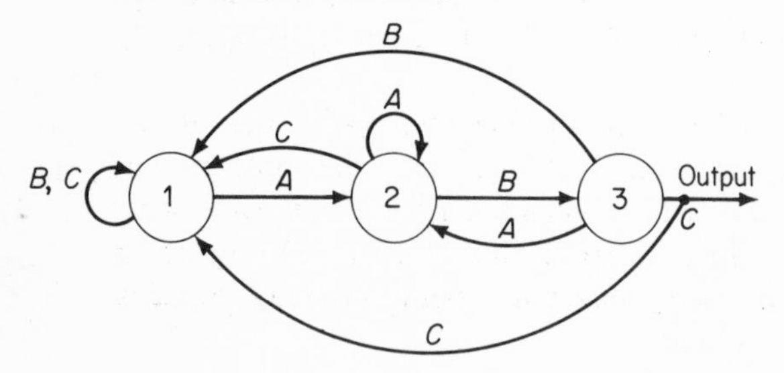

FIGURE 3-2. The State Diagram for a Coded Sequential Switch.

Shown in Fig. 3-2 is the state diagram for the required sequential switch. Proper sequential inputs of *A*, *B*, and *C* change the internal states from **1** to **2**, then to **3**, and back to **1** again, in addition to yielding an output. Any out-of-sequence *B* or *C* input returns the switch to the **1** state, whereas an out-of-sequence *A* will return the circuit or leave it in the **2** state. For example, an input sequence of *AABCBABACABCBA* will yield two outputs and return the circuit to the **2** state.

3-3 THE *RS* FLIP-FLOP

The *RS* flip-flop, whose state diagram is shown in Fig. 3-1, has several standard symbols, two of which are given in Fig. 3-3. There are two external inputs, *S* and *R*, and two outputs, *Q* and $\bar{Q}$, which are complements. When the flip-flop is *set* or in the **1** state, the *Q* output is TRUE, and when *reset* or in the **0** state, the $\bar{Q}$ is TRUE. A **1** input at *S* sets the flip-flop, and a **1** input at *R* resets the flip-flop. With zero inputs to both *R* and *S*, the flip-flop rests; i.e., the circuit retains its current state. A simultaneous **1** input to both *R* and *S* is not logically defined and is not normally permitted. Actually, in most circuit configurations simultaneous **1** inputs will force both the *Q* and $\bar{Q}$ outputs to **0**. This is logically contradictory and could cause errors in subsequent functions. If simultaneous **1**'s are applied, the next state of the flip-flop will be determined by the input that is returned to **0** first. If both inputs are returned to **0** simultaneously the output is theoretically indeterminate, whereas in practice inequalities in component parameters will bias the outcome.

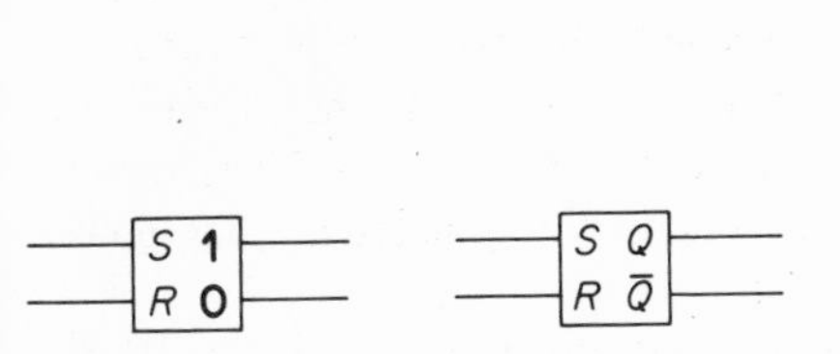

FIGURE 3-3. Standard Symbols for the *RS* Flip-Flop.

FIGURE 3-4. *RS* Flip-Flop Construction Using NOR Gate Logic.

The *RS* flip-flop can be designed by simply interconnecting a pair of two-input NOR gates, with appropriate feedback, as shown in Fig. 3-4.

Observe that with the flip-flop in the reset state ($\bar{Q} = \mathbf{1}$) and with $R = S = \mathbf{0}$ at the inputs, a **1** is fed back to gate *B*, forcing *Q* to **0**. Likewise, a **0** is fed back to the input of gate *A* and, in combination with **0** at the *S* input, holds $\bar{Q}$ at **1**. Also, by the symmetrical nature of the circuit we see that a set flip-flop will remain set when $R = S = \mathbf{0}$. This basic property of resting in either state enables the flip-flop to store one bit of information. Next let us consider the operation with inputs of $S = \mathbf{1}$ and $R = \mathbf{0}$. Regardless of the previous state, $\bar{Q}$ is forced to **0** when the *S* input to NOR gate *A* is **1**. The two inputs to gate *B* are now both **0**; therefore, the output at *Q* is **1** and the flip-flop is set. With the *Q* output at **1**, $\bar{Q}$ is kept at **0** by means of the feedback from *Q* to gate *A*, and the flip-flop remains set. Likewise, by circuit symmetry the flip-flop will reset with inputs of $S = \mathbf{0}$ and $R = \mathbf{1}$ (if previously set). If already reset, inputs of $S = \mathbf{0}$ and $R = \mathbf{1}$ will have no effect; i.e., the flip-flop remains reset.

A modified form of the *RS* flip-flop can be designed with NAND logic, as shown in Fig. 3-5. Operation of the circuit is similar to the NOR gate flip-flop, except that the circuit rests when the inputs at *a* and *b* are HIGH. Therefore, the inputs are labeled $\bar{S}$ and $\bar{R}$ to conform to the logical definitions of the *RS* binary element.

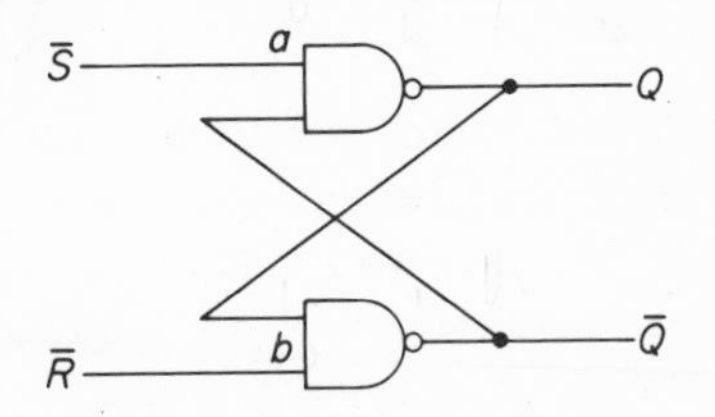

FIGURE 3-5. Flip-Flop Designed with NAND Logic.

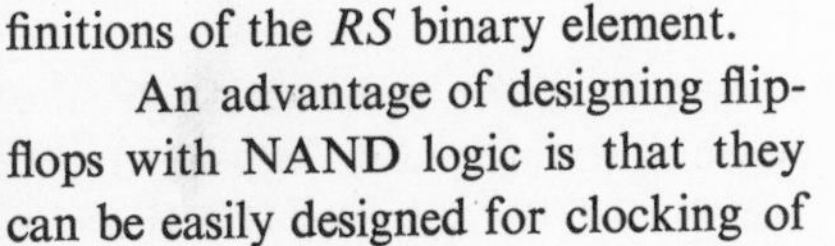

An advantage of designing flip-flops with NAND logic is that they can be easily designed for clocking of the inputs. Clocked operation, known also as synchronous operation, is essential for the proper operation of complex digital systems.

3-4 CLOCKING, SYNCHRONOUS OPERATION

All digital circuits have a finite response time, referred to as the propagation delay, that is associated with the time required for the output to respond to changes in the input conditions. Propagation delays in digital circuits vary from 2 to 3 nsec for high-speed emitter coupled logic (ECL) integrated circuit gates to tens or hundreds of nanoseconds for more complex functions and slower types of integrated circuits such as DTL. Also, the propagation time varies from unit to unit and is a function of many other variables such as operating temperature, fan-out, interconnection wiring practice, and power supply voltage.

Consider now a simple combinational function, $F = A(\bar{B} + C)$. If, initially, $A = B = C = \mathbf{0}$, the function is $F = \mathbf{0} \cdot (\mathbf{1} + \mathbf{0}) = \mathbf{0}$.

Now let the three input variables be inverted simultaneously at time t_x; then $F(t \geqslant t_x) = \mathbf{1} \cdot (\mathbf{0} + \mathbf{1}) = \mathbf{1}$. The three input variables, however, may be

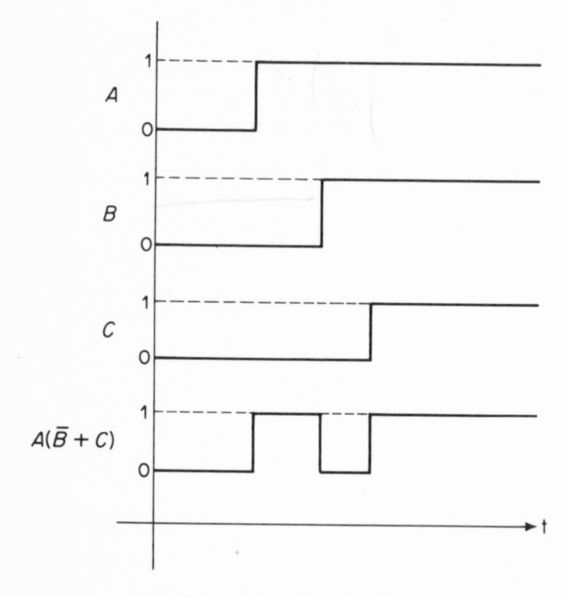

FIGURE 3-6. Effect of Differences in Arrival Time on the Output of a Combinational Circuit.

derived from circuits with different delay factors and therefore have different arrival times. Assume e.g., that A goes HIGH first, followed by B, and then by C. This is shown in the timing diagram of Fig. 3-6. The result is an erratic output waveform for $A(\bar{B} + C)$ that could cause computational errors or other difficulties in subsequent circuits.

Synchronous operation is achieved by gating the propagation of the variables from stage to stage with a sequence of timing pulses derived from a master oscillator (clock). The clock is normally LOW and goes HIGH for τ sec, usually at the start of each clock period (P). The period of the clock pulses is referred to as the bit-time and the various bit-times may be numbered, i.e., bit-time n, bit-time $n + 1$, etc. This is indicated in Fig. 3-7(a).

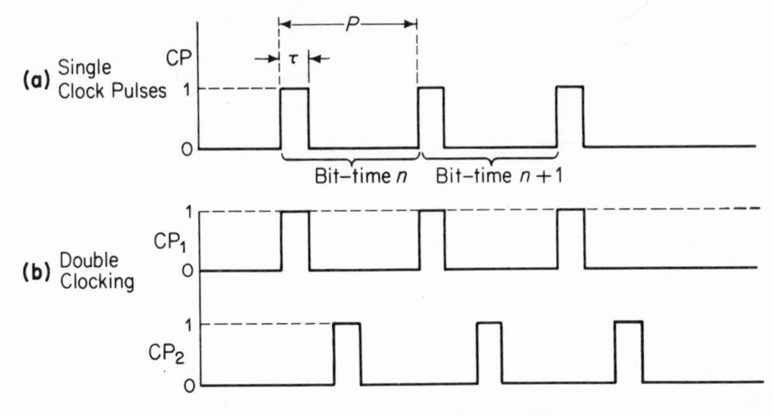

FIGURE 3-7. Typical Clock Pulses for Timing.

In a properly designed system input data to a circuit is not permitted to change when the clock is HIGH. The inputs to a circuit during bit-time n then will be *enabled* by the clock pulse at the start of bit-time $n + 1$. This will determine the outputs during bit-time $n + 1$ when the clock is LOW. Input variables and output functions at different bit-times are indicated by subscripts; e.g., in a clocked *RS* flip-flop, if $R^n = S^n = \mathbf{0}$, $Q^{n+1} = Q^n$; and if $R^{n+2} = \mathbf{1}$, and $S^{n+2} = \mathbf{0}$, then $Q^{n+3} = \mathbf{1}$.

Also, in many sequential circuits *double clocking* is required to prevent the feedback paths from causing *race-around* conditions. In Sec. 3-7 the use of double clocking in the *JK* flip-flop is described.

3-5 THE CLOCKED *RS* FLIP-FLOP

By using four NAND gates the clocked *RS* flip-flop shown in Fig. 3-8 may be developed. When the clock is LOW, $\bar{S}$ and $\bar{R}$ are **1**'s and the flip-flop rests. On application of a clock pulse, gates A and B are enabled and the state of the flip-flop is set or reset as a function of the R and S inputs at

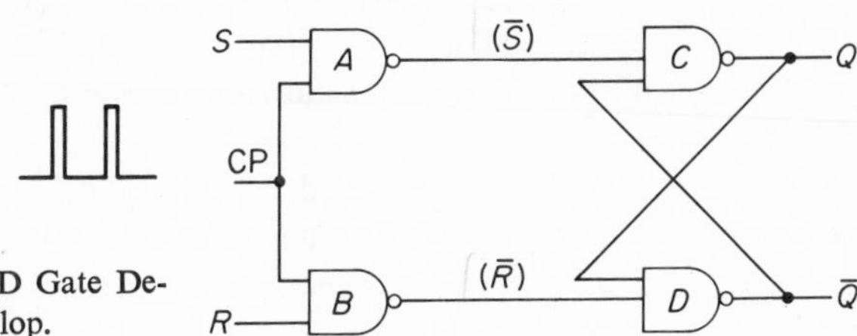

FIGURE 3-8. A NAND Gate Design of the *RS* Flip-Flop.

that time. This determines the state of Q and $\bar{Q}$ during the remainder of the bit-time. The operation of the clocked RS flip-flop may now be described precisely by means of the table of Fig. 3-9.

Input During Bit–Time n		Output at Bit–Time $(n+1)$
R^n	S^n	Q^{n+1}
0	0	Q^n
0	1	1
1	0	0
1	1	Undefined

FIGURE 3-9. Truth Table of Clocked RS Flip-Flop.

Inspection of the table in Fig. 3-9 shows that Q^{n+1} (the output state of Q during bit-time $n + 1$) is a **1** when S^n is **1** or when R^n is **0** and Q^n is **1**. This may be expressed by the following Boolean function:

$$Q^{n+1} = S^n + \overline{R^n} \cdot Q^n \qquad (3\text{-}1)$$

Eq. (3-1) is called the *difference equation* for the function. Logic circuits that satisfy all of the difference equations are a necessary condition for the synthesis of a sequential circuit. Although the foregoing was a relatively simple example, the procedure is general and we will use it extensively in the design of more complex sequential circuits.

3-6 CIRCUIT DESIGN OF THE RS FLIP-FLOP

With the present availability of competitively priced integrated circuits, the trend in the design of new digital systems is to use integrated circuits. However, in many situations the circuit designer is still required to use or understand discrete component design techniques. Some of the factors that the design engineer must consider in choosing between discrete components and integrated circuits are:

1. The availability of suitable power supplies. This is especially true when making modifications to existing equipments where the power supplies are fixed.
2. Compatibility of input and output logic levels at the interface with the new digital circuits.
3. Timing considerations, such as speed of response and choosing between asynchronous and clocked systems.

We will first examine some of the design details of a discrete component RS flip-flop circuit and then the considerations in using a commercially available integrated circuit flip-flop.

Discrete Component *RS* Flip-Flop

Referring to Fig. 3-4 we see that the *RS* flip-flop can be implemented by simply interconnecting two NOR gates. The design of the flip-flop reduces then to the design of a suitable two-input NOR gate, such as is shown in Fig. 3-10.

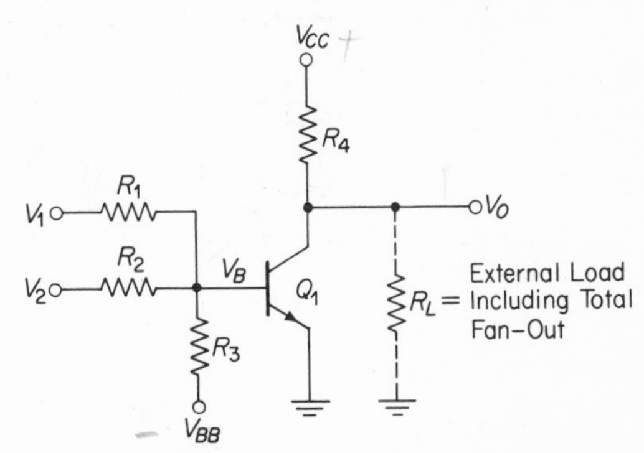

FIGURE 3-10. Two-Input NOR Gate Discrete Component Design.

The circuit in Fig. 3-10 uses positive logic with V_{CC} positive, V_{BB} negative, and an npn transistor. To operate properly transistor Q_1 should remain cut off when the inputs V_1 and V_2 are both at logical **0**. Likewise, when either V_1 or V_2 (or both) is at logical **1**, Q_1 should be in saturation. When Q_1 is in cutoff, the voltage division between R_L and R_4, in conjunction with V_{CC}, determines V_0; therefore, the allowable voltage range of logical **0** and **1** must be defined and adhered to. In a typical application, with $V_{CC} = +12$ V, logical **0** is defined as 0 to +1.0 V and logical **1** as +6 to +12 V. Details of the design of similar circuits are presented in Sec. 2-2.

By interconnecting two discrete component NOR gates, as in Fig. 3-11, an *RS* flip-flop may be developed.

With worst-case conditions (i.e., including any combination of minimum to maximum spread in transistor parameters, resistor values, power supply variations, and external loading at Q and $\bar{Q}$) the flip-flop should set, reset, and rest as logically required and consistent with the defined voltage ranges of **1** and **0**.

The external loads may be either *source loads* or *sink loads*. Circuits such as OR and NOR gates (Fig. 2-8) are called source loads when they are driven by a source of current. AND and NAND gates (Fig. 2-10) are usually sink loads.

The circuit of Fig. 3-11 will now be analyzed to verify its proper functioning with the nominal component values shown. The 1N914 is a small-signal silicon diode, suitable for fast switching applications. The diode may be considered an ideal open circuit when reverse biased. In the forward direction the diode is (for small signals, forward conduction ≤ 10 mA) approximately a zero-resistance device in series with a constant voltage drop of $V_\gamma \approx +0.7$ V.

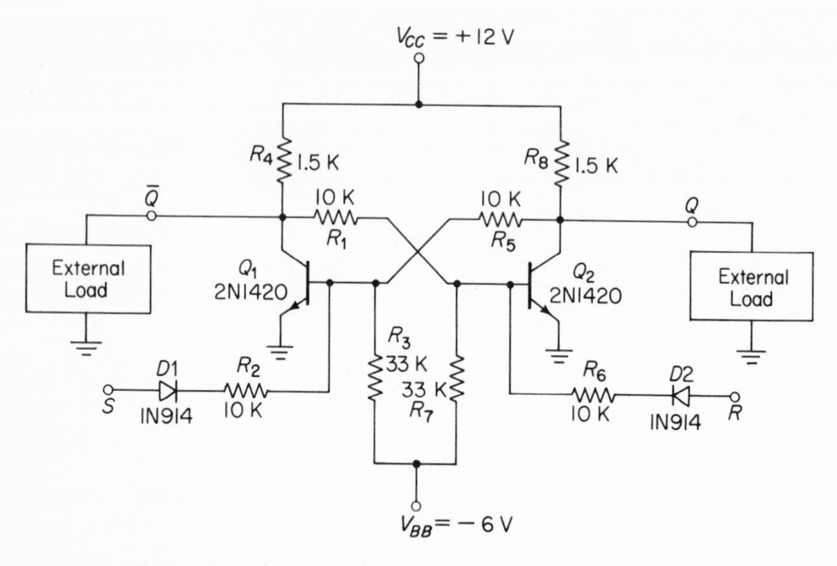

FIGURE 3-11. Discrete Component *RS* Flip-Flop Implementation.

The 2N1420 is a silicon transistor with $h_{FE} = 50$ minimum. In saturation, the output impedance is approximately 50 Ω. Also in saturation, $V_{CE} \approx 0.4$ V and $V_{BE} \approx 0.7$ V. For base currents greater than 5 mA, the saturation input resistance should be considered in addition to the fixed voltage drop of +0.7 V.

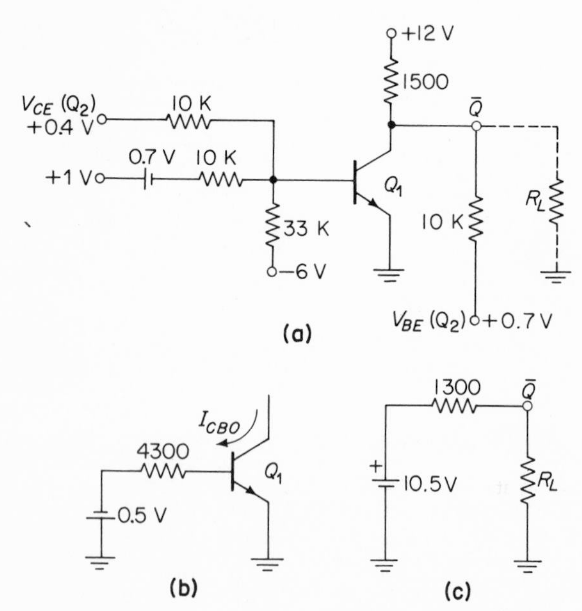

FIGURE 3-12. Input-Output Circuits of Q_1 with Flip-Flop Reset.

Stability of the device with $R = S = \mathbf{0}$ will now be determined. Since the circuit is symmetrical it is not restrictive to assume that the binary is in the **0** state. Verification that the circuit is stable and resting is then equally applicable to the **1** state. In the **0** state Q_1 should be in cutoff and Q_2 in saturation. To show this the significant inputs and outputs of Q_1 are redrawn in Fig. 3-12(a). With the $\bar{Q}$ output (Q_1) at logical **1**, the worst-case load is a source load, as indicated by R_L to ground.

The Thévenin equivalent of the input to the base of Q_1 is shown in Fig. 3-12(b). Note that the base-emitter junction of Q_1 remains reverse biased, with a worst-case **0** input of $+1$ V, provided the collector-to-base leakage current I_{CBO} is less than $0.5/4.3\text{ K} = 115\ \mu\text{A}$. With $I_{CBO} \leq 0.1\ \mu\text{A}$ at $+25°\text{C}$, and assuming the leakage current doubles every $+10°\text{C}$, the transistor will remain cut off at junction temperatures $\leq +125°\text{C}$.

The Thévenin equivalent of the output circuit is given in Fig. 3-12(c). By the definition of logical $\mathbf{1} = +6$ V minimum, the maximum allowable load is then

$$R_L \geq \left(\frac{6}{10.5-6}\right) 1{,}300 = 1.74\text{ K}$$

If the flip-flop is to drive gates with similar input impedances, then a unit load may be defined as 10K. The fan-out of the flip-flop is therefore five.

To determine the ON state of Q_2, the significant inputs and outputs of Q_2 are redrawn in Fig. 3-13. Note that the worst-case inputs are $+6$ V from Q_1 and 0 V at V_{in}.

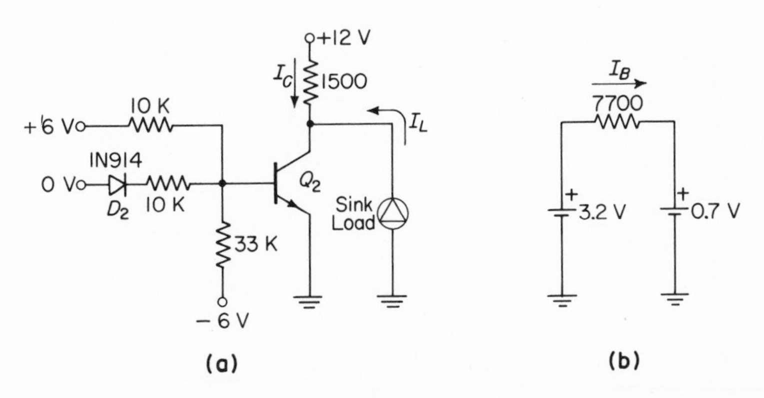

FIGURE 3-13. Input-Output Circuit of Q_2 with Flip-Flop Reset.

With $V_{BE} \geq 0$ V, D_2 is reverse biased and the Thévenin equivalent of the input to the base is as shown in Fig. 3-13(b). I_B is therefore

$$I_B = \frac{3.2 - 0.7}{7,700} = 325\ \mu\text{A} \quad \text{and} \quad I'_C(\text{sat}) = \frac{12 - 0.4}{1,500} = 7.75\ \text{mA}$$

If we again define a unit load as 10 K, then with a +12-V supply a unit sink load will be 1.2 mA. With a fan-out of six, the total saturation current of Q_2 is

$$I'_C(\text{sat}) = [7.75 + 6(1.2)] \times 10^{-3} \approx 15\ \text{mA}$$

The minimum h_{FE} required for Q_2 is therefore

$$(h_{FE})_{\text{min}} = \frac{15}{0.325} = 46$$

The safety margin with the required $(h_{FE})_{\text{min}}$ and the manufacturer's stated $(h_{FE})_{\text{min}}$ of 50 is in reality marginal. The h_{FE} of a transistor normally will vary with operating temperature and different values of I_C. Aging of the semiconductor also produces changes in its initial parameters. Finally, the tolerance of the resistors and their aging and temperature characteristics must be considered. A check should be made (see Prob. 4) with the 10- and 1.5-K resistors on the high side of their tolerance and the 33-K resistor on the low side.

The bistable characteristic of an *RS* flip-flop can be determined from the symmetry of the circuit. The analysis for the set state is identical to the reset analysis just completed. A significant aspect of the *RS* flip-flop is that it is unstable in the linear operating region with both Q_1 and Q_2 conducting. There is positive feedback from the collector of Q_2 to the base of Q_1 and from the collector of Q_1 to the base of Q_2. With no external inputs, when power is first applied to the flip-flop, the circuit instantly stabilizes in either the **0** or **1** state. Inequalities and variations in component values will bias the outcome, and a given flip-flop almost always will turn on in the same state. In many digital systems, indiscriminate turn-on of this type is undesirable and the turn-on is predetermined by applying controlling inputs to *R* or *S*.

Setting and resetting of the flip-flop can be accomplished by either turning the ON transistor OFF or the OFF transistor ON. For example, when the circuit of Fig. 3-11 is reset, it could be set (Q_1 turned ON) by applying a positive signal to the *S* input. This causes Q_1 to start conducting, which in turn reduces the drive to Q_2. The collector voltage on Q_2 then rises, forcing Q_1 to conduct harder. The circuit is said to *flip* from one state to the other—a regenerative transition process that occurs very rapidly. In a well-designed flip-flop with relatively high-frequency transistors ($f_T \geq 50$ MHz), the circuit can be reliably set and reset at a 5-MHz rate.

The reset flip-flop also can be set by applying a negative input to Q_2, causing Q_2 to stop conducting. The feedback then will turn Q_1 ON and force Q_2 into cutoff. When using positive logic, the negative pulses are usually generated by differentiating the input pulse and using the negative-going trailing edge for triggering. A typical differentiating circuit with a positive pulse blocking diode is shown in Fig. 3-14 with input-output waveforms.

Setting and resetting in this manner is considered preferable because it provides better noise immunity. Note that a transistor in saturation has a much lower gain (h_{FE}), requiring more input drive to turn it OFF than is needed to turn it ON when in cutoff; therefore, it is not as susceptible to low level noise pickup. Shown in Fig. 3-15 is an *RS* flip-flop, designed with pnp transistors. Included are the differentiating networks at the inputs. Note that the *R* and *S* inputs are reversed from the circuit of Fig. 3-11 because the input differentiating networks and diodes reverse the polarity of the input pulses. Also, since triggering of the flip-flop occurs when the input returns to zero,

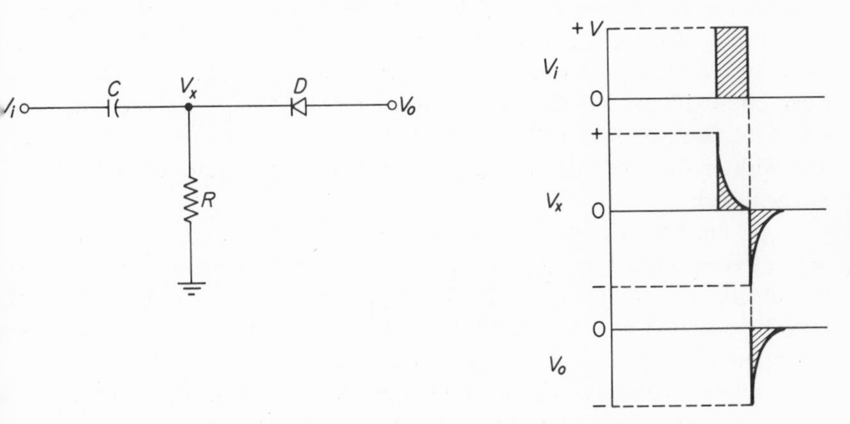

FIGURE 3-14. Differentiating Circuit with Blocking Diode.

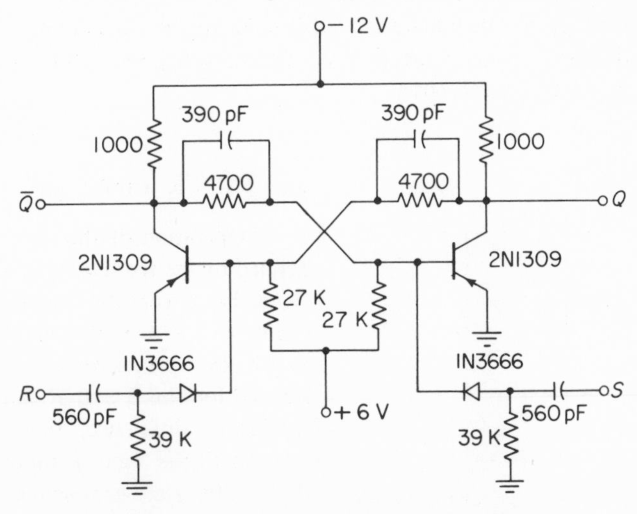

FIGURE 3-15. PNP Transistor *RS* Flip-Flop.

transitions occur at the end of an input pulse. Triggering in this manner is referred to as *trailing edge logic*. This leads to an important consideration in the design of digital systems, i.e., consistency in the choice of interfacing circuits. In most systems we would encounter difficulties if one circuit operated on the leading edge of a pulse and another on the trailing edge.

Integrated Circuit *RS* Flip-Flop

When designing with integrated circuits (IC's) we have several approaches available in specifying a suitable *RS* flip-flop. It often is convenient to simply interconnect a pair of IC NAND or NOR gates as shown in Figs. 3-4 or 3-5. Suitable two-input gates are readily available in different logic types, packaged in multiples of up to four in a single unit. A circuit designed with diode-transistor logic (DTL) gates will typically have a response time of 30–50 nsec and a fan-out of 10 and will dissipate in the vicinity of 20 mW of power. By using transistor-transistor logic (TTL) gates, switching time is reduced to 10–20 nsec, and with emitter-coupled logic (ECL) gates, switching times of under 5 nsec are achieved.

IC flip-flops, designed exclusively for *RS* operation, are not always available in every logic type because of limited usage. It is more convenient for manufacturers to supply other flip-flop types, such as the *JK* flip-flop, which may be operated as a simple *RS* type. Both synchronous and asynchronous operations are possible and are discussed in Sec. 3-7.

When a particular circuit is used repetitively and in large quantities, as in the arithmetic unit of a digital computer, it often becomes economically feasible to have multiple functions designed on a single IC chip. This design concept, called large-scale integration (LSI) holds tremendous promise in reducing size, eliminating interconnection wiring, reducing switching speeds, and eventually reducing overall costs.

3-7 THE *JK* FLIP-FLOP

Operation of the *JK* flip-flop is defined by the truth table of Fig. 3-16. Note that except for the input condition $J = K = \mathbf{1}$, the *JK* is identical to the *RS* flip-flop with J and K substituted for the S and R inputs. The significant difference, then, is that when both the J and K inputs are **1**'s the flip-flop state complements.

J	K	Q^{n+1}
0	0	Q^n
1	0	1
0	1	0
1	1	$\bar{Q}^n$

FIGURE 3-16. Truth Table for the *JK* Flip-Flop.

The *JK* flip-flop may be contructed with discrete components, integrated circuit logic gates, or on a single integrated circuit chip. There are numerous types commercially available, mainly for synchronous (clocked) operation. We will now develop some typical designs using logic gates as building blocks.

The Clocked *JK* Integrated Circuit Flip-Flop

A clocked integrated circuit *JK* flip-flop may be constructed with four NAND gates and two time delays (Δ) as shown in Fig. 3-17. The Δ delays are required, as will be seen, to prevent a *race-around* condition from occurring. Duration of the clock pulses (τ) and their period (P) are both critical.

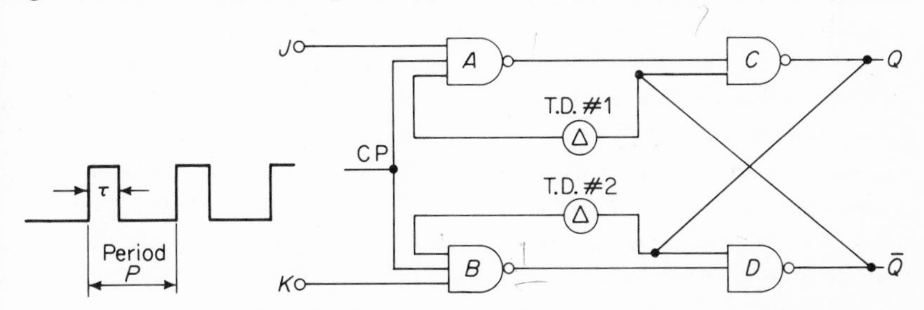

FIGURE 3-17. Clocked *JK* Flip-Flop Constructed with NAND Logic.

The required relationship between propagation delays in the NAND gates and time delays (Δ) will be developed. Finally, in a properly designed system, changes in the J and K inputs do not occur when the clock is HIGH. In circuits where asynchronous inputs occur, separate gating should be provided to inhibit the J and K inputs for the duration of the clock pulse. Changes in J and K occurring during bit time n (when the clock is LOW) then will change the state of the flip-flop, after clock pulse $n + 1$, in accordance with the truth table of Fig. 3-16.

Consider now the operation with inputs of $J = K = \mathbf{0}$. Both gates A and B are inhibited, forcing their outputs to **1**. This satisfies the rest conditions for the $\bar{R}$ and $\bar{S}$ inputs of the NAND gate *RS* flip-flop of Fig. 3-5. Therefore, the flip-flop will store its current state, set or reset, indefinitely; i.e., $Q^n = Q^{n+1}$ for $J = K = \mathbf{0}$.

The operation with inputs of $J = \mathbf{1}$ and $K = \mathbf{0}$ depends on the previous state of the flip-flop. If the flip-flop is already SET, Q is HIGH and $\bar{Q}$ is LOW. After an elapsed time of Δ sec, gate B can be enabled because of the **1** fed back from Q, and gate A is inhibited with the feedback of a **0** from $\bar{Q}$. However, with $J = \mathbf{1}$, $K = \mathbf{0}$, gate B also is inhibited due to the **0** at the K input, so that the flip-flop remains in the SET state. If, on the other hand, the flip-flop is RESET, a **1** fed back from $\bar{Q}$ enables gate A and an enabling clock pulse coupled with a **1** at J forces the output of gate A to **0**. This causes Q to change to **1** and $\bar{Q}$ to **0**, setting the flip-flop. We see, then, that $Q^{n+1} = \mathbf{1}$ for $J =$ and $K = \mathbf{0}$ during bit-time n, regardless of the previous state of the flip-flop. Also, by the symmetry of the circuit, it is apparent that $Q^{n+1} = \mathbf{0}$ for $J =$ and $K = \mathbf{1}$ during bit-time n. Any change in the state of the flip-flop during a clock pulse will now be fed back to the input gates before the next clock pulse appears. This leads to relationship (3-2).

$$\tau < \Delta < P \qquad (3\text{-}2)$$

Finally, let us consider the operation when the inputs are $J = K = \mathbf{1}$. Again, assuming that the flip-flop were SET, $Q = \mathbf{1}$, $\bar{Q} = \mathbf{0}$, gate A is inhibited and gate B is enabled. A clock pulse will then cause the output of gate B to go LOW, forcing $\bar{Q}$ to **1** and Q to **0**. Likewise, if the flip-flop were RESET gate B is inhibited and gate A enabled. Observe that the feedback to gates A and B is acting as a steering network, when both J and K are HIGH, allowing only the input that changes the state of the flip-flop to propagate through. The need for the time delays TD-1 and TD-2 should now be apparent. If the delay (Δ) is longer than the duration of the clock pulse (τ), the feedback from Q and $\bar{Q}$ to gates A and B will not change during the duration of the clock pulse and the flip-flop will function properly. Without the time delays the outputs would race around to the inputs, delayed only by the propagation time of two gates, causing the flip-flop to keep changing state as long as the clock is high. The flip-flop, with Eq. (3-2) satisfied, will then change its state with each clock pulse, yielding the required response, $Q^{n+1} = \bar{Q}^n$ for $J = K = \mathbf{1}$.

Type-*T* Flip-Flop

If the J and K inputs of the *JK* flip-flop circuit of Fig. 3-17 are tied together, the circuit will change state with each clock pulse when the input is HIGH. This operation is referred to as toggling. The single input is called the T input, and the flip-flop is referred to as a toggle or type-T flip-flop.

Dual Rank Flip-Flop

Although the circuit shown in Fig. 3-15 fulfills all the logic requirements for proper operation of the *JK* flip-flop, the two time delays in the feedback paths present difficulties of a practical nature in their manufacture. Delay networks are not easily designed or fabricated into integrated circuits. To overcome this difficulty a double-clocked flip-flop, known as the *dual rank flip-flop*, is used. The dual rank flip-flop is seemingly a complex device. In the discrete component version it would require approximately twice the number of components to construct as a single rank flip-flop. This illustrates an interesting aspect of integrated circuit technology, namely, that adding the equivalent of approximately 20 additional transistors, diodes, and resistors to a circuit may not significantly affect the product's cost. Once the masks and manufacturing processes for a given device have been perfected, the major cost factors in a mass produced unit are the number of leads, packaging, testing, and marketing of the device.

A dual rank *JK* flip-flop can be designed with eight NAND gates, as illustrated in Fig. 3-18. The circuit contains, in effect, two separate clocked *RS* flip-flops in tandem. Feedback for steering purposes is derived from the output of the second rank and supplied to the input gates of the first rank. The first rank is often referred to as the *master* and the second as the *slave.*

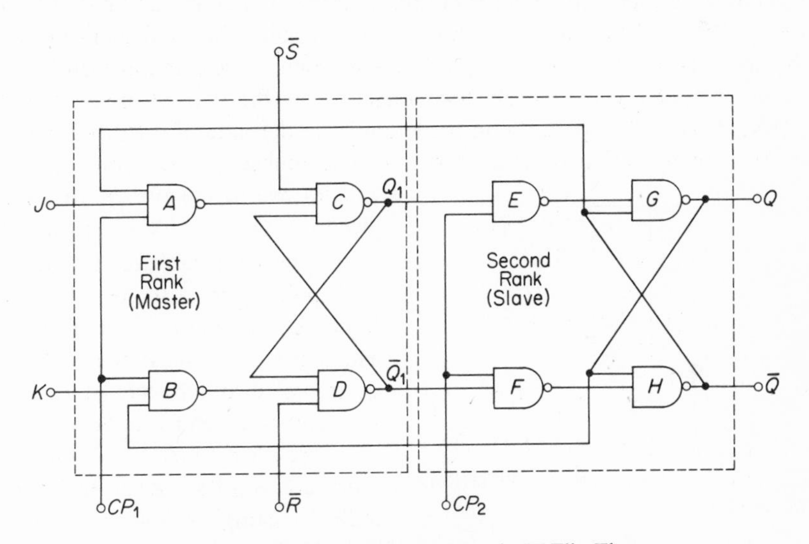

FIGURE 3-18. NAND Logic Dual Rank *JK* Flip-Flop.

To analyze the operation of the dual rank *JK*, we will first assume that the separate set and reset inputs are LOW, i.e., $\bar{S} = \bar{R} = \mathbf{1}$. With CP_1 LOW, gates *A* and *B* are inhibited, their outputs are **1**'s, and gates *C* and *D* form a stable flip-flop that is either SET or RESET. If SET, $Q_1 = \mathbf{1}$ and $\bar{Q} = \mathbf{0}$. Then, with the enabling of CP_2, the output of gate *E* will be a **0** and the output of gate *F* a **1**. This in turn will force the output of gate *G* HIGH, and the second rank will be SET. When the CP_2 line goes LOW, gates *E* and *F* are inhibited and the second rank remains SET. Again, by circuit symmetry, we see that if $Q_1 = \mathbf{0}$ and $\bar{Q}_1 = \mathbf{1}$, application of a CP_2 pulse will reset the second rank. By this mechanism we see that the second rank is in effect a slave to the state of the first rank.

Operation of the first rank is similar to that of the second rank, except that the feedback from Q and $\bar{Q}$ steers the inputs to gates *A* and *B*. When Q

is HIGH, gate B is enabled, permitting a K entry to cause subsequent resetting and when $\bar{Q}$ is HIGH, gate A is enabled to permit setting by a **1** input at J therefore, when CP_1 is enabled, the first rank will rest if $J = K = \mathbf{0}$, toggle if $J = K = \mathbf{1}$, and set or reset (or remain set or reset) if $J = \mathbf{1}$ and $K = \mathbf{0}$ or $J = \mathbf{0}$ and $K = \mathbf{1}$.

By sequentially enabling first CP_1 and then CP_2, inputs at J and K first control the state of the first rank and then transfer this state to the second rank. *Race around* is avoided when $J = K = \mathbf{1}$ by not enabling both clock inputs simultaneously. In most integrated circuits CP_2 is derived internally by inverting CP_1 with a single input NAND or NOR gate. This is illustrated in the next circuit we will analyze (Fig. 3-20). When CP_1 and CP_2 are derived from a single input, the input is normally LOW; therefore, CP_1 is LOW and CP_2 is HIGH. With the application of a clock pulse, the leading edge enables the first rank and the trailing edge the second rank. $\bar{S}$ and $\bar{R}$ can then be used to set or reset asynchronously both ranks of the flip-flop when the clock is LOW. This is of importance in predetermining the state of the flip-flop when starting an operation or when power is first applied to the circuit. Also, it enables one to use the JK as an RS flip-flop, provided CP_2 is enabled.

$\bar{R}$	$\bar{S}$	J	K	CP	Q^{n+1}
1	1	0	0	0	Q^n
0	1	0	0	0	0
1	0	0	0	0	1
1	1	0	0	1	Q^n
1	1	1	0	1	1
1	1	0	1	1	0
1	1	1	1	1	$\bar{Q}^n$

FIGURE 3-19. Truth Table for Dual Rank JK Flip-Flop.

The truth table for a dual rank JK flip-flop, with a single clock pulse input, is given in Fig. 3-19. The CP is assumed to be a pulse that is **1** for a short period at the start of each bit time and then returns to **0**.

Another version of the dual rank JK flip-flop is given in Fig. 3-20. Clocking of both ranks is accomplished by a single clock pulse at CP. The leading edge enables gate A and B, permitting the first rank (gates C and D) to be set or reset. The trailing edge of the clock pulse causes both gates A and B to go LOW and the output of NOR gate X to go HIGH. This enables the second rank and permits non-synchronous operation by means of the P_j and P_k inputs; the reader should verify this by showing that when the clock input is LOW the operation with the P_j and P_k inputs is identical to that of the simple RS flip-flop. If we do not wish to restrict the timing of preset and prereset inputs to when the clock is LOW, then by setting $P_k = \mathbf{1}$ and $K = \mathbf{0}$ the flip flop will always prereset, and with $P_j = \mathbf{1}$ and $J = \mathbf{0}$ the flip-flop will always preset. Also, by tying J and K together, the flip-flop will function as a toggle.

Several versions of the clocked IC flip-flop are available from different manufacturers. Some are suitable only for a specific application; however, most manufacturers have emphasized the universal approach in an attempt to

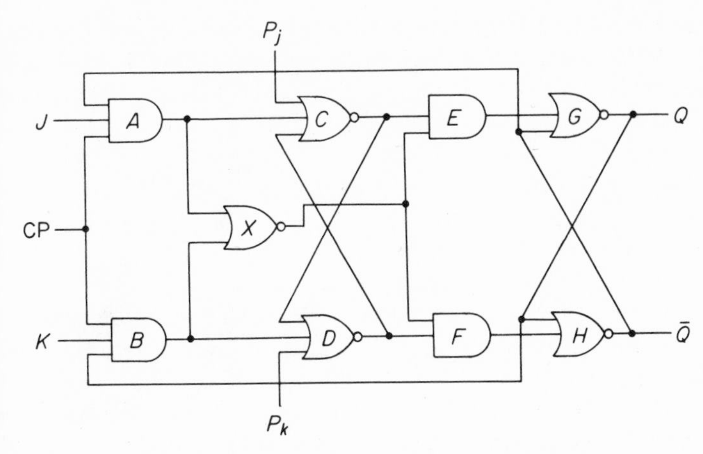

FIGURE 3-20. Alternate Dual Rank *JK* Flip-Flop.

tandardize and reduce the number of different products. The flip-flop shown y its symbol and truth table in Fig. 3-21 is a common type that may be used s a *JK* or as a clocked (gated) *RS* flip-flop.

For *JK* operation, S_2 is connected to $\bar{Q}$ and C_2 to Q. Then S_1 becomes he J input and C_1 the K input. This may be verified by inspection of the truth able of Fig. 3-21. When S_1 is connected to S_2 and C_1 to C_2, the circuit func-

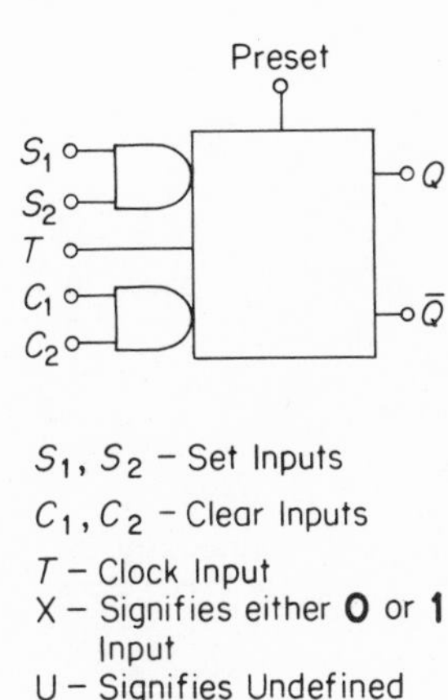

FIGURE 3-21. IC Clocked Flip-Flop with Truth Table.

Bit Time n				$n+1$
S_1	S_2	C_1	C_2	Q^{n+1}
0	X	0	X	Q^n
0	X	X	0	Q^n
X	0	0	X	Q^n
X	0	X	0	Q^n
0	X	1	1	0
X	0	1	1	0
1	1	0	X	1
1	1	X	0	1
1	1	1	1	U

tions as a clocked *RS* flip-flop. Also, S_1 and C_1 may be used as R and S inputs gated by the S_2 and C_2 inputs.

The type D is another variation of the gated flip-flop that is available and used primarily in shift registers and counters. Both these applications are discussed in Chap. 6.

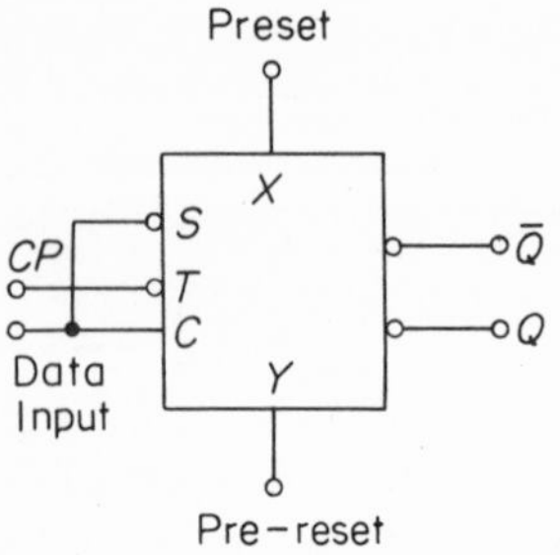

Direct Inputs Clock HIGH			
Y	X	Q	$\bar{Q}$
0	0	NC	NC
0	1	1	0
1	0	0	1
1	1	U	U

Clocked Inputs		
n	$n+1$	
Data	Q	$\bar{Q}$
1	0	1
0	1	0

NC – Signifies No Change
U – Signifies Undefined Condition

FIGURE 3-22. Type D Clocked Flip-Flop, Symbol, and Truth Tables.

The symbol and truth tables for the type D flip-flop are given in Fig. 3-22. Note that when the input is HIGH the clear input is **1** and when it is LOW the set input is **1**. Also, the clock input is inverted so that a bit at the data input is gated into the flip-flop when the clock changes from HIGH to LOW.

Metal-Oxide-Silicon Field-Effect Transistors (MOS-FETs)

Flip-flops, as well as other logic gates, are commercially available as monolithic integrated circuits designed by *MOS technology*. MOS-FETs are generally high impedance units that operate at relatively high-power supply and logic voltage levels. Typically, a power supply voltage of -25 to -30 V is used and logic levels (negative logic) are typically 0 to -3 V for logical **0** and -9 to -12 V for logical **1**.

An advantage of MOS technology is the relative ease with which complex devices can be economically produced as monolithic integrated circuits.[5] Insulated-gate field-effect transistors require fewer processing steps to fabricate into large-scale integrated circuits than equivalent bipolar transistor circuits. This leads to higher yield factors and lower costs. Complementary symmetry direct-coupled circuits have been developed that have low power dissipation and require only transistors and connecting leads to design complete circuits.

Unfortunately, logic levels of MOS-FETs are not compatible with most bipolar transistor integrated circuits and therefore interface circuits are necessary to couple the two types.

3-8 DISCRETE COMPONENT *JK*- AND *T*-TYPE FLIP-FLOPS

While integrated circuit flip-flops, other than *RS* types, are mainly operated synchronously and directly coupled, their discrete component counterparts are usually capacitively coupled and operated asynchronously. Discrete component capacitors are practical to use and, unlike integrated units, the total number of semiconductor junctions used is a significant cost consideration. Therefore, the discrete component design trend has been toward the use of two transistors and the fewest number of diodes and passive components possible. External gating of inputs could provide synchronous operation; however, the basic flip-flop is unclocked.

An *RS* flip-flop could be operated satisfactorily as a *JK*-or *T*-type if suitable steering is provided at the inputs to accommodate simultaneous **1** inputs at *S* and *R*. This is accomplished by enabling only the *R* input when the flip-flop is set and the *S* input when the flip-flop is reset. Steering is provided, as shown in Fig. 3-23, by connecting the returns of the input differentiating network resistors to Q and $\bar{Q}$.

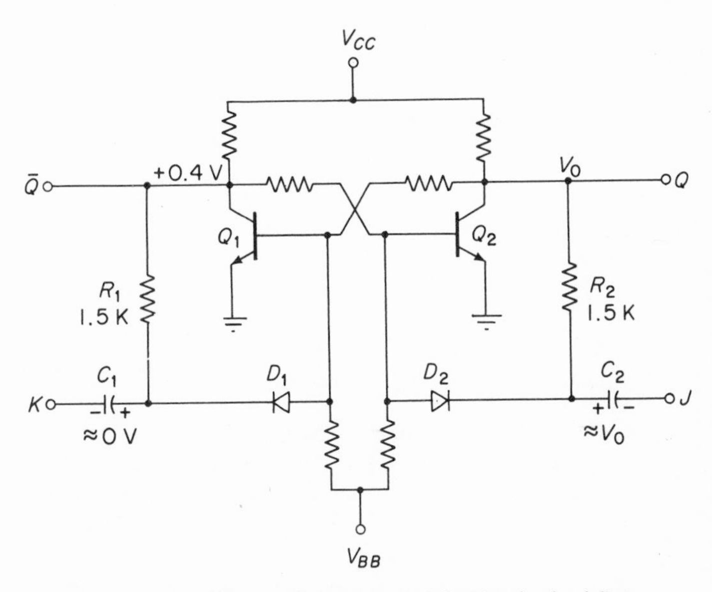

FIGURE 3-23. Discrete Component *JK* Flip-Flop in the **1** State.

To verify that the flip-flop will operate satisfactorily as a *JK*-type, let us assume the unit has been set and the two inputs, *J* and *K*, returned to **0**. Then, as indicated in Fig. 3-23, C_1 and C_2 will charge to approximately 0 V and V_0 volts, respectively. After the capacitors are fully charged, if a positive-going pulse (logical **1**) is applied simultaneously to the *J* and *K* inputs, the potential on C_2 will block the input at *J*, but the input at *K* will be differentiated by C_1 and R_1. The negative slope (trailing edge of input pulse) will then pass through D_1 and cut Q_1 off. This resets the flip-flop and causes C_2 to discharge to ≈ 0 V and C_1 to charge to $\approx V_0$. A second input (**1**) to *J* and *K* will now be steered to Q_2 as required. Connecting *J* and *K* together changes the *JK* to a type-*T* flip-flop. Except for the steering mechanism, design and operation of the flip-flop is identical to that of the *RS* flip-flop.

REFERENCES

1. Phister, M., Jr., *Logical Design of Digital Computers*, Chaps. 2 and 5, John Wiley & Sons, New York, 1967.
2. Maley, G. A., and Earle, J., *The Logical Design of Transistor Digital Computers*, Prentice-Hall, Englewood Cliffs, New Jersey, 1963.
3. Millman, J., and Taub, H., *Pulse, Digital, and Switching Waveforms*, Chap. 10, McGraw-Hill Book Company, New York, 1965.
4. Strauss, L., *Wave Generation and Shaping*, Chap. 9, McGraw-Hill Book Company, New York, 1960.
5. Wallmark, J. T., and Johnson, H., *Field-Effect Transistors*, Sec. 12, Prentice-Hall, Englewood Cliffs, New Jersey, 1966.

PROBLEMS

1. The Q output of a flip-flop is sometimes referred to as the **1** output and the $\bar{Q}$ as the **0** output. Explain the difference between the **1** output in this sense and a **1** binary state. What is the state of the **1** output of a flip-flop when the RESET input goes HIGH?
2. An *RS* flip-flop, by definition, will store its last state indefinitely when $R = S = \mathbf{0}$. What would happen if there was a power supply interruption with a circuit constructed from solid-state active elements? Could a flip-flop be constructed with NOR or NAND logic that retains its current state during a power shutdown?
3. Draw a state diagram for the NAND gate flip-flop shown in Fig. 3-8.
4. The *RS* flip-flop of Fig. 3-11 is designed with resistors that have a $\pm 5\%$ tolerance. Because of aging and temperature effects the tolerance can be expected to change to $\pm 10\%$. What is the worst-case combination of resistor tolerances in regard to stability with $R = S = \mathbf{0}$? Does the circuit function properly with this combination of tolerances (assuming fanout of 5)?
5. Redesign the *RS* flip-flop of Fig. 3-11 using pnp germanium-type 2N3323 transistors. Power supply polarities may be reversed as required. Pertinent characteristics of the transistors are $V_{CE} = 35$ V (max), $h_{FE} = 30$ (min), $I_{CBO} = 2\ \mu A/25°C$. Determine the following:
   (a) Maximum fan-out with source loads if a unit load is 10 K.

(b) The maximum safe operating temperature, assuming I_{CBO} doubles with every $+10°C$ rise.

6. The *RS* flip-flop of Prob. 4 is to operate in a system utilizing trailing edge logic. Design suitable input networks for the *R* and *S* inputs to permit this.
7. Explain the effect of attempting to preset the *JK* flip-flop of Fig. 3-18 when CP_1 is HIGH. Design an external logic circuit that would prevent this condition from occurring.
8. Verify the functioning of the *JK* flip-flop of Fig. 3-20 as a toggle type by tying the *J* and *K* inputs together. Trace through the state of each flip-flop, assuming the flip-flop is initially RESET.
9. Design a type-*D* flip-flop (Fig. 3-22) using either all NAND or all NOR gates.
10. Verify that the flip-flop of Fig. 3-23 will toggle when the *J* and *K* inputs are tied together.

CHAPTER 4

NUMBER SYSTEMS AND ARITHMETIC OPERATIONS

A major internal function of digital computers is the performance of arithmetic operations. Solutions to many problems are obtained by reducing more complex relationships to the basic arithmetic operations of iterative addition and subtraction. Multiplication and division also may be performed in this manner. This chapter discusses these operations and illustrates several methods of implementing them. Basic to an understanding of the digital techniques used is a knowledge of number systems and conversion between numbers of different radices; therefore, a discussion of number systems precedes the material on circuit implementation.

4-1 INTRODUCTION TO NUMBER SYSTEMS

The *denary* number system, commonly called the decimal or Arabic number system, is used almost universally in our everyday lives. Most people take it for granted and perform numerical operations quite readily without being concerned about the structure of the numbers they are using.

Decimal numbers are structured with 11 symbols, 10 used to represent the different integer quantities from 0 through 9 and the eleventh, the *radix point*, used to assign positional weighting. All quantities are derived by assigning different weights to the position in which the symbols 0 through 9 appear relative to the radix (decimal) point. The significant advantage of the decimal system over other number systems, such as the Roman numeral system, is the logical construction with the use of a radix point and not merely the quantity of symbols used. A natural extension, then, of decimal numbers is the generalized structure of weighted position notation shown in Eq. (4-1).

$$\cdots a_2 a_1 a_0 . a_{-1} \cdots = \cdots a_2 R^2 + a_1 R + a_0 + a_{-1} R^{-1} + \cdots \quad (4\text{-}1)$$

R, the radix or *base* of the number system, is usually an integer greater than one, and the a_i's are digits between zero and the quantity $(R - 1)$. The progression of increasing positive exponents continues as far as required to the left of the decimal point and increasing negative exponents to the right of the decimal point. The 10 most commonly used symbols for the decimal system are 0, 1, 2, 3, 4, 5, 6, 7, 8, and 9.

In the *octanary* system (known by the familiar name of *octal*) R is equal to 8; therefore, only eight symbols are required and the symbols 0, 1, 2, 3, 4, 5, 6, and 7 are used. Consider now the meaning of a number such as 423.2. Without defining the radix used, the number is meaningless since the same symbols could be used in octal, decimal, or in many other number systems. In the average man's everyday computations radix 10 is universally implied; however, in the strictest sense, all that could be inferred from the number 423.2 is that the radix is 5 or greater. Even this inference is based on the assumption that conventional numerical values were used for the symbols. If, then, any ambiguity could exist, as is the case with literature on digital technology, the radix should be stated or indicated by use of a subscript. For example, the number $(423.2)_5$ has the decimal value

$$\begin{aligned}(423.2)_5 &= (4 \times 5^2 + 2 \times 5^1 + 3 \times 5^0 + 2 \times 5^{-1})_{10} \\ &= (100 + 10 + 3 + \tfrac{2}{5})_{10} \\ &= (113.4)_{10}\end{aligned}$$

4-2 THE BINARY NUMBER SYSTEM

The bistable property of electronic switching circuits, such as flip-flops, relays, and magnetic cores makes the binary base very useful in digital data processing and computation. The binary number system by definition, has a radix of two and uses the digits **0** and **1**. A number in this system might appear, e.g., as 11011.011. The expansion of 11011.011 in radix 10 is

$$\begin{aligned}11011.011 &= \Big(1 \times 2^4 + 1 \times 2^3 + 0 \times 2^2 + 1 \times 2^1 \\ &\qquad + 1 \times 2^0 + \frac{0}{2} + \frac{1}{2^2} + \frac{1}{2^3}\Big)_{10} \\ &= (16 + 8 + 2 + 1 + 0.25 + 0.125)_{10} \\ &= (27.375)_{10}\end{aligned}$$

The radix 2 is implied and not written as a subscript in most digital systems literature. A number such as 0110010, when appearing in this book without a subscript, most probably would be $(0110010)_2$. Only when the context in which the number appears makes the radix unclear will the subscript be added. Also, in writing binary numbers it is good practice to fill in all the significant positions even if they are zeros. For example, if a digital circuit is performing an operation with 6 significant digits, then the decimal number 7 would be written 000111 and not 111.

A limited number of radices other than binary are of special interest and are used in digital computations and circuits. The most commonly used are the *quaternary* ($R = 4$), *quinary* ($R = 5$), *octal* ($R = 8$), *duodecimal* ($R = 12$), and *hexadecimal* ($R = 16$). With radices greater than 10, additional numerical symbols are needed, and usually the small letters of the alphabet starting with a are used, i.e., $a = 10, b = 11, c = 12, d = 13$, etc. To illustrate, the hexadecimal number $(b5)_{16}$ has the decimal equivalent

$$(b5)_{16} = (11 \times 16^1 + 5 \times 16^0)_{10} = (181)_{10}$$

Even a limited discussion on different radices used in digital computer technology would be incomplete without the introduction of *negative radix* notation. One could represent all numbers, positive and negative, and perform all the arithmetic operations with a number system based on a negative radix. Consider, e.g., the radix (-10). The number $(28)_{10}$ would be written as $(188)_{-10}$ and $(-88)_{10}$ as $(92)_{-10}$. A little practice will enable the reader to convert back and forth from a positive radix number to any negative radix. For example,

$$\begin{aligned}(188)_{-10} &= [1 \times (-10)^2 + 8 \times (-10)^1 + 8 \times (-10)^0]_{10} \\ &= (100 - 80 + 8)_{10} = (28)_{10}\end{aligned}$$

A characteristic of negative radices is that both positive and negative numbers appear without a polarity sign. This simplifies the notation of mixed numbers. However, the real promise of negative radices is that circuits designed to perform arithmetic operations with negative radices lend themselves to modularized construction. The use of negative radices in computer design has appeared in the literature in recent years and may play a significant role in future computer designs.[1,2]

Number System Conversion

As seen in Sec. 4-1, the reference position within a number in any number system is the radix point. The digits to the left of the radix point describe a whole number and the digit positions increase with positive powers of the radix, whereas digits to the right of the radix point describe fractional numbers and the digit positions decrease with negative powers of the radix. Since proceeding from the reference point is characterized by multiplication on one side and division on the other, it is convenient to handle the conversion of whole and fractional parts of mixed numbers separately.

A rather trivial operation on a whole number will provide insight into the development of a technique for the conversion of numbers. The operation is successive division of the number by the radix and noting that each remainder or *spillover* is a significant digit of the number, starting with the least significant. The division is continued until the quotient is a digit less than the radix. The next division produces a zero quotient and a remainder equal to the dividend. This is illustrated by taking a decimal number and dividing it by the radix 10:

$$
\begin{array}{r|l}
10 & 7653 \\
10 & 765 + 3 \times 10^0 \\
10 & 76 + 5 \times 10^1 \\
10 & 7 + 6 \times 10^2 \\
 & 0 + 7 \times 10^3
\end{array}
$$

The number is easily reformed by proper positioning of the remainders.

In a similar manner, to convert a number in one base to a number in another base, divide the given number by the new radix noting the remainders, continuously, until the quotient is zero. Arrange the remainders, first to last, as the least significant digit (LSD) to the most significant digit (MSD) of the new number with the appropriate new number symbols. As an illustration, the decimal number 293 is converted to binary.

$$
\begin{array}{r|l}
2 & 293 \\
2 & 146 + 1 \text{ (LSD)} \\
2 & 73 + 0 \\
2 & 36 + 1 \\
2 & 18 + 0 \\
2 & 9 + 0 \\
2 & 4 + 1 \\
2 & 2 + 0 \\
2 & 1 + 0 \\
 & 0 + 1 \text{ (MSD)}
\end{array}
$$

The binary number is 100100101. It is interesting to note that the reverse of this procedure could be used to convert a binary number to decimal.

In converting a decimal number to a number of a new radix the division by the radix is performed in the decimal system. Similarly, in conversion of a number in one base to a number of another base the division of the radix is carried out in the number system of the original number. Since division in other than the decimal system is awkward (unless one is familiar with it) it is more desirable to convert the number to a decimal number and then to convert it to the number with the new radix. As an example, the octal number 527 is converted to binary as follows:

$$(527)_8 = 5 \times 8^2 + 2 \times 8^1 + 7 \times 8^0 = (343)_{10}$$

$$\begin{array}{r|l}
2 & 343 \\
2 & 171 + 1 \\
2 & 85 + 1 \\
2 & 42 + 1 \\
2 & 21 + 0 \\
2 & 10 + 1 \\
2 & 5 + 0 \\
2 & 2 + 1 \\
2 & 1 + 0 \\
 & 0 + 1
\end{array}$$

The binary number is 101010111.

There is a simple relationship among binary numbers and numbers with a radix of powers of two. This is readily seen by partitioning the binary number in equal groups corresponding to the number of binary digits required to represent the digits of the new number system. For example, in the octal system with digits 0 through 7, each digit can be represented with three binary digits; therefore the binary number is partitioned in groups of three. The binary number 101010111 is partitioned as follows:

$$101/010/111 \qquad 5/2/7$$

The corresponding digits in the octal system also are shown. The octal number is then 527. Conversion from octal to binary is simply carried out in the reverse process. Because of the ease of this conversion from octal to binary and back to octal, the octal number system is widely used by computer programmers and engineers. It is convenient to use for the recording of large numbers rather than writing long strings of ones and zeros as with binary numbers and unlike the decimal system, the conversion can be performed by inspection.

In a manner similar to dividing a whole number by the new radix, the digits of a fractional number are obtained by multiplying the number to the right of the radix point by the new radix. The overflow to the left, across the radix point, corresponds to the digits in the new number system. This is illustrated in the decimal system as follows:

	0.5379×10
(MSD) 5	0.379×10
3	0.79×10
7	0.9×10
(LSD) 9	0.0

As another example, the decimal number 0.8125 is converted to binary:

	0.8125×2
(MSD) 1	0.6250×2
1	0.2500×2
0	0.5000×2
(LSD) 1	0.0000

The binary number is 0.1101. Also, in converting fractional numbers from one number system to another, it may be convenient first to convert the number to a decimal number (because of the ease of multiplying in the familiar decimal system) and then to convert to the number system with the new radix. As an example the octal number 0.5 is converted to binary:

$$(0.5)_8 = 5 \times \tfrac{1}{8} = (0.625)_{10}$$

Then,

	0.625×2
(MSD) 1	0.250×2
0	0.500×2
(LSD) 1	0.000

The binary number is 0.101. The relationship between octal and binary, or between a radix of any power of two and binary, in fractional numbers is the same as for whole numbers.

The conversion processes described above will fail if conversion is attempted across a base point. This is expected because movement of the base point alters the value of the number by the radix factor, which is different in any two systems.

To convert a mixed number between two number systems, the recommended procedure is to convert the whole and fractional parts separately and then add the results.

4-3 BINARY CODED DECIMAL NUMBERS

Digital instruments, such as voltmeters and frequency counters, and many computer systems often have their write-in and readout expressed in decimal numbers for the convenience of the operator. In order to build digital circuits that use two-state devices and are compatible with decimal requirements, it is either necessary to use conversion logic or to encode the decimal digits with binary bits. The latter method, called *binary coded decimal* (abbreviated BCD) utilizes a minimum of four binary digits for each significant decimal digit. Since only 10 of the 16 possible permutations of each four-bit representation are used, BCD systems do not use the symbols with maximum efficiency.

Decimal Digit	BCD				
	8-4-2-1	2-4-2-1	7-4-2-1	Excess-3	Biquinary 5043210
0	0000	0000	1100	0011	01 00001
1	0001	0001	0001	0100	01 00010
2	0010	0010	0010	0101	01 00100
3	0011	0011	0011	0110	01 01000
4	0100	0100	0100	0111	01 10000
5	0101	1011	0101	1000	10 00001
6	0110	1100	0110	1001	10 00010
7	0111	1101	1000	1010	10 00100
8	1000	1110	1001	1011	10 01000
9	1001	1111	1010	1100	10 10000

FIGURE 4-1. Commonly Used BCD Codes.

The 8-4-2-1 code is the most straightforward BCD representation, with the weight of each digit corresponding to the same value as in the binary number system. In Fig. 4-1 several commonly used BCD codes are given with their decimal equivalents.

To illustrate, the number $(5490.3)_{10}$ is written in the 8-4-2-1 code as

0101 0100 1001 0000 . 0011

The ease with which the 8-4-2-1 BCD number can be read as a decimal number is obvious. Also, the electronic circuitry that is required to convert each four-digit word in BCD to a 10-wire output for use in driving numerical indicators is relatively simple. Other BCD codes are used to fulfill special requirements. The biquinary code's symmetry lends itself to digital waveform synthesizers; the 7-4-2-1 code is used in exact-count codes for error detection (see Chap. 8); whereas the excess-3 code simplifies the carry logic in parallel addition (see Prob. 4-12).

4-4 BINARY ADDITION

The fundamental arithmetic operation in digital computers is the addition of multidigit binary numbers. For example, multiplication can be accomplished by programming a computer to perform successive additions. Likewise, subtraction of positive numbers is usually performed by complementing the subtrahend and adding rather than by direct logical means. Without attempting to exhaust all of the algorithms or circuits used to perform binary arithmetic, a number of fundamental approaches will be presented.

The Half-Adder

The operation of adding two positive binary digits of the same significant value is accomplished with a *half-adder*. Consider the addition of the binary digits A, the augend, and B, the addend (*and* refers to the arithmetic operation and not the Boolean AND function). The sum of A and B, where the digits A and B can each have the values of **0** or **1**, is either 00, 01, or 10. The digit of the same significant value as A and B is called the sum digit (S), and the next significant digit is the carry digit (C).

The following Boolean expressions are now evident by inspection of the truth table for the half-adder shown in Fig. 4-2.

A	B	S	C
0	0	0	0
0	1	1	0
1	0	1	0
1	1	0	1

FIGURE 4-2. Truth Table for the Half-Adder.

$$S = \bar{A}B + B\bar{A} = A \oplus B \tag{4-2}$$

$$C = A \cdot B \tag{4-3}$$

The half-adder is usually drawn as shown in Fig. 4-3(a) with the letters HA and four leads, although all three symbols shown in Fig. 4-3 are used.

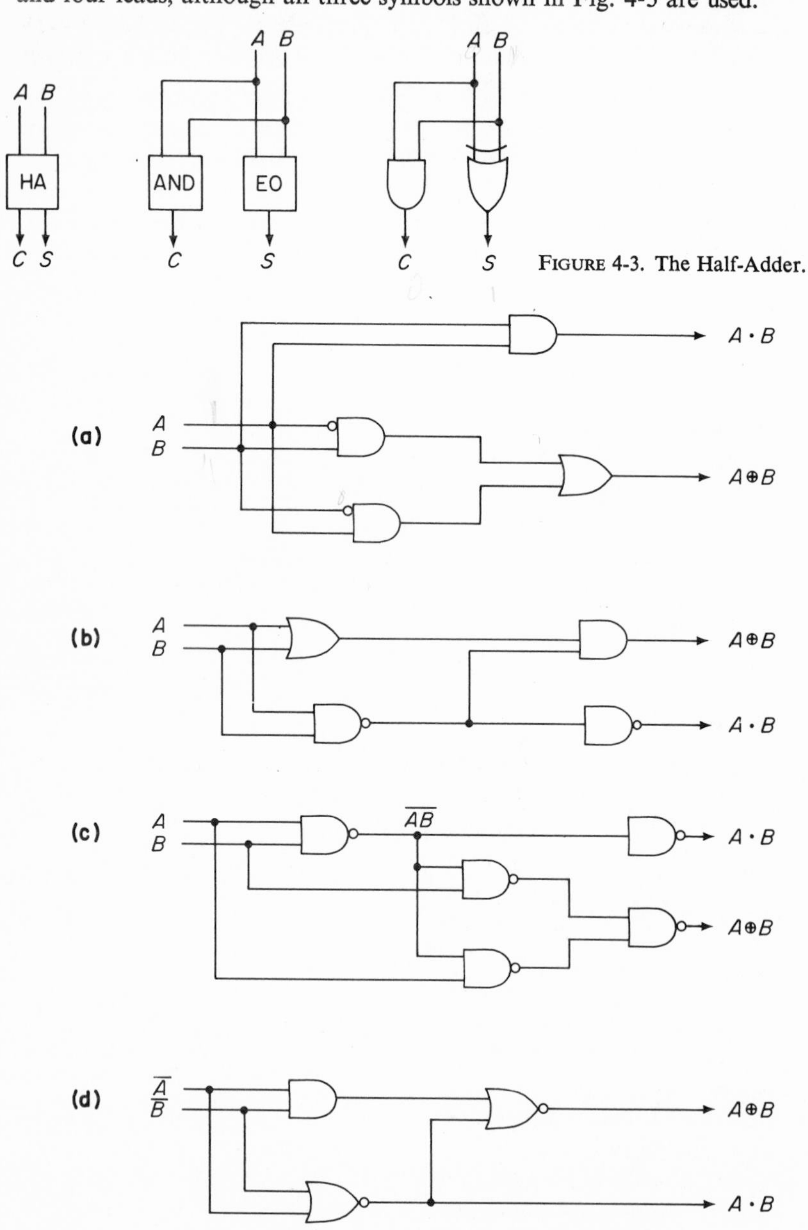

FIGURE 4-3. The Half-Adder.

FIGURE 4-4. Four Implementations of the Half-Adder.

Figure 4-4 shows four of the many possible logic gate implementations of the half-adder. The circuits are synthesized directly from the definition of $A \cdot B$ and the various equivalent expressions for $A \oplus B$. Configurations (a) and (b) in Fig. 4-4 are derived directly from the definitions of exclusive OR, as expressed in Eqs. (4-4) and (4-5). Although the configuration of Fig. 4-4(c)

$$A \oplus B = A\bar{B} + B\bar{A} \tag{4-4}$$

$$A \oplus B = (A + B)(\overline{A \cdot B}) \tag{4-5}$$

appears wasteful in that it requires five gates instead of four to synthesize, it is particularly convenient when using integrated circuits since it can be constructed from all NAND gates. The appropriate expression for the exclusive OR used for the NAND configuration is found by use of DeMorgan's theorem as follows:

$$\begin{aligned} A \oplus B &= A\bar{B} + B\bar{A} \\ &= \overline{\overline{A\bar{B}} \cdot \overline{B\bar{A}}} \\ &= \overline{\overline{(A\bar{B} + A\bar{A})} \cdot \overline{(B\bar{A} + B\bar{B})}} \quad \text{(where } A\bar{A} = B\bar{B} = 0\text{)} \\ &= \overline{\overline{(A \cdot \overline{AB})} \cdot \overline{(B \cdot \overline{AB})}} \qquad (4\text{-}6) \end{aligned}$$

Finally, the half-adder may be designed using only three gates, as seen in Fig. 4-4(d).

The Full-Adder

The complete operation for the addition of two binary numbers requires, for each bit of like significance, a circuit capable of adding three inputs. These are augend, the addend, and the *carry* from the previous bit position. As described in the truth table of Fig. 4-2, the half-adder makes no provision for a carry from the previous bit. A circuit for the nth order then must accept three inputs, A_n, B_n, and C_{n-1}, and provide two outputs, S_n and C_n. The truth table for this operation is shown in Fig. 4-5.

A_n	B_n	C_{n-1}	S_n	C_n
0	0	0	0	0
0	0	1	1	0
0	1	0	1	0
1	0	0	1	0
0	1	1	0	1
1	1	0	0	1
1	0	1	0	1
1	1	1	1	1

FIGURE 4-5. Truth Table for a Full-Adder.

Inspection of the truth table for the full-adder yields the following relationships for C_n and S_n:

$$C_n = \bar{A}_n B_n C_{n-1} + A_n B_n \bar{C}_{n-1} + A_n \bar{B}_n C_{n-1} + A_n B_n C_{n-1}$$
$$C_n = A_n B_n + C_{n-1}(A_n \oplus B_n) \tag{4-7}$$
$$S_n = \bar{A}_n \bar{B}_n C_{n-1} + \bar{A}_n B_n \bar{C}_{n-1} + A_n \bar{B}_n \bar{C}_{n-1} + A_n B_n C_{n-1} \tag{4-8}$$

Mapping of Eq. (4-8) on a three-variable map yields no simplification of the expression; however, an expansion of $A \oplus B \oplus C$ yields

$$\begin{aligned} A \oplus B \oplus C &= (A\bar{B} + B\bar{A}) \oplus C \\ &= (A\bar{B} + B\bar{A})\cdot\bar{C} + \overline{(A\bar{B} + B\bar{A})}\cdot C \\ &= A\bar{B}\bar{C} + \bar{A}B\bar{C} + (\bar{A} + B)(\bar{B} + A)\cdot C \end{aligned}$$
$$A \oplus B \oplus C = A\bar{B}\bar{C} + \bar{A}B\bar{C} + \bar{A}\bar{B}C + ABC \tag{4-9}$$

which is equivalent to the expression for S_n in Eq. (4-8); therefore,

$$S_n = A_n \oplus B_n \oplus C_{n-1} \tag{4-10}$$

The full-adder shown in Fig. 4-6(a) is now synthesized directly from Eqs. (4-7) and (4-10). When redrawn as in Fig. 4-6(b) the full-adder is seen to consist of two half-adders and an OR gate for combining the two subcarries.

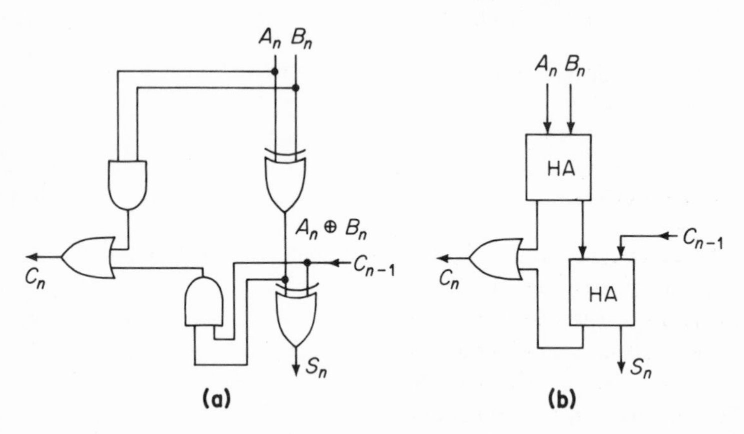

FIGURE 4-6. The Full-Adder.

The circuit of Fig. 4-6(a) is only one of many implementations of the full-adder from which the logic designer may choose. Final choice of circuit configuration is generally determined by cost, availability of components, and required speed of response. Another consideration is that the negation of A_n and B_n are often available, which enables the designer to use the configuration of Fig. 4-4(d). Furthermore, the present availability of entire full-adders on a single integrated circuit chip makes the design of full-adders or half-adders with discrete components, or even individual IC gates, uneconomical except where logic level or power supply incompatibilities exist.

Binary Parallel Adder

Consider now the task of adding two n-digit binary numbers. Assuming that all the digits are available simultaneously (on separate input leads), the addition can be performed in parallel. Usually the least significant bit has no carry input; if so, a half-adder will be sufficient for the first stage. All other orders will require full-adders, resulting in the circuit of Fig. 4-7. A characteristic of the parallel adder of Fig. 4-7 is that the carries *ripple through* each stage, starting with the least significant bit. Since each stage has a propagation delay associated with it, the outputs on the lines from S_1 through S_{n+1} will be subject to change until the carry ripple-through is completed; therefore, the outputs should be gated and clocked out after the adder has settled down.

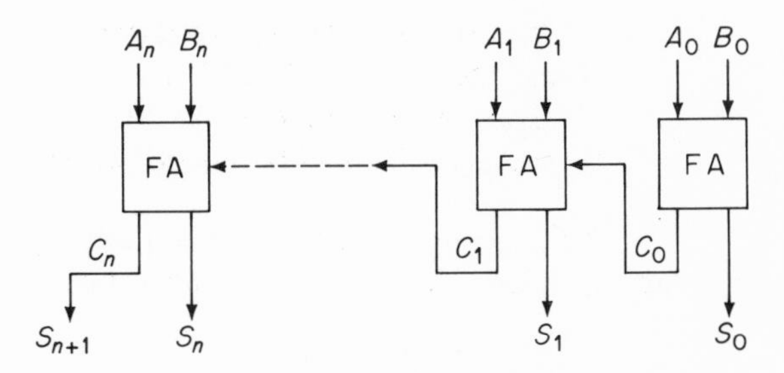

FIGURE 4-7. An n-Stage Parallel Adder.

The carry ripple-time imposes a serious limitation on the speed at which computers can perform arithmetic operations. As seen in Fig. 4-6(a), a carry propagates through two gates in rippling through a full-adder stage. Therefore, an n-bit adder will require $2n$ propagation delays to settle. An alternate approach that reduces this delay is formulated by logically determining the carry to each stage simultaneously. The simultaneous carry parallel adder, also referred to as a *look-ahead carry*, may be designed to perform the required addition with a total of four gate delays. This is achieved at the expense of added circuit complexity.

The Simultaneous-Carry Parallel Adder

Equation (4-7) for C_n shows that a carry is generated in the nth order and propagated to the next higher order when $A_n \cdot B_n$ is true or when $A_n \oplus B_n$ is true simultaneously with a carry from the previous stage. This relationship may be expanded as follows:

$$C_n = A_nB_n + (A_n \oplus B_n)C_{n-1}$$

$$C_n = A_nB_n + (A_n \oplus B_n)A_{n-1}B_{n-1} + (A_n \oplus B_n)(A_{n-1} \oplus B_{n-1})C_{n-2}$$

$$C_n = A_nB_n + (A_n \oplus B_n)A_{n-1}B_{n-1}$$
$$+ (A_n \oplus B_n)(A_{n-1} \oplus B_{n-1})A_{n-2}B_{n-2}$$
$$+ (A_n \oplus B_n)(A_{n-1} \oplus B_{n-1})(A_{n-2} \oplus B_{n-2})C_{n-3}$$

etc.

Inspection of the expansion on C_n shows that the carry may be determined for any n by

1. Performing all the exclusive OR operations indicated. The exclusive OR function is required for determining the sum digits so that this step does not add any complexity other than the increased fan-out imposed on individual circuits (see Fig. 4-8).
2. Perform the indicated AND and OR operations. These functions are performed by standard IC gates, provided the number of inputs is not excessive. When more than 10 inputs are required, *gate expanders* are necessary. This adds to the complexity and propagation delays.

To illustrate the procedure we will design a 4-bit simultaneous-carry parallel adder. The carries are found as follows:

$$C_0 = A_0B_0 + (A_0 \oplus B_0)C_{-1} \tag{4-11}$$

$$C_1 = A_1B_1 + (A_1 \oplus B_1)C_0$$
$$C_1 = A_1B_1 + (A_1 \oplus B_1)A_0B_0 + (A_1 \oplus B_1)(A_0 \oplus B_0)C_{-1} \tag{4-12}$$

$$C_2 = A_2B_2 + (A_2 \oplus B_2)C_1$$
$$C_2 = A_2B_2 + (A_2 \oplus B_2)A_1B_1 + (A_2 \oplus B_2)(A_1 \oplus B_1)A_0B_0$$
$$+ (A_2 \oplus B_2)(A_1 \oplus B_1)(A_0 \oplus B_0)C_{-1} \tag{4-13}$$

$$C_3 = A_3B_3 + (A_3 \oplus B_3)C_2$$
$$C_3 = A_3B_3 + (A_3 \oplus B_3)A_2B_2 + (A_3 \oplus B_3)(A_2 \oplus B_2)A_1B_1 \tag{4-14}$$
$$+ (A_3 \oplus B_3)(A_2 \oplus B_2)(A_1 \oplus B_1)A_0B_0$$
$$+ (A_3 \oplus B_3)(A_2 \oplus B_2)(A_1 \oplus B_1)(A_0 \oplus B_0)C_{-1}$$

The simultaneous-carry parallel counter is implemented by direct application of Eqs. (4-11)–(4-14). The resultant circuit for the first three stages is shown in Fig. 4-8. The same procedure can be followed for greater numbers of stages; however, circuit complexity and component fan-in and fan-out limitations arise. A compromise may then be made in the approach. For example, if a 12-bit parallel adder is required, the circuit could be partitioned into three

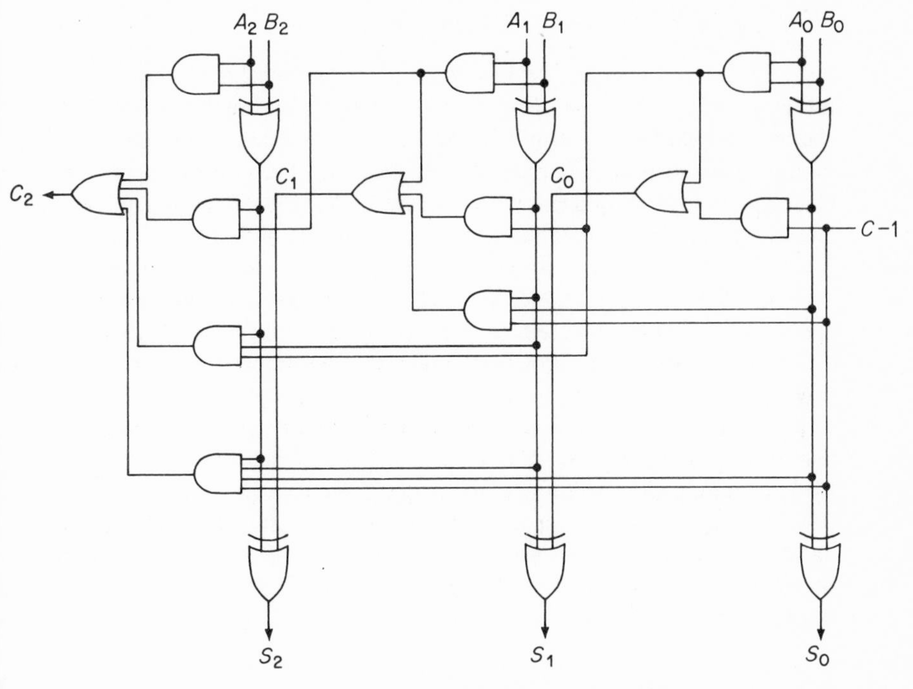

FIGURE 4-8. Three-Stage Simultaneous-Carry Adder.

groups of four-bit simultaneous-carry adders. The carry is then either rippled through from group to group or may be generated simultaneously on a group level.

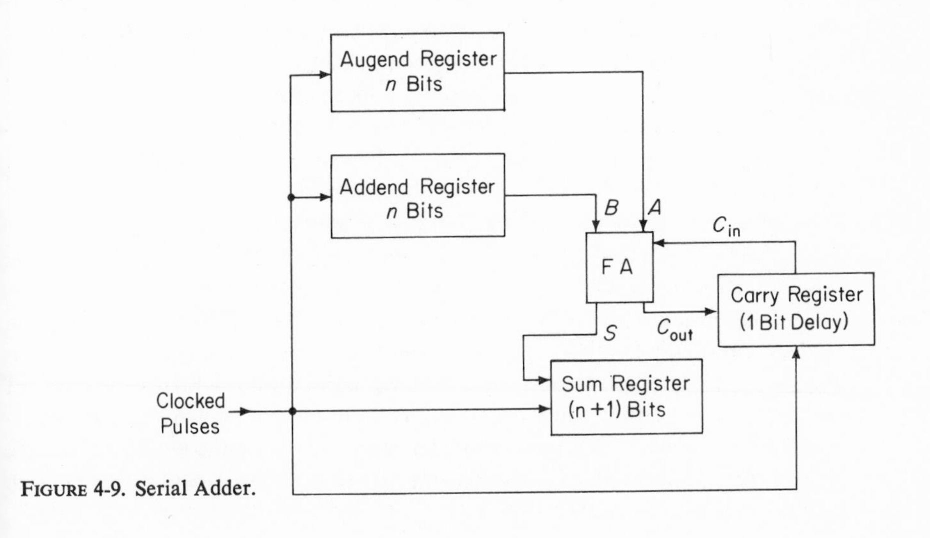

FIGURE 4-9. Serial Adder.

Serial Addition

In the serial-adder two binary numbers are synchronously clocked, one bit position at a time, starting with the least significant bit. The inputs are summed simultaneously in a full-adder with the carry from the preceding order. The sum output is then clocked out, and the carry is stored for one clock period to synchronize it with the next higher order summation. A typical circuit that performs serial addition is shown in Fig. 4-9. To perform addition the numbers A and B are first loaded into the augend and addend registers. The sum register and carry storage register are simultaneously cleared. Then, by applying $(n + 1)$ clock pulses, the sum is entered into the sum register. Readout is either in parallel form from the sum register, serially by clocking the sum register, or serially at the S output of the full-adder during the addition process. In the circuit of Fig. 4-9 it should be noted that the augend, addend, and sum registers are auxiliary circuits with the basic addition being performed by a single full-adder and a one-bit carry storage register. In effect, then, component count has been reduced at the expense of operation time. Note that the number of full-adders is reduced by n whereas the operation time is increased by a factor of $n + 1$.

Accumulation

In the parallel-adder method of summing multidigit binary numbers only the addition of two numbers was considered. When several numbers are to be summed, use of the parallel adder becomes unwieldy and alternate means such as addition by accumulation are generally used. Consider the task of adding three numbers, A, B, and C. The approach with parallel adders would be to take a partial sum (say A plus B) and then to add the partial sum to C. This would require either two sets of parallel adders or a sum register to store the partial sum of A and B. In the latter method, with appropriate timing and gating circuitry, the partial sum is recirculated into the same parallel adder and summed with C. Aside from the circuit complexity, the process is relatively slow when applied to adding long columns of numbers.

Accumulators offer an alternate approach to parallel adders and are widely used to perform arithmetic operations in digital computers. The basic accumulation process is shown in block diagram form in Fig. 4-10. Assume that N multidigit numbers are to be summed. If each of the numbers is an n-digit number, then a total of $n \times N$ leads enter gating logic (A) for parallel fed numbers and n leads for serial fed numbers. With parallel feed—i.e., when all N numbers are present simultaneously—gating pulses are used to clock the numbers 1 through N consecutively into the addend register. From the addend register the numbers are gated into the initially cleared accumulator

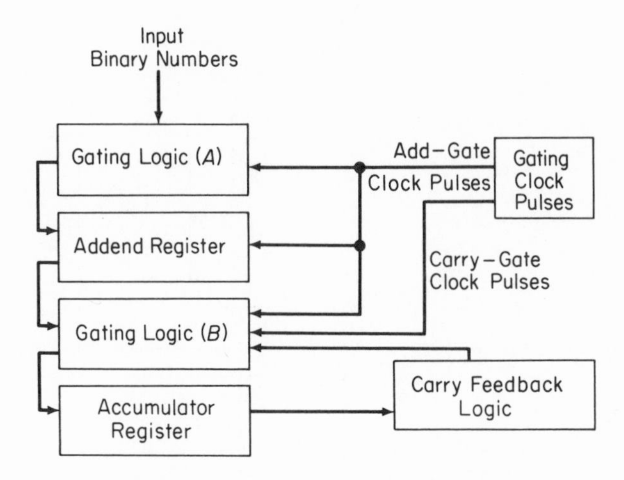

FIGURE 4-10. Basic Accumulation Process.

register. After the first entry, carries are generated, as with the full-adder, and these also must be logically gated into the accumulator. As in the parallel adder, the carry logic can be set up to either ripple through the carry or logically predetermine the carry to each stage simultaneously. The latter approach again is faster but more costly in circuit complexity and total component count.

As each new number is gated in, along with the appropriate carries, the accumulator elements change state and retain the cumulative sum. The number of input numbers is limited only by the total storage capacity of the accumulator register.

One stage of a circuit that performs these functions is shown in Fig. 4-11. Double clocking is used, as in the block diagram of Fig. 4-10, to prevent errors arising from the simultaneous arrival of inputs from the augend register and carries from previous order. Operation of the single stage is as follows: On application of an add-gate clock pulse, a **1** from the addend register will enable AND gate (**1**), OR gate (**2**), and toggle the flip-flop. The add-gate clock pulse is followed by a carry-gate clock pulse. If the flip-flop is in the **0** state ($\bar{Q} = 1$) and the input from the addend register is also a **1**, AND gate (**3**) is enabled. A carry is then generated as required since the conditions signify the flip-flop was toggled from **1** to **0**. If the flip-flop is in the **1** state ($Q = 1$) and a carry is present from the previous stage, then AND gate (**5**) is enabled, generating a carry to the next higher order. Note that a possible race problem exists with this last function if the flip-flop toggles before the carry is generated. However, the propagation delay in OR gate (**2**) and the *JK* flip-flop is generally sufficient to prevent it from happening. Otherwise, additional clocking, such as using a clocked *JK* flip-flop, can be incorporated.

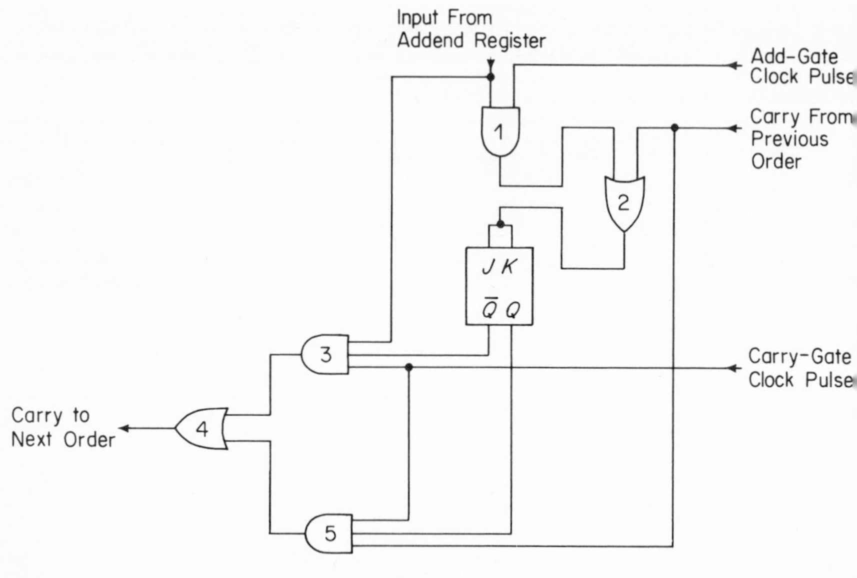

FIGURE 4-11. A Single Stage of an Accumulator Circuit.

4-5 BINARY SUBTRACTION

The development of circuitry to perform binary subtraction of positive, multidigit numbers parallels very closely the approaches presented on binary addition in the preceding section. Rather than detailing the design of all these circuits again, only the full-subtractor development is presented. Section 4-12 discusses *complementing* and the reduction of binary subtraction (of positive numbers) and addition (of negative numbers) to simple binary addition.

A_n	B_n	P_{n-1}	D_n	P_n
0	0	0	0	0
0	0	1	1	1
0	1	0	1	1
1	0	0	1	0
0	1	1	0	1
1	0	1	0	0
1	1	0	0	0
1	1	1	1	1

FIGURE 4-12. Truth Table for Binary Full-Subtractor.

Binary Full-Subtraction. Subtraction of positive, multidigit numbers is performed between two binary numbers, A (the minuend) and B (the subtrahend) on a digit by digit basis starting with the least significant order. The truth table for the nth significant digit is given in Fig. 4-12.

D_n is the difference digit and P_n is the borrow from the next higher der. P_{n-1} is the borrow to the next lower order and must be subtracted om the minuend. Inspection of the truth table of Fig. 4-12 yields the following relationships for D_n and P_n:

$$D_n = P_{n-1} \oplus A_n \oplus B_n \tag{4-15}$$

$$P_n = \bar{A}_n B_n + \overline{A_n \oplus B_n} \cdot P_{n-1} \tag{4-16}$$

A circuit that fulfills these relationships and performs the full-subtraction peration is shown in Fig. 4-13. The full-subtractor may now be used in configurations similar to that of Figs. 4-7–4-9 to achieve the subtraction of multiigit numbers by the same methods used for addition.

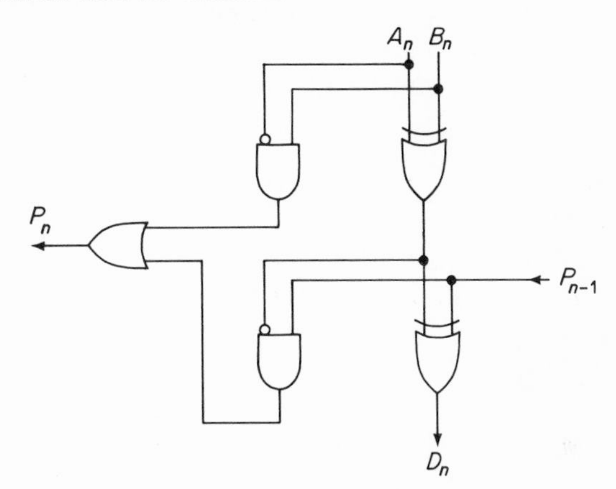

IGURE 4-13. The Full-Subtracor.

Since the difference between two numbers may be a negative quantity, he computer must be organized to label and recognize negative numbers. This s normally accomplished by adding an additional digit to each n-digit number. simple code such as **0** for negative numbers and **1** for positive numbers then vill signify the correct polarity. A computer may now be designed to add or ubtract any combination of positive or negative numbers in what is known as *sign plus magnitude system.* Unfortunately, such a method is quite complex n the required gating logic and results in redundant circuitry with both subractors and adders performing essentially the same functions. In most compuers subtraction and addition of mixed polarity numbers is accomplished by omplementing as required and then performing the single operation of ddition.

Subtraction by Complementing

The process of subtraction by complementing may be illustrated by an example using decimal numbers. Consider the following alternate method for

performing the operation 623 − 246 = 377; Instead of using the convention
borrow-subtract method, let us first find the 10's complement of 246 and ad
this to the minuend (623). The 10's complement of an n-digit decimal numbe
is obtained by subtracting the number from 10^n as follows:

$$\begin{array}{rl} 1{,}000 & (n = 3;\ 10^n = 1000) \\ -246 & \\ \hline 754 & (\text{10's complement of } 246) \end{array}$$

The minuend is then added to the complemented subtrahend, 623 + 754 =
1,377. The most significant digit in the sum is a **1** and must be dropped to ac
count for the 1,000 *borrowed* in the complementing process. The remaind
(377) is then the correct difference. If the minuend had been smaller than th
subtrahend, the difference would be a negative number. The proper polar
ty is indicated by the value of the most significant digit; i.e., a **1** indicat
positive polarity as in the above example, and a **0** indicates negative polarit
When the polarity is negative, the difference is obtained by again complemen
ing the sum and assigning a negative sign. To illustrate, consider the exampl
258 − 623 = −365

$$1{,}000 - 623 = 377 \text{ (10's complement)}$$

$$258 + 377 = 0635 \text{ (sum)}$$

The fourth digit of the sum is **0**; therefore,

$$1{,}000 - 635 = 365 \text{ (complement of sum)}$$

and the correct difference is −365.

The complementing procedure appears to be unduly complicated, es
pecially if applied as illustrated to the decimal number system. With binar
numbers, however, the process of complementing is relatively simple, and th
substitution of addition for subtraction may result in a major savings in hard
ware. Most computers therefore are designed to perform subtraction by eithe
the 1's or 2's complement. Both these methods will now be discussed.

Two's Complement Subtraction. Subtraction of binary numbers by 2'
complement is performed as follows:

STEP 1. Obtain the 2's complement of the subtrahend. If an n-digit number system
is being used, then the 2's complement is found by subtracting the subtra
hend from 2^n. For example, in a computer using binary numbers with
significant digits, the number $(14)_{10}$ is 01110. The 2's complement is then

$$\begin{array}{r} 100000 \\ -01110 \\ \hline 10010 \end{array}$$

A simpler method for obtaining the 2's complement of a binary number is to invert each digit and to add +1. The 2's complement of $(18)_{10}$ may be found as follows:

$$
\begin{array}{rl}
10010 & (18)_{10} \\
\hline
01101 & \text{(complement of each digit)} \\
+00001 & \text{(add 1)} \\
\hline
01110 & \text{(2's complement)}
\end{array}
$$

Using this method the circuitry is relatively simple. The complement of individual digits is often available (i.e., the $\bar{Q}$ output of flip-flops) or may be obtained with inexpensive inverters. The addition of 1 to the complemented digits is at most an extra step on existing adders.

TEP 2. Perform the indicated addition of the minuend and the 2's complement of the subtrahend. To illustrate, consider the operation 18 − 14 equal to +4 with binary numbers:

$$
\begin{array}{rrl}
 & 10010 & (18)_{10} \\
 & +10010 & \text{(2's complement of 14)} \\
\hline
\boxed{1} & 00100 & \text{(sum)}
\end{array}
$$

Since the sixth digit in the above sum is a **1**, the difference is positive. Dropping the sixth digit will account for the 100000, or $(32)_{10}$, that was originally borrowed and added to $(-14)_{10}$ when the 2's complement was derived; therefore, the resultant difference of $(18 - 14)_{10}$ is +00100.

TEP 3. If the borrowed number (2^n) cannot be accounted for (i.e., if the sixth digit of the sum in the example of STEP 2 had been a **0**), then the difference is negative. The sum then must be complemented again and a negative sign assigned to the answer. To illustrate, consider the operation $(6 - 14)_{10}$ equal to $(-8)_{10}$:

$$
\begin{array}{rrl}
 & 00110 & (6)_{10} \\
 & 10010 & \text{(2's complement of 14)} \\
\hline
\boxed{0} & 11000 & \text{(sum)}
\end{array}
$$

The n + 1 digit is **0**; therefore, the difference is negative and equal to the negative of the 2's complement of 11000, or −01000.

One's Complement Subtraction. The 1's complement of a binary number 'n digits) is found by subtracting from $2^n - 1$. This turns out to be the complement of each digit. For example, the 1's complement of 0010111 is 1101000. Subtraction may now be performed as follows:

STEP 1. Obtain the 1's complement of the subtrahend.

STEP 2. Add the minuend to the subtrahend's complement

STEP 3. If the sum in Step 2 has a **1** in the $n + 1$ digit position, the difference i *positive* and the correct difference may be found by adding **1** to the sum.

If the sum has a **0** in the $n + 1$ digit position, the difference is *nega tive* and the correct difference is found by taking the 1's complement of th sum.

This procedure is illustrated by the following examples:

(a) $(34 - 23 = 11)_{10}$

	100010	$(34)_{10}$
	010111	$(23)_{10}$
	101000	(1's complement of subtrahend)
	+100010	(add minuend)
1	001010	(7th digit is 1)
	+1	(add 1)
	001011	(difference = +11)

(b) $(17 - 23 = -6)_{10}$

	10001	$(17)_{10}$
	01000	(1's complement of 23)
0	11001	(sum, 6th digit is 0)
	−00110	(complement sum with negative sign)

4-6 BINARY MULTIPLICATION

Multiplication is the third arithmetic operation we shall discuss. The operation consists of forming the product of two numbers, the multiplicand and the multiplier. A product of single-digit binary numbers may be found by application of the logic defined by the truth table of Fig. 4-14. Note that $P = A \cdot B$.

A	B	P
0	0	0
0	1	0
1	0	0
1	1	1

FIGURE 4-14. Multiplication Truth Table.

The multiplication of multidigit numbers, however, is considerably more involved than Fig. 4-14 would indicate. One fundamental process is the forming of a product by the iterative addition of the multiplicand. It is relatively simple to implement but too time consuming for most computer systems. More elaborate methods have been developed that effect reductions in computation time. These require more complex circuitry and generally greater expense. Choice of an algorithm, then, involves making the best compromise between complexity and operational speed.

Multiplication by Iterative Addition

Multiplication by repeated addition is illustrated with decimal and binary numbers by the following example:

$$23 \times 3 = 69 \qquad 10111 \times 11 = 1000101$$

$$\begin{array}{r} 23 \\ 23 \\ +23 \\ \hline 69 \end{array} \qquad \begin{array}{r} 10111 \\ 10111 \\ +10111 \\ \hline 1000101 \end{array}$$

The circuit of Fig. 4-15 implements this method of multiplication with an accumulator, *binary down-counter*, and suitable control logic. The design of binary counters with flip-flops is analyzed in detail in Chap. 6. It will be adequate for this discussion to understand the operation of down-counters on a functional, block diagram basis. The n-bit down-counter used in the multiplication circuit of Fig. 4-15 may have any number from 0 to $2^n - 1$ entered by means of preset inputs. The device then will count (down) with each application of an input pulse. The zero state of the counter is then logically

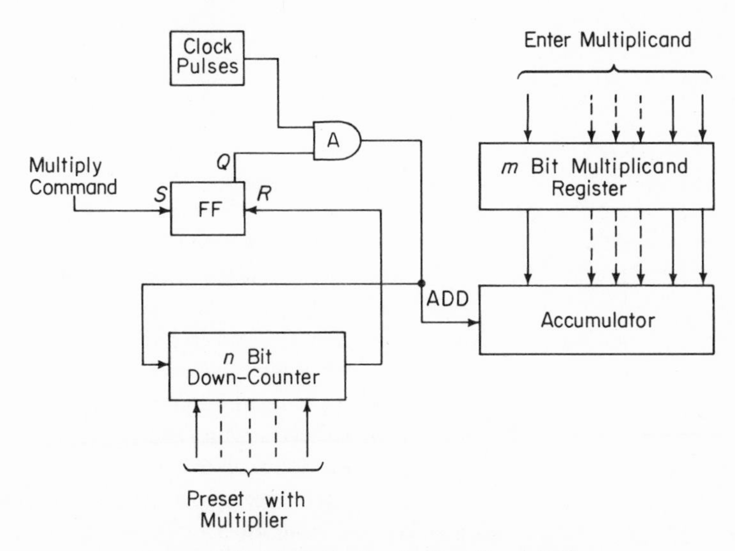

FIGURE 4-15. Multiplication by Iterative Addition.

determined (i.e., by NORing the Q output of each flip-flop in the counter) and an output pulse is produced.

Multiplication is initiated by a start command setting the RS flip-flop. Prior to the start of multiplication, the accumulator is reset, the down-counter is preset with the multiplier, and the multiplicand is present on the m input leads and entered into the multiplicand register. The SET FF will now enable AND gate A and allow clock pulses to pass through. With each clock pulse applied, the multiplicand is summed into the accumulator. Also, the count in the down-counter is reduced by **1**. When the counter is completely reset, an output pulse appears that resets the flip-flop. This disables gate A, stops further additions to the accumulator, and the product is stored in the accumulator.

Multiplication by Shifting

Multiplication by shifting offers significant improvement in speed over the method of iterative addition. Most computers are designed to perform multiplication in this manner. Further improvements are effected by clever variations that take advantage of **0**'s or repetitive **1**'s in the multiplier.

The shifting method of multiplication may be illustrated by the following example of long-hand multiplication in binary form:

$$
\begin{array}{rl}
101101 & = (45)_{10} \\
\times\ \ 11001 & = (25)_{10} \\
\hline
101101 & \\
000000 & \\
000000 & \\
101101 & \\
101101 & \\
\hline
10001100101 & = (1125)_{10}
\end{array}
$$

Note that the product is the sum of five partial products, one corresponding to each of the five digits in the multiplier. When a particular multiplier is **0**, the partial product is also **0**. When the digit is **1**, the partial product is the multiplicand appropriately shifted. For example, multiplication by the fourth significant digit (a **1**) is equivalent to multiplying by 2^3; hence the partial product is the multiplicand shifted to the left by three digit positions.

Design of the shifting multiplier is facilitated by use of a shift register. The operation and design of shift registers is discussed in detail in Chap. 8. The following brief explanation of the operating characteristics, however, will suffice to understand the application in a shift multiplier.

The shift register consists of an array of flip-flops that may be preset to store binary numbers. In addition, the flip-flops are wired so that on each application of a clock (shift) pulse the binary state of each stage is shifted one bit to the right (or left). A four-stage shift register (shift to right) is shown in Fig. 4-16. After presetting, each new shift pulse will set or reset FF_3, depending

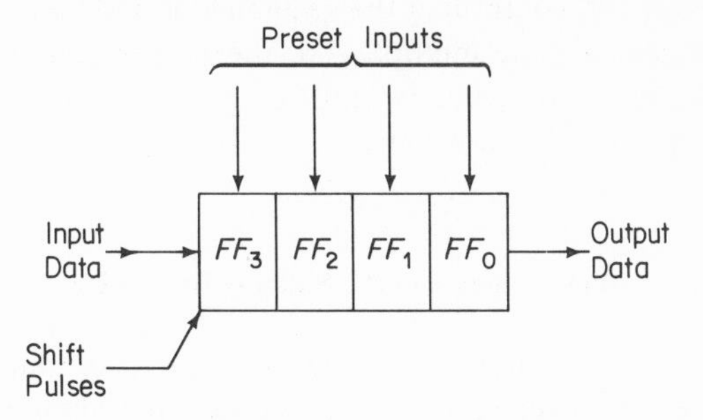

FIGURE 4-16. Four-Bit Shift-Register.

upon the state of the input data. FF_3 in turn will determine the state of FF_2; FF_2, the state of FF_1; and FF_1, the state of FF_0. The state of FF_0 is read out with each shift pulse applied. In recent years multistage shift registers have been developed on single-integrated circuit chips, utilizing MOS-FET techniques. Most of these devices, however, cannot be preset because of limitations in the number of connections that can be made.

The shifting multiplier can be designed in a variety of forms. One such implementation is shown in Fig. 4-17. The n-bit multiplier is entered into a n-bit

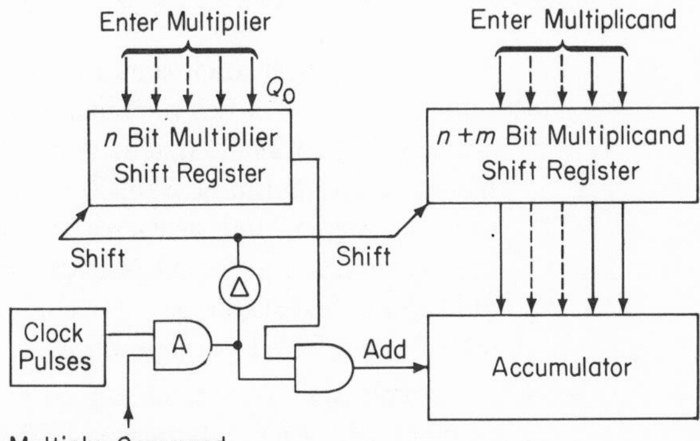

FIGURE 4-17. Shifting Multiplier.

shift register (shift to right) and the m-bit multiplicand into a $n + m$-bit shift register (shift to left). When the multiply command is HIGH, the clock pulses are enabled by AND gate A. If the least significant bit in the multiplier register is a **1** (Q_0 is HIGH), then the accumulator will add in the multiplicand from the multiplicand register. After a short delay (Δ) the multiplier is shifted one bit to the right and the multiplicand one bit to the left. The second clock pulse then will repeat the process. If the second significant bit of the multiplier is a **1**, the multiplicand ($\times 2^1$) is entered into the accumulator. If the multiplier bit is **0**, the multiplier and multiplicand are shifted without addition.

The delay (Δ) allows the accumulation to be executed before shifting the data bits in the registers. Double clocking may be used instead of the delay, with CP_1 controlling the accumulator and CP_2 the shift registers. Also, instead of shifting the multiplicand, the proper relative position to the accumulator may be maintained by shifting the partial products in the accumulator to the right, one bit at a time.

Improvements in the Shifting Multiplier

If a computer is designed to perform multiplication by the circuit configuration shown in Fig. 4-17, then the time required to execute each multiply command is given by Eq. (4-17).

$$t_m = n(t_s + t_a) \tag{4-17}$$

where t_m = total multiplication time, t_s = time required for registers to shift, t_a = time for each addition in the accumulator, and n = number of multiplier bits.

Equation (4-17) represents the maximum time required to perform a given multiplication of two numbers. However, this time may be reduced significantly by several means. Four of these methods will now be discussed.

1. The total multiplication time may be reduced by stopping the multiplication process when the remaining significant bits in the multiplier are all **0**'s. Consider a system that is designed for 10-bit multipliers. If a particular multiplier were 0001011011, then the multiply command could be halted after seven partial products are formed. Also, since the multiplicand and multiplier can be interchanged, the number requiring the fewest number of accumulations could be logically determined and selected as the multiplier.
2. A variable shift command can be generated that shifts the registers automatically when a **0** appears in the multiplier. This avoids the unnecessary operation of summing all **0**'s into the accumulator. On the average, then, the number of accumulation steps are reduced by one-half.
3. A string of **1**'s in the multiplier can be reduced to one addition to, and one subtraction from, the accumulator. Consider the fact that a number consisting of p **1**'s is equal to $2^p - 1$; i.e., the number 11111 $= 2^5 - 1 = (31)_{10}$. Then, if the next five digits in the multiplier register were all **1**'s, instead of forming five partial products (additions

in the accumulator), the multiplicand is first subtracted and then shifted five times to the left and added to the accumulator.

4. The multiplier bits are considered in pairs. If the first pair of multiplier bits is 00, no addition is required and the multiplier is shifted two bits. If the pair is 01 or 10, one appropriate addition to the accumulator is performed. Lastly, if the pair is 11, an addition of three times the multiplicand is required. This is accomplished by the alternate method of subtracting the multiplicand one time from the partial product and carrying a **1** to the next pair of digits. This may be temporarily stored in a flip-flop that represents the carry digit.

   The next pair of multiplier digits are also either 00, 01, 10, or 11. With the carry added to the new pair, the sum becomes 00, 01, 10, 11, or 100. The first four sum pairs are treated as the first pair of digits. However, if the sum is 100, a new carry is generated without performing a subtraction from the partial product. This process is summarized in the table of Fig. 4-18.

Multiplier Pair	Old Carry	Required Operation	New Carry
0 0	0	Shift Twice	0
0 1	0	Add, Shift Twice	0
1 0	0	Shift, Add, Shift	0
1 1	0	Subtract, Shift Twice	1
0 0	1	Add, Shift Twice	0
0 1	1	Shift, Add, Shift	0
1 0	1	Subtract, Shift Twice	1
1 1	1	Shift Twice	1

IGURE 4-18. Multiplication by ultiplier Bit-Pairs.

-7 BINARY DIVISION

Arithmetic division can be performed in a digital computer by several lgorithms, depending upon requirements for speed, complexity (cost), and acuracy. Note that accuracy was not a basic consideration when selecting a ethod for addition, subtraction, or multiplication. Inadequate noise margins r faulty design and fabrication techniques will cause errors; however, the basic lgorithms will yield absolute accuracy when operating on integers. Fractional umbers may require *rounding-off*, but again any desired degree of accuracy ay be attained by providing sufficient bit capacity to the right of the radix oint. However, division may be performed by approximation techniques that re inexact but very fast.

Division by Iterative Subtraction

Division by repeated subtraction is analogous to the method describe for binary multiplication in Sec. 4-6. In this method the *divisor* is repeatedl subtracted from the *dividend* until the *remainder* is smaller than the diviso The circuit could be implemented either with an adder designed to perform 1 or 2's complement subtraction or an accumulator designed to perform subtra tion directly. This latter method is illustrated in Fig. 4-19.

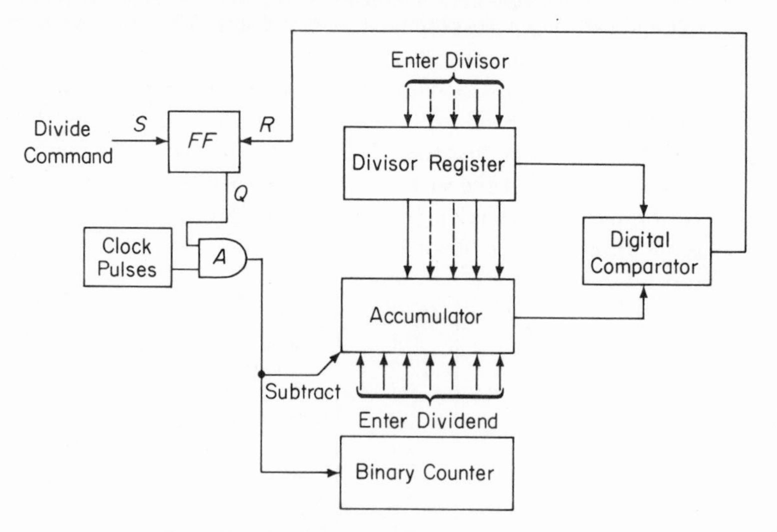

FIGURE 4-19. Division by Iterative Subtraction.

Before initiating the divide command, the divisor is entered into th divisor register and the dividend into the accumulator. The divide comman then sets the *FF* and enables AND gate *A*. With each application of a cloc pulse, the sum in the accumulator is reduced by the number in the divisor re gister. When the remaining sum is less than the divisor, the output of th digital comparator goes HIGH and resets the *FF*. The count in the binar counter then represents the quotient and the sum in the accumulator rep resents the remainder.

The method of iterative subtraction offers both simplicity and accurac when performing division. The procedure is suitable for small, desk-top cal culators and in special purpose computers where the relatively slow speed i tolerable. Clearly, the clock period must be adequate to allow the accumulato to perform subtraction and the comparator to function. The total computatio time therefore is equal to the product of the clock period and the value of th quotient.

Digital comparators may be designed in a variety of forms. One typica ircuit is the parallel comparator shown in Fig. 4-20. Two n-digit numbers, A nd D, are compared on a bit by bit basis, starting with the most significant it. If the bits of like significance are equal, ($\overline{A_x \oplus D_x} = \mathbf{1}$), the next lower evel comparison is enabled. With $A \geq D$, all the inputs to the output NAND ate are HIGH and the output is **0**. At the highest level at which the A and) digits are unequal, if $\overline{\bar{A}_x D_x} = \mathbf{0}$, then $D > A$ and the output goes HIGH. Comparison speed is limited by two propagation delays per stage, and this may e significant when long numbers are used.

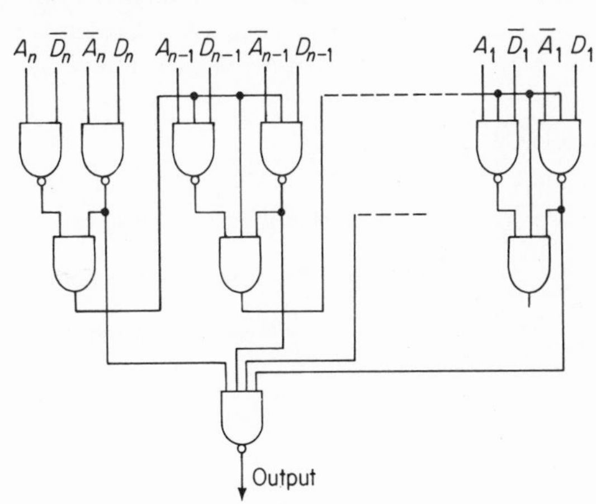

FIGURE 4-20. Parallel Digital Comparator.

Restoring Division by Shifting the Divisor

Restoring division is performed in a computer by shifting the divisor in manner closely paralleling multiplication by shifting. The method also s similar to long-hand division as seen n the example shown in Fig. 4-21.

```
          1 0 1 1 1
1 0 1 | 1 1 1 0 0 1 1
       -1 0 1
          1 0 0
         -0 0 0
          1 0 0 0
           -1 0 1
              1 1 1
             -1 0 1
                1 0 1
               -1 0 1
                0 0 0
```

FIGURE 4-21. Restoring Division.

In restoring division, trial subractions of the divisor from the partial dividend are made. If the divisor can be subtracted (without yielding a negative difference), the subtraction is performed and a **1** is entered into the quotient. If the subtraction cannot be made, a **0** is entered into the quotient and the partial dividend is not altered. The divisor is then shifted one bit to the right and the procedure is repeated.

Nonrestoring Division

Nonrestoring division is performed by always taking the difference b tween the divisor and partial dividend, regardless of whether the difference positive or negative. When the partial dividend is positive, the divisor is sul tracted and a **1** is entered into the corresponding quotient bit. This is identic to the procedure followed in restoring division. However, if he tpartial div dend is negative, the difference is obtained by adding the divisor to the parti dividend. A **0** then is entered into the quotient.

The nonrestoring procedure for division is illustrated in the example Fig. 4-22. The first partial dividend is found by subtracting the divisor fro

```
       10001
110 ⟌1100111
    -110
    +000111  (First Partial Dividend is Positive)
     -110
    -101001  (Second Partial Dividend is Negative)
      +110
     -10001  (Third Partial Dividend is Negative)
       +110
      -0101  Etc.
       +110
       +001  Etc.
```

FIGURE 4-22. Nonrestoring Div sion.

the whole dividend. Since the difference is positive, the quotient bit is a **1**. Th second partial dividend, however, is negative. Therefore, the quotient bit entere is a **0**. The third partial dividend is now formed by adding the divisor to th negative difference. The procedure is repeated until the remainder is **0** or unt all the allowed-for quotient bits are filled. Depending upon the particular com putational format used, fractional accuracy may or may not be provided fo When fractional accuracy is not necessary, the remainder is ignored except fo its sign. A positive or **0** remainder signifies a **1** in the lowest order quotien bit. When fractions are to be represented by one or more bit positions, th dividend must provide an adequate number of bit positions to the right of th

```
        10111
101 ⟌ 1110011
     +011
   [1]0100011  (+)        101
      +011                010
   [0]111011   (-)         +1
      +101                011 = 2's Complement
   [1]001111   (+)              of Divisor
        +011
      [1]0101  (+)
         +011
       [1]000  (+)
```

FIGURE 4-23. Nonrestoring Divi sion with 2's Complement Sub traction.

binary point. With either system fractional accuracy may be achieved by the programmer scaling the divisor and dividend prior to entering the problem. Scaling is discussed in Sec. 4-8.

The main advantage of the nonrestoring algorithm is that it avoids the need for comparing the divisor and partial dividend. The procedure for determining the polarity of the partial dividend and either adding or subtracting the divisor may be performed with relative simplicity by 1's or 2's complement subtraction. This is illustrated in Fig. 4-23, using the problem solved in Fig. 4-21. by restoring division.

Note that the first difference is obtained by adding the 2's complement of the divisor to the dividend. Since the appropriate bit in the difference (signified by dotted enclosure) is **1**, the quotient entry also is **1** and the procedure is repeated. The second partial dividend is negative, therefore the quotient entry is a **0** and the following step is appropriately an addition of the divisor. This method reduces the division process to one of addition, with the appropriate gating and shifting circuitry. The process requires no new circuit concepts to synthesize. Finally, procedures for skipping across **0**'s in the partial dividend (**1**'s in two's-complement form) and using divisor multiple bits logic to speed up the operation may be used. The development of these procedures parallels closely the material on fast multiplication.

4-8 SCALING AND FLOATING-POINT ARITHMETIC

The word-size capacity of a computer's arithmetic unit is a limiting factor in the computational accuracy attainable. To achieve the maximum accuracy from a system with fixed word lengths, numbers are scaled to yield the largest number of significant digits possible in each operation.

Scaling may be performed by the programmer in a manner similar to determining the decimal point when using a slide rule. In a system based upon fixed binary point notation, where the weighing factor for each bit position is a fixed power of the binary base, scaling must be used to handle wide ranges of numbers. The position of the binary point is predetermined for each number before entering the computer and is followed throughout the computation. When performing a series of operations the scale may be changed several times during the total operation by shifting numbers in registers. This procedure is both cumbersome and time-consuming.

The scaling operation may be performed automatically in the computer by using a *floating-point* format. In floating-point notation the magnitude of each number is represented by two groups of bits. This is illustrated in Eq. (4-19) where the number N is expressed in radix R' by a mantissa a and characteristic b.

$$N = aR^b \tag{4-19}$$

The similarity of Eq. (4-19) to expressing numbers by logarithms should be apparent. The usual practice in floating-point notation is to use radix 2 and to normalize numbers so that the mantissa is less than 1 and greater than $\frac{1}{2}$. This is illustrated by the following examples:

$$1011000 = 0.1011 \times R^{+111}$$

$$0.000101 = 0.1010 \times R^{-011}$$

An example of floating-point format is given in Fig. 4-24. The characteristic is four bits long and the mantissa is seven bits long, with one bit in each term reserved for polarity notation.

Mantissa							Characteristic			
$\pm$	m_6	m_5	m_4	m_3	m_2	m_1	$\pm$	c_3	c_2	c_1

FIGURE 4-24. Floating-Point Word Format.

A computer may be programmed to shift numbers, automatically right or left, until the highest significant **1** in a number is in the most significant bit position of the mantissa. This would be the m_6 position of the Fig. 4-24 format. The number of shifts are either added to or subtracted from the characteristic, depending upon the direction of shifting.

When performing multiplication the mantissas are multiplied and the characteristics are added. The product is then reshifted to the left, if required, so that the most significant bit in the mantissa is a **1**. Likewise, in division, the mantissas are divided and the characteristics are subtracted. Division continues until all the dividend mantissa bit positions are filled, yielding the smallest remainder the computer's word length mechanization will permit. Further accuracy could be achieved only by using longer word formats.

When performing addition or subtraction, the procedure is altered slightly. In addition, the larger number is first scaled so that the most significant mantissa bit is a **1**. The smaller number is then scaled to the same characteristic. The mantissas are then added and the characteristics are transfered to the sum. Subtraction also may be performed in this format, either directly or by complementing.

4-9 FAST DIVISION BY REPEATED MULTIPLICATION

An algorithm for obtaining an approximation of the quotient will now be discussed. It is based on developing a converging series from appropriate multiplication. The method is generally restricted to systems using floating-point notation since the divisor is required to be in normalized (fractional) form.

Consider the operation of Eq. (4-20) where the quotient Q of two numbers, A and B, is to be determined. The divisor B is normalized so that $\frac{1}{2} \leq B < 1$. The reciprocal of B is therefore a number greater than 1 and less than

$$Q = \frac{A}{B} \tag{4-20}$$

or equal to 2. Furthermore, a simple decoder circuit, called a *look-up table*, could be permanently wired to yield a numerical value for the reciprocal of B based on the n most significant bits of B. For example, if $B = 0.101$, the reciprocal is equal to 1.6. If the look-up table were designed with 8 or more entries, i.e., to decode at least the first three digits of B, then the reciprocal would be determined exactly. However, for the general case B could be expected to have more significant digits than any practical look-up table could store. Therefore the reciprocal obtained is an approximation of B, or

$$R = \frac{1}{B} \pm \delta$$

How good the approximation is depends, obviously, on the extent of the look-up table and the particular value of B.

If both A and B are now multiplied by R, the quotient Q becomes

$$Q = \frac{AR}{BR} = \frac{AR}{1 \pm B\delta} \tag{4-21}$$

Furthermore, if both numerator and denominator in the relationship of Eq. (4-21) are multiplied by $(1 \mp B\delta)$, then

$$Q = \frac{AR(1 \mp B\delta)}{1 - (B\delta)^2} \tag{4-22}$$

For $(B\delta)$ small, $(B\delta)^2 \to 0$ and the quotient

$$Q \approx AR(1 \mp B\delta) \tag{4-23}$$

The process can be repeated by multiplying the numerator and denominator in Eq. (4-22) by $1 + (B\delta)^2$, $1 + (B\delta)^4$, etc. until any degree of accuracy required is obtained. The required multiplication factors are obtained as follows:

1. R is obtained from the look-up table
2. $(1 \pm B\delta) = BR$

3. $(1 \mp B\delta) = 2 - (1 \pm B\delta) =$ complement of $1 \pm B\delta$
4. $[1 - (B\delta)^2] = (1 \pm B\delta)(1 \mp B\delta)$
5. $[1 + (B\delta)^2] = 2 - [1 - (B\delta)^2] =$ complement of $[1 - (B\delta)^2]$

etc.

The procedure is particularly suited for systems that have very fast multiplication units and where exact solutions are not required.

REFERENCES

1. Wadel, L. B., "Negative Base Number Systems," *IRE Trans. Electron. Comput.* (correspondence), EC-6: 123, June 1957.
2. DeRegt, M. P., "Negative Radix Arithmetic," *Comput. Des.*, 6(5), May 1967, and 6(6), June 1967.
3. Phister, M., Jr., *Logical Design of Digital Computers*, Chaps. 2 and 9, John Wiley & Sons, New York 1958.
4. Mac Sorley, O. L., "High Speed Arithmetic in Binary Computers," *Proc. IRE*, 49(1), January 1961.
5. Richards, R. K., *Arithmetic Operations in Digital Computers*, D. Van Nostrand Company, Princeton, New Jersey, 1955.
6. Allen, M. W., "A Decimal Addition-Subtraction Unit," *Proc. IRE*, Part B, Suppl. 1, 103: 138-145, April 1956.
7. Tocher, K. D. and Lehman, M., "A Fast Parallel Arithmetic Unit," *Proc. IEEE*, Part B, Suppl. 3, p. 03: 520-527, April 1956.
8. Baron, R. C., Piccirilli, A. T., Wallace, D. L., and Ward, R. L., "Concepts of Digital Logic and Computer Operations," Doc. no. 71-280A, Computer Control Company, April 1965.
9. Flores, I., *The Logic of Computer Arithmetic*, Prentice-Hall, Englewood Cliffs, New Jersey, 1963.

PROBLEMS

1. Convert the following numbers to different radices as indicated:

   (a) $(1001.11)_2$ to decimal

   (b) $(983)_{10}$ to binary

   (c) $(324)_5$ to octal

   (d) $(587)_{10}$ to hexadecimal

(e) $(3ca)_{16}$ to binary

(f) $(93.625)_{10}$ to binary

2. Perform the following arithmetic operations:

   (a) Find the octal sum of $(376.43)_8$ and $(27.50)_8$

   (b) Find the binary sum of $(26.25)_{10}$ and $(213)_4$

3. Develop a procedure (set of rules) to convert a binary number to a decimal that uses the reverse of the divide-by-two decimal to binary conversion process. (This technique is called the "double-dabble" process, where the word "dabble" denotes double and add.)

4. In a manner similar to that of Prob. 3, develop a process in the octal base to convert an octal number to a decimal. Using this process, convert $(376)_8$ to a decimal.

5. Find the 2's complement of the following binary numbers.

   (a) 011010 (b) 100111001

   (c) 1000000 (d) 1100010

6. A circuit is to be designed that takes the 2's complement of serially fed binary numbers, starting with the least significant order. Devise a set of rules for the conversion. (Note that the digits remain the same until after the first **1**.)

7. Design a circuit with suitable logic gates to perform the complementing described in Prob. 6.

8. Perform the following arithmetic operations in binary form using 2's complements where necessary:

   (a) $(26)_{10} + (12)_{10} - (29)_{10}$

   (b) $(001001)_2 - (101000)_2$

   (c) $(-14)_{10} + (20)_8 + (114)_4 - (01101)_2$

9. Design a circuit that converts a four-wire input (8-4-2-1) BCD code to the (7-4-2-1) BCD code. Show all work including reduction of expressions by mapping. Use AND, OR, and NOT gates as required.

10. The circuit of Fig. 4-4(c) shows a half-adder designed with NAND logic. This was synthesized from the definition of the half-adder and Eq. (4-6) for the exclusive-OR.

    (a) Derive an equivalent expression for the exclusive-OR that lends itself to synthesis with NOR logic.

    (b) Design a half-adder with all NOR gates. What is the minimum number of gates required?

11. Design a ripple-through adder that adds numbers in 8-4-2-1 BCD code. Write the Boolean expression for the carry from one BCD group to the next higher group and synthesize the decoding logic required to generate the carry.

12. Repeat Prob. 11 using numbers in an *excess-3* code.

13. In a manner similar to the development of the simultaneous carry adder of Fig. 4-8, design a three-stage simultaneous carry subtractor.

14. A computer, as one of its functions, is required to add in parallel a pair of three-significant-digit binary numbers. One of the numbers, A, is always positive (inputs A_0, A_1, and A_2). The second number, B, may be negative or positive (inputs B_0, B_1, B_2, and B_3) where the fourth digit denotes the sign of B; i.e., B_3 is a **1** for B positive and **0** for B negative. If $|A| > |B|$, design a circuit that will add A and B. You may use half-adders, *RS* flip-flops, and AND, OR, NAND, NOR, or NOT gates as required.

15. Design an iterative addition-type multiplier, in block diagram form, that will form the product of two numbers, A and B, where A is a positive number and B is a negative number in 2's complement notation. Use an accumulator, as in Fig. 4-15, and assume the accumulator is programmed to add or subtract on command.

16. Redesign the shifting multiplier of Fig. 4-17 to perform the same function by using a ripple-through adder instead of an accumulator. An additional storage register may be used. Show the basic gating logic and timing relationships.

17. In the shifting multiplier circuit of Fig. 4-17, replace the clock by a double-clocked system that eliminates the need for the delay Δ. Draw a timing diagram that shows the relationship of the clock pulses (CP_1 and CP_2) and define the minimum elapse time between CP_1 and CP_2 in terms of the circuit functions.

18. Prepare a functional block diagram of a system that will perform restoring division. Describe critical timing relationships and show if double clocking is required. You may use an accumulator that is programmed to perform subtraction on command.

19. Perform the operation $295 \div 13$ in long-hand binary form, using both restoring and nonrestoring division. Repeat the nonrestoring division using 2's complement subtraction as required. Assume fixed binary point notation with the division carried out to four quotient digits beyond the radix point.

20. Develop a technique for nonrestoring division using 1's complement subtraction. Repeat Prob. 19 with this format.

21. Convert the numbers $A = (235.75)_{10}$ and $B = (12.625)_{10}$ to binary form and scale to suitable floating-point notation. Then perform the operations $A + B$, $A - B$, $A \times B$, and $A \div B$.

CHAPTER 5

SELECTED TIMING AND SWITCHING CIRCUITS

Circuits normally catagorized as *analog* are often used in the design of digital systems. Aside from obvious applications such as regulated power supplies, analog circuits are frequently required to interface digital functions to their sources of input data or to provide outputs to visual and audio displays. Conversion of data from analog to digital or digital to analog form is therefore an essential subject and is discussed in detail in Chap. 7.

This chapter analyzes the operating characteristics and circuit implementations of several analog and quasi-digital circuits. Most of these circuits are used because they produce acceptable results at lower cost or with less complexity than the equivalent purely digital approach. The designer must bear in mind, however, that the development of new devices, especially through advances in integrated circuit technology, are providing options and continuously changing the economic balance between the different approaches.

5-1 INTEGRATED CIRCUITS—THE OPERATIONAL AMPLIFIER AND THE VOLTAGE COMPARATOR

The *operational amplifier* is a high-gain, directly coupled device with relatively high input impedance and low output impedance. The amplifier is intended for use in feedback circuitry where, with sufficient gain, the characteristics of the feedback loop fully determine the overall performance.

The *voltage comparator* (also referred to as a *differential comparator*) is a device very ismilar to the operational amplifier. The voltage comparator typically has less open loop gain (less than 60 dB), lower output drive capability, and faster response than a typical operational amplifier.

Both devices are readily available in a variety of monolithic integrated circuit designs. The usual circuit implementation consists of one or two direct-coupled differential amplifier stages that provide voltage gain, followed by an output stage that restores the dc level and provides power amplification with low output impedance. Unfortunately, manufacturers have not standardized on a minimum number of common, interchangeable designs. Differences in specifications, power supply requirements, pin connections, etc. have led to unwarranted variations and insignificant differences that cause considerable confusion to the design engineer. Adding to the user's difficulties are variations in the methods of specification. Fortunately, with most feedback and switching applications, wide variations in characteristics can be tolerated and the overall performance of properly designed circuits determined.

Some of the more significant characteristics of typical operational amplifiers and voltage comparators follow.

Voltage Gain

As stated above, the input stage of either device is a differential amplifier. Two input connections are usually provided: one is the inverting input (− terminal) and the other is the noninverting input (+ terminal). The voltage gain is defined as the output signal divided by the differential in put signal. Measurements are made at low frequency or dc, with a given output load, and usually at room temperature. Manufacturers will specify either minimum gain, typical gain, or both.

Monolithic operational amplifiers are available with stated voltage gains of more than 100,000; however, for most switching applications voltage gains of greater than 5,000 are usually unnecessary. In the circuits that we will consider, very high open loop gain adds little to performance and makes the devices more sensitive to drift caused by temperature gradients within the amplifier.

Offset Voltage

Monolithic amplifiers are usually superior to discrete component operational amplifiers (nonchopper stabilized types) due to inherent component matching and physical proximity of the input transistor junctions. Typical offset voltage drifts are in the 5 to 50 $\mu V/^{\circ}C$ range. It should be noted, however, that these ratings are for steady-state conditions. Transient thermal gradients due to changes in signal levels cause more severe offsets. For example, a temperature differential of 0.1 °C between the junctions of the input pair of transistors will result in an offset of approximately 240 μV. Also, initial offset voltages in the 1 mV range are not uncommon. In more critical applications initial offset may be balanced out with an external biasing adjustment. However, the designer should be aware that external biasing will change the balance in the collector currents drawn by the input transistor pair and have an additional effect on the total drift rate.

Input Impedance

Specifications on input impedance range typically from 20,000 Ω to 5 MΩ for monolithic designs and are in the 10^{10} Ω range for discrete component MOS-FET devices. Requirements for input impedances greater than 10^6 Ω are encountered in sample-and-hold circuits (see Chap. 7) and in analog-to-digital converters when the analog signal is derived from a very high impedance source.

Output Swing

Manufacturer's specifications on output capabilities of operational amplifiers are in many instances less than optimum. The maximum output voltage swing (typically ± 5 to ± 15 V) and a maximum output current swing (typically ± 2 to ± 20 mA) are normally specified. However, the maximum voltage output and peak current output may not be available simultaneously. Also, full output may not be available over the entire stated frequency range of the amplifier.

Frequency Response

Ideally, an operational amplifier will have a flat frequency response from dc to some corner frequency (at which the gain is down 3 dB). The gain then will proceed with uniform attenuation of 20 dB/decade (6 dB/octave) to at least 10 dB below unity gain. This will ensure good stability when the amplifier is connected with negative feedback for reduced gain. Depending upon the operational amplifier chosen, this ideal characteristic is not inherent in the device and external compensation networks may be required. These are very simple, usually one or two capacitors and resistors, connected between terminals of the amplifier.

Whereas the operational amplifier response is given as a *gain-bandwidth product*, the voltage comparator response is defined in terms of switching speed. For example, a high-speed voltage comparator will switch from HIGH to LOW output in less than 100 nsec.

Slew Rate

The slew rate of an amplifier is defined as the time rate of change of output when the input is overdriven. Maximum slew rates in the 2 to 100 V/μsec range are available. The achievable rate in a given circuit may be considerably less than the manufacturer's figures because of external circuit impedance levels and stray capacitances.

5-2 OPERATIONAL AMPLIFIER CIRCUITS

The operational amplifier is frequently used as a summing amplifier in digital-to-analog converters. A basic part of this application is the use of appropiate feedback to achieve a desired transfer function. In Fig. 5-1 the fundamental feedback arrangement for the operational amplifier is shown.

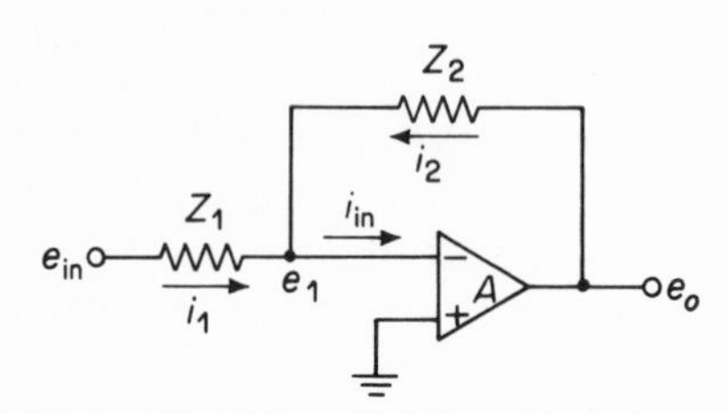

FIGURE 5-1. Operational Amplifier with Feedback Elements.

With the voltage gain A in the 10^3–10^6 range, and $|e_0| \leq 15$ V,

$$|e_1| = \frac{|e_0|}{A} \approx 0 \text{ V} \tag{5-1}$$

Also, if the input impedance is high relative to Z_1 and Z_2 (for frequencies of interest), then the input current is negligible and Eq. (5-2) is valid.

$$i_1 + i_2 \approx 0$$
$$i_1 = -i_2 \tag{5-2}$$

Linear Amplification

In the circuit of Fig. 5-1, replacing Z_1 by R_1 and Z_2 by R_2 and applying Eq. (5-2), Fig. 5-2 is derived. Refering to Fig. 5-2, the gain of the resulting linear amplifier is calculated as follows:

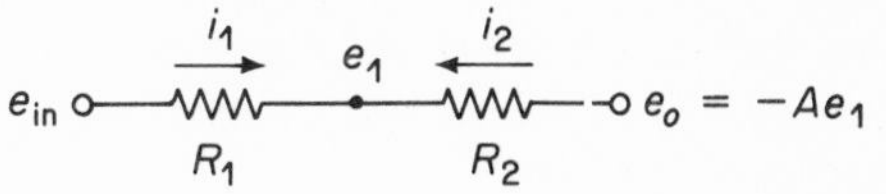

FIGURE 5-2. Equivalent Circuit of Operational Amplifier with Resistive Feedback Elements.

e_{in} — R_1 (i_1 →) — e_1 — R_2 (← i_2) — $e_o = -Ae_1$

$$i_1 = -i_2$$

$$\frac{e_{in} - e_1}{R_1} = -\frac{e_0 - e_1}{R_2} \tag{5-3}$$

$$e_{in}R_2 + e_0R_1 = e_1(R_1 + R_2) \tag{5-4}$$

Substituting $-(e_0/A)$ for e_1 in Eq. (5-4) and reducing, Eq. (5-5) is derived,

$$\frac{e_0}{e_{in}} = \frac{-AR_2}{R_1(1 + A) + R_2} \tag{5-5}$$

which may be rewritten as Eq. (5-6).

$$\frac{e_0}{e_{in}} = \frac{-R_2}{R_1[(1 + A)/A] + (R_2/A)} \tag{5-6}$$

For A very large,

$$\frac{1 + A}{A} \rightarrow 1 \quad \text{and} \quad \frac{R_2}{A} \ll R_1$$

Therefore, the closed-loop transfer function (gain of the linear circuit) reduces to

$$A_v = \frac{e_0}{e_{in}} \approx -\frac{R_2}{R_1} \tag{5-7}$$

Eq. (5-7) is an approximation and is equivalent to assuming that $e_1 = 0$. The gain of the circuit is relatively insensitive to wide variations in A, and the maximum error or variation in gain is completely predicted by specifying a minimum value of A.

The stability of the circuit also must be considered. The feedback circuit could oscillate if the open-loop response of the operational amplifier did not have proper high frequency characteristics. As stated in Sec. 5-1, the open-loop gain of the amplifier should be flat to some fixed frequency and then roll off with a slope of -6 db/octave. Some operational amplifiers have this response designed into the device, whereas others require external compensation networks to achieve the required response. Information regarding external compensation networks is usually provided in manufacturers' data sheets.

Summing Amplifier

A linear summation of several input voltages can be achieved by replacing Z_1 with n resistors, as shown in Fig. 5-3. Assuming that $e_x \approx 0$, and Z_{in} is very high,

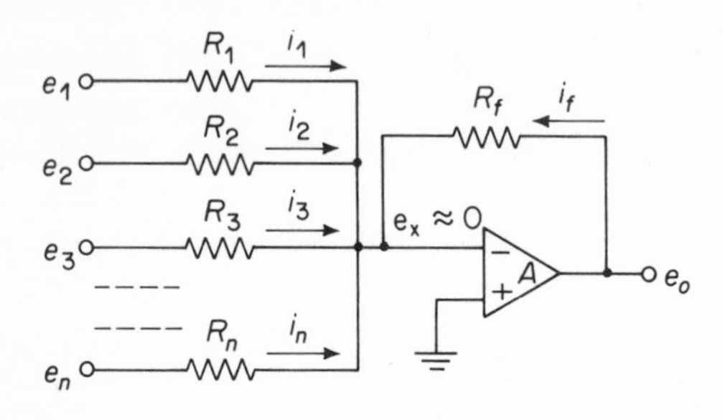

FIGURE 5-3. Operational Summing Amplifier Circuit.

$$i_1 + i_2 + i_3 + \cdots + i_n = -i_f \tag{5-8}$$

$$\frac{e_1}{R_1} + \frac{e_2}{R_2} + \frac{e_3}{R_3} + \cdots + \frac{e_n}{R_n} = -\frac{e_0}{R_f} \tag{5-9}$$

$$e_0 = -R_f\left(\frac{e_1}{R_1} + \frac{e_2}{R_2} + \cdots + \frac{e_n}{R_n}\right) \tag{5-10}$$

If the resistors $R_f, R_1, R_2, \ldots, R_n$ are all equal in value, then Eq. (5-10) becomes

$$e_0 = -(e_1 + e_2 + e_3 + \cdots + e_n) \tag{5-11}$$

Also, a weighted summation may be obtained by proper choice of resistors. This is one method of achieving digital-to-analog conversion and is discussed in Chap. 7.

5-3 SCHMITT TRIGGER CIRCUIT

The Schmitt trigger is an important switching circuit that is widely used in digital systems as an amplitude comparator. The device has two output states that are functions of the amplitude of the input excitation. Therefore, the Schmitt trigger functions as a one-bit analog-to-digital convertor. Switching between the two output states is very rapid due to internal regenerative feedback that aids the transition from one state to the other. In this manner the Schmitt trigger is similar to the *RS* flip-flop; however, the two circuits differ in several of their operating characteristics. Triggering of the device is achieved at a single input terminal. Also, the output does not retain its output state when the input excitation is removed; therefore (except for a small hysteresis region), the device does not exhibit the properties of a memory element.

The input-output relationship of the Schmitt trigger is illustrated in Fig. 5-4. Observe that when the input excitation exceeds the threshold V_{T1},

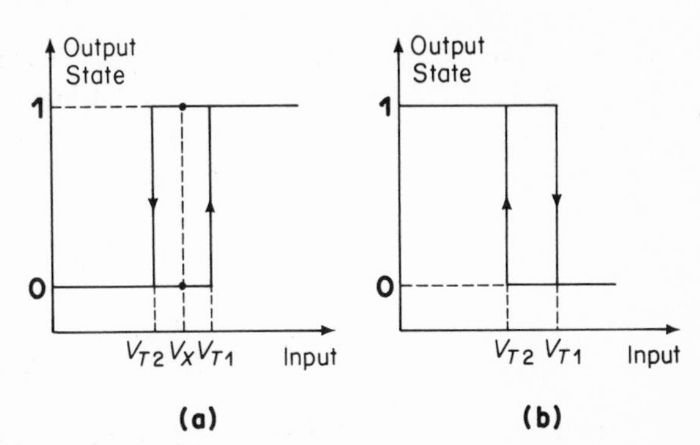

FIGURE 5-4. Output Response of a Schmitt Trigger to Input Excitation Level.

the output switches to the **1** state. Then, as the input is reduced below the threshold V_{T2}, the output returns to the **0** state. As indicated, the device exhibits hysteresis in the region between V_{T1} and V_{T2}. Note that if the input is set at V_x the output will remain indefinitely in its current logical state of either **1** or **0**. The circuit also can be designed to yield inverted output states, as shown in Fig. 5-4(b).

Although the Schmitt trigger retains its current state when the input excitation level is between V_{T1} and V_{T2}, the device is not normally used as a memory element. Rather, the ability to determine when the input signal is above or below given levels makes the circuit useful as a threshold device. In an amplitude shift keying (ASK) system, for example, the device may be used to determine when a received signal is above or below the threshold separating the **1** and **0** regions.

Circuit Design of the Schmitt Trigger

A Schmitt trigger may be designed either with discrete components or by applying appropriate feedback from the output to the input of an integrated circuit operational amplifier. The latter approach is simple to implement, yields excellent results, and clearly illustrates the operation of the device; therefore, we will start by analyzing the operational amplifier implementation.

In order to facilitate the illustration of this application, some of the pertinent characteristics of a representative operational amplifier are given in Fig. 5-5. Actually, the Schmitt trigger design is relatively insensitive to different amplifier specifications, and the basic characteristics listed are used merely as typical values.

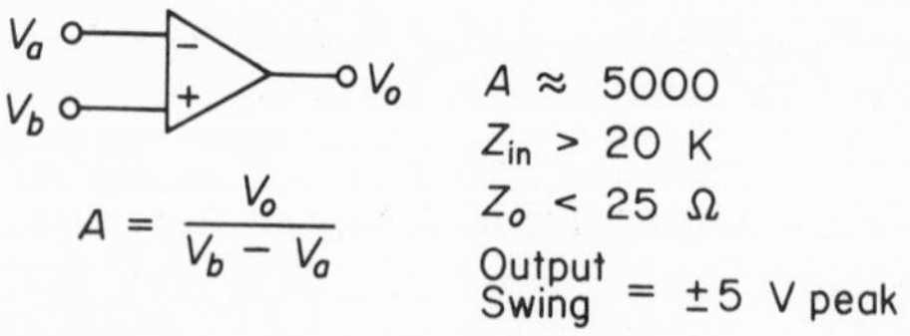

$$A = \frac{V_o}{V_b - V_a}$$

$A \approx 5000$
$Z_{in} > 20$ K
$Z_o < 25\ \Omega$
Output Swing $= \pm 5$ V peak

FIGURE 5-5. Typical Characteristics of an Operational Amplifier.

Consider the circuit of Fig. 5-6(a). The value of R_1 is chosen to be very large compared to the amplifier's output impedance, and the parallel combination of R_1 and R_2 is small compared to the input impedance. The feedback voltage (V_{fb}) is then given by Eq. (5-12). V_{TH} is an externally applied voltage

$$V_{fb} = V_{TH} + \frac{(V_0 - V_{TH})R_2}{(R_1 + R_2)}$$

$$V_{fb} = \frac{V_{TH}R_1}{R_1 + R_2} + \frac{V_0 R_2}{R_1 + R_2} \tag{5-12}$$

that sets the threshold level of the Schmitt trigger. The feedback from output to input is positive and, assuming the source impedance of V_{TH} is low, the feedback loop gain is equal to $AR_2/R_1 + R_2$. With positive feedback and the *feedback loop gain* > 1, the circuit is unstable in its linear operating region; therefore, the output is forced into saturation at either $+5$ or -5 V.

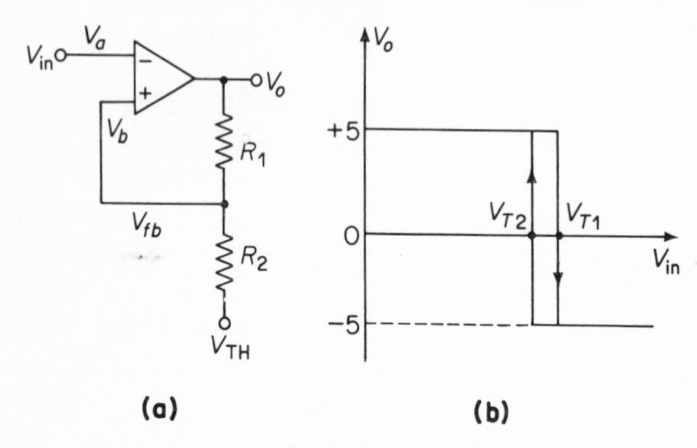

FIGURE 5-6. (a) Schmitt Trigger Circuit and (b) Input-Output Response.

When the output is in saturation at $+5$ V the voltage fed back is given by Eq. (5-13). When in saturation at -5 V, the voltage fed back is given by

$$V_{fb1} = \frac{V_{TH}R_1}{R_1 + R_2} + \frac{5R_2}{R_1 + R_2} \tag{5-13}$$

$$V_{fb2} = \frac{V_{TH}R_1}{R_1 + R_2} - \frac{5R_2}{R_1 + R_2} \tag{5-14}$$

Eq. (5-14). The circuit of Fig. 5-6(a) will now be analyzed using the component values shown in Fig. 5-7(a). Since $R_1 = 10$ K is much larger than Z_0

and Z_{in} is much larger than the source impedance of the feedback voltage,

$$\text{Feedback loop gain} \approx \frac{R_2 A}{R_1 + R_2} \geq \frac{50(5{,}000)}{50 + 10{,}000} = 25$$

This exceeds unity by an adequate margin, and proper regenerative switching between output states is assured. Applying Eqs. (5-13) and (5-14) and using a $V_{TH} = 2.0$ V, V_{fb1} and V_{fb2} are obtained:

$$V_{fb1} = \frac{2.0(10{,}000)}{50 + 10{,}000} + \frac{5(50)}{50 + 10{,}000} \approx 2.025 \text{ V}$$

$$V_{fb2} = \frac{2.0(10{,}000)}{50 + 10{,}000} - \frac{5(50)}{50 + 10{,}000} \approx 1.975 \text{ V}$$

Assuming that the input is set at a value smaller than V_{TH} (e.g., let $V_{in} = -2$ V), V_0 is forced into saturation at $+5$ V, provided Eq. (5-15) remains satisfied.

$$(V_{fb1} - V_{in})(A) = (2.025 - V_{in})(5{,}000) \geq 5$$
$$V_{in} \leq 2.024 \text{ V} \tag{5-15}$$

If V_{in} is raised above $+2.024$ V the output comes out of saturation, initiating regenerative switching and forcing the output to -5 V, provided Eq. (5-16) is satisfied.

$$(V_{fb2} - V_{in})A = (1.975 - V_{in})(5{,}000) \leq -5$$
$$V_{in} \geq 1.976 \text{ V} \tag{5-16}$$

The limiting values of V_{in} from Eqs. (5-15) and (5-16) therefore define V_{T1} and V_{T2}. This results in the input characteristic shown in Fig. 5-7(b).

The small difference between V_{T1} and V_{T2} (in this example, 48 mV) is equal to the hysteresis. The hysteresis may be reduced by decreasing the feedback loop gain. When the loop gain equals unity, the hysteresis is reduced to zero. Operation at near zero hysteresis is difficult to achieve because of component tolerances and variations in amplifier gain. Also, as the feedback loop gain is lowered, regenerative action is reduced and the switching time increases. When operating with the loop gain at less than unity, the circuit ceases to function as a Schmitt trigger. Then, as shown in Fig. 5-7(d), the device becomes a differential amplifier with a linear operating region near the value of the threshold setting V_{TH}.

Elimination of all regenerative feedback (by setting R_2 to zero and opening R_1) results in narrowing the linear operating region to a 2-mV spread

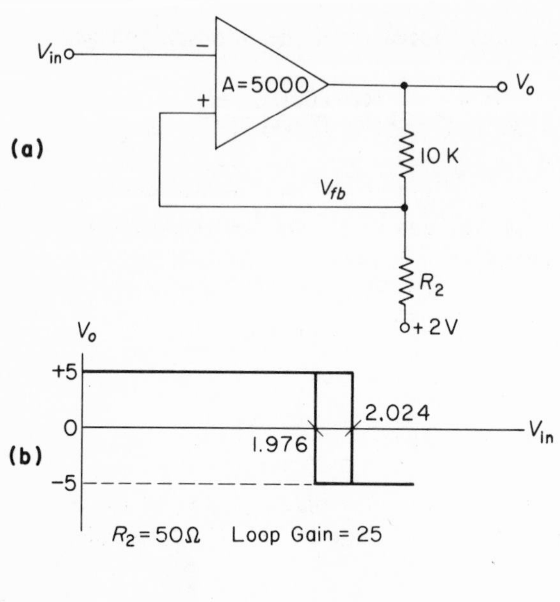

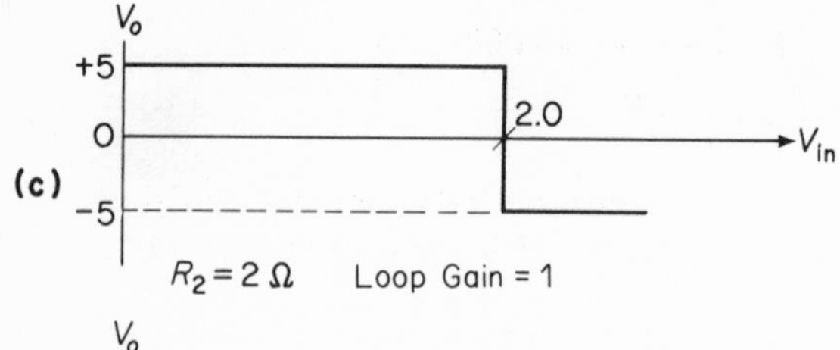

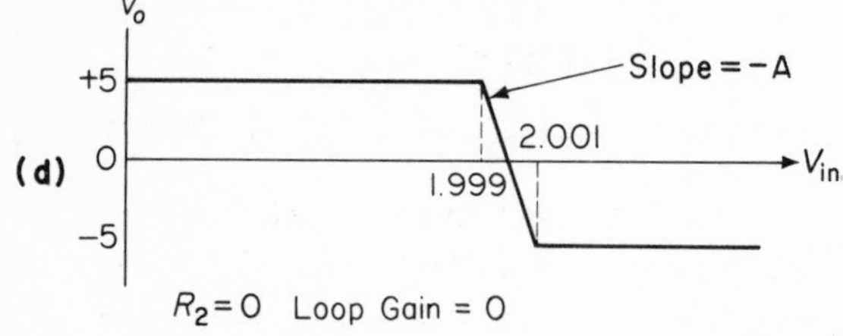

FIGURE 5-7. Schmitt Trigger Circuit and the Effect of Feedback Loop Gain on Input-Output Characteristics.

centered around V_{TH}. In this configuration use of a high-speed *differential comparator* is preferable to the operational amplifier.

Discrete Component Design

A discrete component design of the Schmitt trigger, using transistors for amplifiers and emitter-coupled feedback, is shown in Fig. 5-8. The component values indicated correspond to the design used in the analysis.

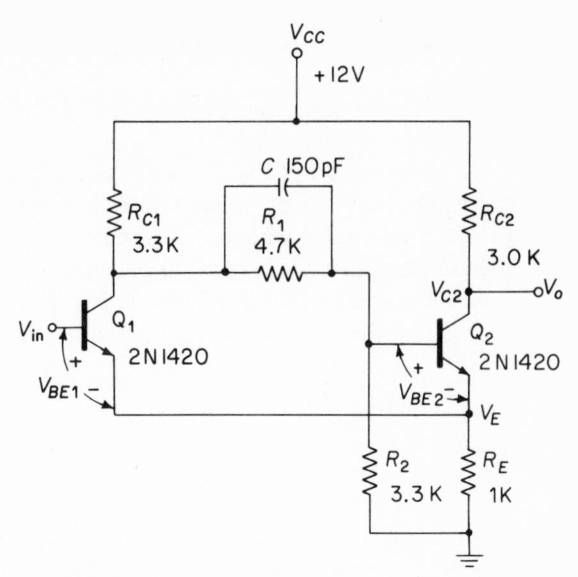

FIGURE 5-8. Emitter-Coupled Schmitt Trigger Design.

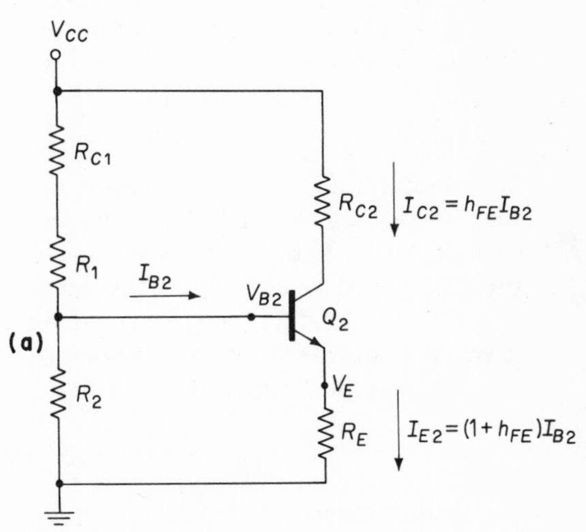

(b) $R_O = \dfrac{R_2(R_1 + R_{C1})}{R_2 + R_1 + R_{C1}}$

V_{BE2}

V_E

I_{B2}

$V' = \dfrac{R_2 V_{CC}}{R_1 + R_2 + R_{C1}}$

$R'_E = R_E(1 + h_{FE})$

FIGURE 5-9. (a) The Equivalent Circuit of Q_2 with Q_1 in Cutoff, and (b) the Thévenin Equivalent of the Base Drive to Q_2.

When feedback loop gain is greater than unity, the input-output characteristic is as shown in Fig. 5-4(a).

In the circuit of Fig. 5-8, Q_1 is in cutoff and Q_2 is conducting whenever the input excitation V_{in} is less than V_{T2}. If the I_{CBO} of Q_1 is assumed negligible (which is reasonable with silicon transistors) and R_{C2} is chosen small enough to prevent saturation of Q_2, then the operating conditions of Q_2 can be calculated from the equivalent circuits shown in Fig. 5-9.

From Fig. 5-9(b), I_{B2} and V_E are determined as follows:

$$I_{B2} = \frac{V' - V_{BE2}}{R_0 + (1 + h_{FE})R_E} \tag{5-17}$$

$$V_E = V' - V_{BE2} - I_{B2}R_0 \tag{5-18}$$

Switching occurs when the input excitation V_{in} rises to the value that permits Q_1 to just start conducting. The regenerative action of the feedback loop then will rapidly force Q_1 to saturate ON and turn Q_2 OFF. V_{T1}, the value of V_{in} at which switching action is triggered, is given by Eq. (5-19).

$$V_{T1} = V_E + V_{\gamma 1} \tag{5-19}$$

In Eq. (5-19), $V_{\gamma 1}$ is the transistor *cut-in voltage*, the base-emitter voltage when Q_1 begins to conduct. This is approximately $+0.6$ V for a silicon transistor at $+25\,^{\circ}\mathrm{C}$. The cut-in voltage also is approximately 0.1 V less than the base-emitter voltage drop V_{BE} of the transistor in conduction. This latter approximation is valid for both silicon and germanium transistors operating with relatively small base drives ($I_B \leq 5$ mA).

V_{T1} may be determined from Eqs. (5-18) and (5-19) as follows:

$$V_{T1} = V' - V_{BE2} + V_{\gamma 1} - I_{B2}R_0$$

If Q_1 and Q_2 are similar transistors, $-V_{BE2} + V_{\gamma 1} \approx -0.1$ V and

$$V_{T1} = V' - 0.1 - I_{B2}R_0 \tag{5-20}$$

The second switching level V_{T2} is determined from the equivalent circuit of Fig. 5-10. Q_2 is cut off, and leakage currenis are assumed negligible ($I_{C2} = I_{B2} = 0$). V'' and R' are the Thévenin equivalent of V_{CC}, R_{C1}, R_1, and R_2.

If the input excitation is reduced to (or below) V_{T2}, Q_2 starts to conduct and Q_1 is cut off. Switching occurs when the base-emitter voltage of Q_2 equals $V_{\gamma 2}$. This yields the following:

$$V_{B2} = V_E + V_{\gamma 2} \tag{5-21}$$

where at the instant of switching

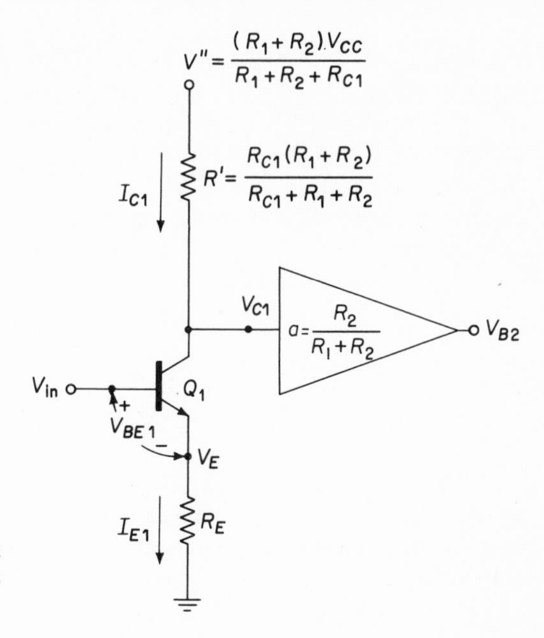

FIGURE 5-10. Equivalent Circuit of Schmitt Trigger When Q_1 Is Conducting.

$$V_E = V_{T2} - V_{BE1} \tag{5-22}$$

and

$$V_{B2} = aV_{C1} = a(V'' - I_{C1}R')$$

$$V_{B2} \approx a\left[V'' - \left(\frac{V_{T2} - V_{BE1}}{R_E}\right)R'\right] \tag{5-23}$$

Substituting Eqs. (5-22) and (5-23) in Eq. (5-21) and simplifying, V_{T2} is calculated as follows:

$$V_{T2} = V_{BE1} + \frac{(aV'' - V_{\gamma 2})R_E}{R_E + aR'} \tag{5-24}$$

where

$$a = \frac{R_2}{R_1 + R_2}$$

$$R' = \frac{R_{C1}(R_1 + R_2)}{R_{C1} + R_1 + R_2}$$

$$V'' = \frac{(R_1 + R_2)V_{CC}}{R_1 + R_2 + R_{C1}}$$

The component values shown in Fig. 5-8 will now be used to illustrate the design of the emitter-coupled Schmitt trigger.

$$V' = \frac{R_2 V_{CC}}{R_1 + R_2 + R_{C1}} = \frac{3.3\text{ K}(12)}{3.3\text{ K} + 4.7\text{ K} + 3.3\text{ K}} = 3.5\text{ V}$$

$$R_0 = \frac{R_2(R_1 + R_{C1})}{R_1 + R_2 + R_{C1}} = \frac{3.3\text{ K}(3.3\text{ K} + 4.7\text{ K})}{3.3\text{ K} + 4.7\text{ K} + 3.3\text{ K}} = 2.33\text{ K}$$

Assuming $h_{FE} = 50$ for the 2N1420 transistor,

$$I_{B2} = \frac{V' - V_{BE2}}{R_0 + (1 + h_{FE})R_E} = \frac{3.5 - 0.7}{2.33\text{ K} + 51\text{ K}} = 53\ \mu\text{A}$$

$$\begin{aligned} V_{T1} &= V' - 0.1 - I_{B2}R_0 \\ &= 3.5 - 0.1 - (53 \times 10^{-6})(2.33 \times 10^3) \\ &\approx 3.3\text{ V} \end{aligned}$$

The value of $I_{B2}R_0$ is approximately 0.1 V. This is a reasonable approximation for a wide range of typical emitter-coupled Schmitt trigger designs. The error in V_{T1} due to simply assuming $V_{T1} = V' - 0.2$ V is negligible compared to the variations in V' resulting from the resistance tolerances of R_1, R_2, and R_{C1}. V_{T2} is calculated as follows:

$$V'' = \frac{(R_1 + R_2)V_{CC}}{R_1 + R_2 + R_{C1}} = \frac{12(4.7\text{ K} + 3.3\text{ K})}{3.3\text{ K} + 4.7\text{ K} + 3.3\text{K}} = 8.5\text{ V}$$

$$a = \frac{R_2}{R_1 + R_2} = \frac{3.3\text{ K}}{3.3\text{ K} + 4.7\text{ K}} = 0.41$$

$$R' = \frac{R_{C1}(R_1 + R_2)}{R_{C1} + R_1 + R_2} = \frac{3.3\text{ K}(3.3\text{ K} + 4.7\text{ K})}{3.3\text{ K} + 4.7\text{ K} + 3.3\text{ K}} = 2.34\text{ K}$$

$$\begin{aligned} V_{T2} &= V_{BE1} + \frac{(aV'' - V_{\gamma 2})R_E}{R_E + aR'} \\ &= 0.7 + \frac{(0.41 \times 8.5 - 0.6)1{,}000}{1{,}000 + 0.41(2.34\text{ K})} \\ &= 2.2\text{ V} \end{aligned}$$

Finally, a check is made to show that Q_2 is not in saturation when in the ON state:

$$\begin{aligned} I_{C2} &= I_{B2}h_{FE} \\ &= (53 \times 10^{-6})50 \\ &= 2.65\text{ mA} \end{aligned}$$

$$\begin{aligned} V_{C2} &= V_{CC} - I_{C2}R_{C2} \\ &= 12 - (2.65 \times 10^{-3})(3{,}000) \\ &= 4.1 \text{ V} \end{aligned}$$

and

$$\begin{aligned} V_E &\approx I_{C2}R_E \\ &= (2.65 \times 10^{-3})(1{,}000) \\ &= 2.65 \text{ V} \end{aligned}$$

V_{CE2} therefore is 1.45 V, which ensures linear operation of Q_2.

This discussion of the discrete component design of the Schmitt trigger was presented partly because the device is a classical example of switching circuits that were used in digital systems prior to the introduction of integrated circuits. Further detail such as methods for independently varying V_{T1} and V_{T2} are treated in the literature,[3] and inclusion of this material is not warranted. Actually, the current availability of fast-acting, high-gain integrated circuit differential comparators has obviated the need for a regenerative device in most applications.

5-4 THE MONOSTABLE MULTIVIBRATOR

The monostable multivibrator (also called a one-shot multi) is similar to the flip-flop and Schmitt trigger insofar as all three are closed loop devices with regenerative feedback. In the monostable circuit, an energy storage element, usually a capacitor, is introduced in series with the regenerative loop. This creates a device with one stable state and one quasistable state. The circuit rests in the stable state until triggered by an input excitation signal. Then regenerative action is initiated, switching the device into the quasistable state. The circuit remains in the quasistable state for a fixed elapsed time, determined by the circuit component values, and switches back to the stable state.

Monostable multivibrators are regenerative devices used to generate rectangular output pulses of a predetermined width. The device often is used to generate uniform data and timing pulses in digital systems or as an inexpensive timing circuit. Using components with $\pm 1\%$ tolerance, pulse widths of 1 μsec to several seconds in length are easily produced with accuracies of better than $\pm 5\%$. Pulse width tolerances of $\pm 1\%$ are attainable with the addition of a single adjustment.

Discrete Component Design of the Monostable Multivibrator

A collector-coupled monostable multivibrator is shown in Fig. 5-11

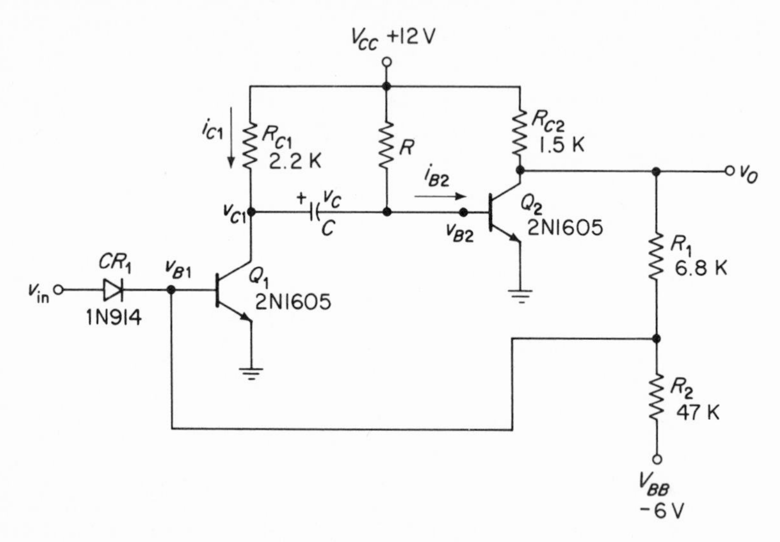

FIGURE 5-11. Collector-Coupled Monostable Multivibrator.

The component values indicated are used to illustrate the analysis numerically. Significant waveforms are given in Fig. 5-12.

Operation in the Stable State. When the monostable multi is in the stable state, Q_2 is biased into saturation by R and V_{CC}. The output v_0 is then equal to $V_{CE2}(\text{sat})$, and Q_1 is cut off by the reverse bias supplied by R_2, R_1, and V_{BB}. In the stable state C charges to $v_C = V_C = V_{CC} - V_{BE2}$, with the polarity as indicated in Fig. 5-11.

Saturation of Q_2 and cut off of Q_1 are calculated as follows:

$$I_{C2}(\text{sat}) = \frac{V_{CC} - V_{CE2}}{R_{C2}} - \frac{V_{CE2} - V_{BB}}{R_1 + R_2} \tag{5-25}$$

$$\approx \frac{V_{CC}}{R_{C2}} = \frac{12\text{ V}}{1.5\text{ K}} = 8\text{ mA}$$

The minimum value of I_{B2} required is

$$I_{B2}(\text{min}) \doteq \frac{I_{C2}(\text{sat})}{h_{FE}(\text{min})} \tag{5-26}$$

$$= \frac{8\text{ mA}}{40} = 200\ \mu\text{A}$$

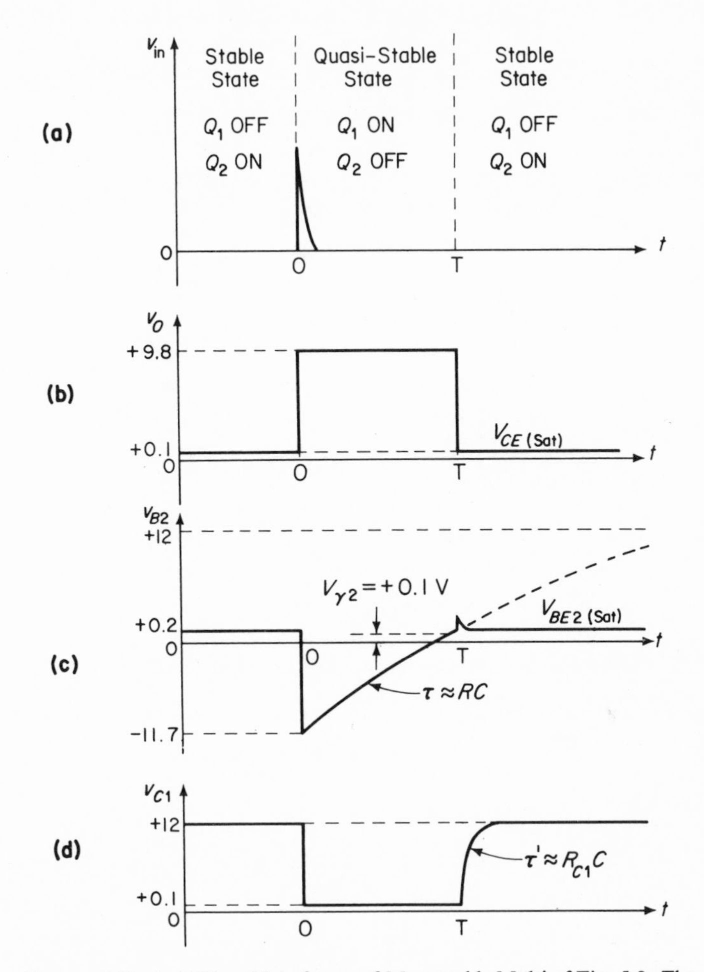

FIGURE 5-12. Switching Waveforms of Monostable Multi of Fig. 5-8. The Exponential Portions Have Time Constants $\tau \approx RC$ and $\tau' \approx R_{C1}C$ as Indicated.

The maximum allowable value of R is

$$R_{\max} = \frac{V_{CC} - V_{BE}(\text{sat})}{I_{B2}(\text{min})} \tag{5-27}$$

$$= \frac{12 - 0.2}{0.200} \approx 60\text{ K}$$

Excluding the leakage current I_{CBO} of transistor Q_1 when in cutoff the base voltage of Q_1 is calculated to be

$$v_{B1} = V_{BE1} = V_{BB} + \frac{R_2(V_{CE2} - V_{BB})}{R_1 + R_2} \quad (5\text{-}28)$$

$$= -6 + \frac{47{,}000(0.1 + 6.0)}{47{,}000 + 6{,}800} = -0.74 \text{ V}$$

The source impedance of V_{BE1} is the parallel combination of R_1 and R_2 (approximately 6 K); therefore, the base-emitter junction of Q_1 will remain reverse biased, provided

$$I_{CBO} \leq \frac{0.74 \text{ V}}{6 \text{ K}} \approx 120\ \mu\text{A}$$

This fixes the maximum junction temperature of Q_1. A transistor's leakage current doubles for approximately every $+10°\text{C}$ rise in junction temperature. If the leakage current $I_{CBO} = 3\ \mu\text{A}$ at $+25°\text{C}$, then Q_1 will remain safely cut off at temperatures as high as $+75°\text{C}$.

Operation in the Quasistable State. A positive input excitation pulse, sufficient in amplitude and duration to start conduction in Q_1, will initiate switching. The drop in collector voltage at Q_1 is then coupled through C to the base of Q_2. If sufficient to take Q_2 out of saturation, the collector voltage at Q_2 rises and feeds back a positive-going waveform to Q_1. The regenerative feedback then causes a rapid transition into the quasistable state, placing Q_1 in saturation and Q_2 in cutoff.

With the collector of Q_1 at saturation potential, Q_2 is kept reverse biased until the capacitor C recharges (through R) sufficiently to raise v_{B2} above $V_{\gamma 2}$. The voltage waveforms, at v_{B2} and across the capacitor (v_c), are given in Fig. 5-12. The charge and discharge time constants τ and τ', are calculated from the equivalent circuits in Fig. 5-13.

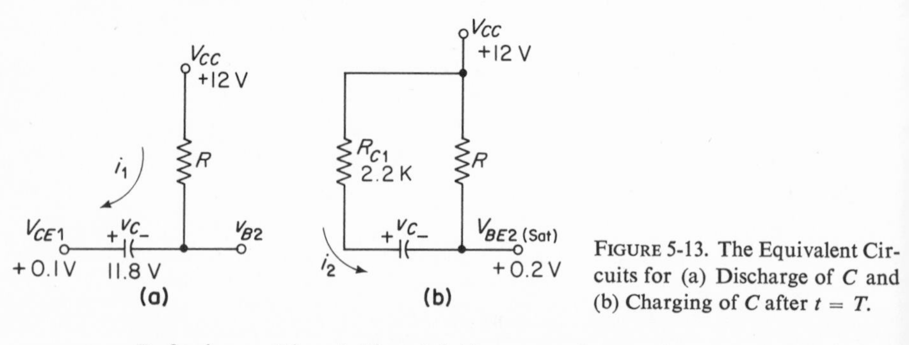

FIGURE 5-13. The Equivalent Circuits for (a) Discharge of C and (b) Charging of C after $t = T$.

Referring to Figs. 5-12 and 5-13, we see that at time $t = 0$, the instant after switching into the quasistable state, the capacitor is charged $v_C = +11.8$ V. Q_1 is in sasuration and V_{CE1} is at approximately $+0.1$ V and $v_{B2} = V_{CE1} - v_C = -11.7$ V. An important consideration here is to not exceed the reverse emitter-base voltage rating of Q_2 (see Prob. 8). The capacitor is now being discharged by i_1, as shown in Fig. 5-13(a). The output impedance of Q_1 in saturation is negligble compared to the value of R; therefore, the discharge time constant is $\tau = RC$. In the region $0 < t \leq T$, v_{B2} is given by Eq. (5-29).

$$v_{B2} = V_{\text{initial}} + (V_{\text{final}} - V_{\text{initial}})(1 - \epsilon^{-(t/RC)}) \tag{5-29}$$

$$v_{B2} = -11.7 + (12 + 11.7)(1 - \epsilon^{(t/RC)})$$

$$v_{B2} = 12 - 23.7\epsilon^{(t/RC)}$$

When v_{B2} equals the cut-in voltage of $Q_2(+0.1\text{ V})$, the circuit switches back to the stable state. This occurs, by definition, at time $t = T$ and may be calculated as follows:

$$v_{B2} = 0.1 = 12 - 23.7\epsilon^{-(T/RC)}$$

$$T = RC \log_\epsilon \frac{23.7}{11.9}$$

$$\approx 0.69RC \tag{5-30}$$

Using the circuit configuration of Fig. 5-11, Eq. (5-30) yields reasonably accurate results with either germanium or silicon transistors. Reducing the power supply voltage to +6 V also will have a negligible effect on T. When using silicon transistors, lower power supply voltages are preferable since transistors wish V_{BEO} ratings of greater than 7.0 V are not usually available in small signal devices.

At time $t = T$, C starts recharging to $v_C = +11.8$ V, as shown in Fig. 5-13(b). If the input resistance of Q_2 is considered small compared to R_{C1} then $\tau' \approx R_{C1}C$. The monostable should not be triggered until C has fully recharged to +11.8 V, otherwise the output pulse will be shorter than 0.69 RC. Therefore, a minimum recovery time of 3-5 τ' should be allowed in the stable state between pulses.

Integrated Circuit Design of the Monostable Multivibrator

Monostable multivibrators are available from integrated circuit manufacturers on a single monolithic chip. Thesa devices have both an internal timing network and provisions for connecting external timing capacitors and resistors. Use of the internal timing yields output pulse widths in the 100–500 nsec range. The output pulse width varies widely with temperature, and variations from unit to unit are large. Variations of 2:1 in pulse width are not uncommon; therefore, the internal components cannot be relied upon for accurate timing. By adding external components, pulse widths can be lengthened into the high millisecond range.

The output pulse amplitude is typically in the 2-3 V range. This is compatible with the input requirements of other integrated circuit devices but is generally insufficient to drive discrete component circuits without further amplification.

Although the monostable function is available in most *families* of integrated circuits, the devices are relatively expensive because of their low usage. A simple monostable multivibrator can be designed by using a few passive components and either a pair of NAND or NOR gates (available in a single package). A typical circuit using two NAND gates is shown in Fig. 5-14. The circuit rests in the stable state with the input HIGH. The output of gate *A* is then LOW, and the output of gate *B* is HIGH.

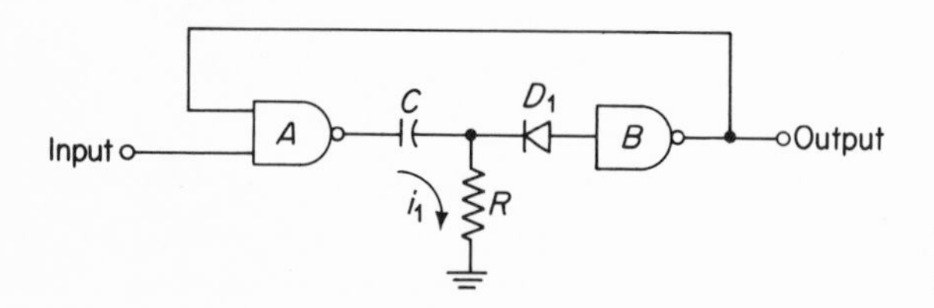

FIGURE 5-14. NAND Gate Design of Monostable Multivibrator.

Switching to the quasistable state is initiated by a negative-going pulse at the input that is equivalent to a **0** input. The output of gate *A* then goes HIGH, causing *C* to charge through *R* by the flow of i_1. This couples a **1** to gate *B*, (assuming *B* is a sink load), which drives the output LOW and keeps the output of gate *A* HIGH. When the voltage drop across *R* drops below the **1** threshold of gate *B*, (approximately 1.4 V), the output of gate *B* goes HIGH and the circuit reverts to the stable state.

Inclusion of the diode CR_1 is optional, it is a method for temperature compensation. At elevated temperatures the switching threshold of the gate decreases, causing the output pulse width to increase. The decreasing cut-in voltage of the diode with temperature tends to keep the pulse width constant.

Typical values of *R* are in the 100 to 500 Ω range. Using a germanium diode for CD_1, $R = 100\ \Omega$, and $C = 10\ \mu\text{F}$, the output pulse width will be approximately 1 msec. Pulse width accuracies of $\pm 20\%$ are attained over a temperature range of 0 to + 75 °C.

5-5 DIGITALLY CONTROLLED PULSE GENERATOR

It is instructive at this time to make a departure from the analog timing approach to the monostable multi design. The generation of rectangular pulses with very precise periods is often required for gating and timing signals. The best short term accuracy by analog means is in the order of $\pm 1\%$. Also, periodic adjustments are necessary to compensate for the aging of components. These limitations can be overcome by using digital logic and an accurate high-frequency clock (the latter normally being available in a clocked digital system).

The circuit shown in Fig. 5-15 is one approach to the design of digitally controlled pulse generators. Pulse width accuracy of this circuit is a function of the clock period T.

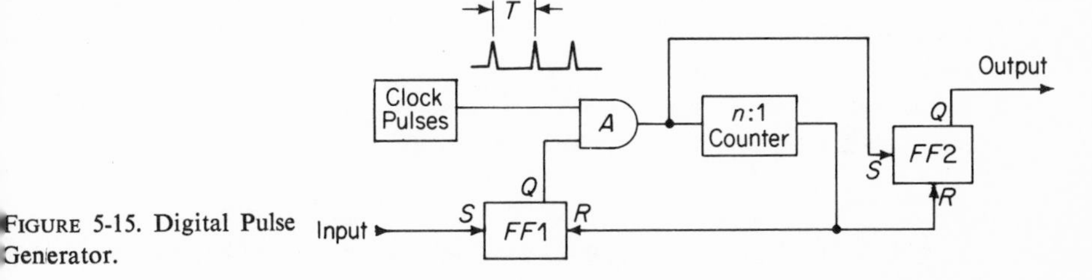

FIGURE 5-15. Digital Pulse Generator.

The digital pulse generator rests in the stable state with $FF1$ and $FF2$ both reset. AND gate A is disabled with the output of $FF1$ LOW, and the clock pulses are blocked from entering the counter. The design of n:1 counters is discussed in Chap. 6. For this discussion it will be sufficient to define the counter as a sequential circuit that yields one output pulse (and clears itself) for every n input pulses.

A **1** input sets $FF1$ and enables gate A. The next n pulses enter the counter. The nth pulse produces an output that resets $FF1$. $FF2$ is set by the first clock pulse that enters the counter and reset by the output of the counter; therefore, the output of $FF2$ is HIGH for exactly $(n - 1)T$ sec. The circuit will function properly if the width of the clock pulse is shorter than the propagation delay in the counter, so that the set input to $FF2$ is not HIGH simultaneously with the reset command.

When very accurate output pulses of variable width are required, the fixed n:1 counter may be replaced by a commercial preset counter. These instruments are available from several manufacturers. They may be preset to count over a range, typically from 00001 to 99999, by turning decimal-coded dials. Most models are supplied with an internal clock that may be used for timing.

5-6 VOLTAGE TIME-BASE GENERATORS; LINEAR SWEEP CIRCUITS

The linear voltage sweep finds widespread usage in the deflection circuits of cathode-ray tube displays and in analog-to-digital converters. Two methods for generating linear sweeps are the Miller integrator circuit and the digitally controlled sweep circuit. The Miller integrator utilizes a high-gain amplifier with negative feedback to increase the effective time constant of an *RC* charging circuit. Either an integrated circuit or discrete component operational amplifier may be used. The digitally controlled sweep circuit makes use of a master clock, digital counter, and digital-to-analog converter. The latter approach achieves greater accuracy at the expense of more circuit complexity.

To continue our discussion of analog circuits used in digital systems we shall analyze two of the more widely used Miller integrator sweep generators.

The Operational Amplifier Sweep Generator

In Sec. 5-2 the transfer function for the circuit of Fig. 5-1 was derived for the special case where Z_1 and Z_2 are purely resistive. This resulted in Eq. (5-5), which is now repeated:

$$\frac{e_o}{e_{\text{in}}} = \frac{-AR_2}{R_1(1 + A) + R_2} \tag{5-5}$$

Eq. (5-5) is based on the assumption that the input current to the operational amplifier is negligible.

If we replace R_1 and R_2 by the complex impedances Z_1 and Z_2, then the transfer function must be written in terms of voltages written in operational form, as in Eq. (5-31).

$$\frac{E_o(s)}{E_{\text{in}}(s)} = \frac{-AZ_2}{Z_1(1 + A) + Z_2} \tag{5-31}$$

In the circuit of Fig. 5-16, Z_1 is replaced by the resistor R and Z_2 by capacitor C. The transfer function, found by substituting R for Z_1 and $1/sC$ for Z_2 in Eq. (5-31), is

$$\frac{E_o(s)}{E_{\text{in}}(s)} = \frac{-A}{1 + (A + 1)RCs} \tag{5-32}$$

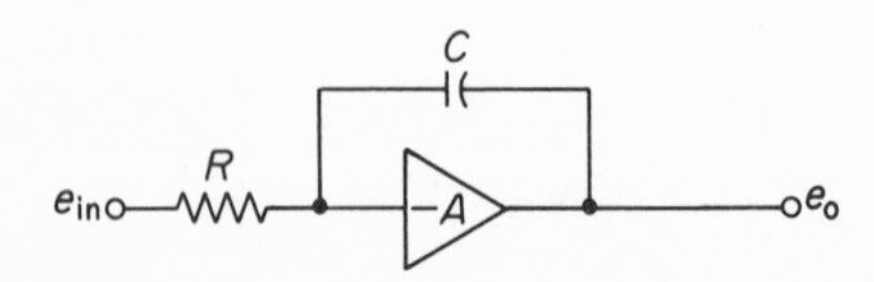

FIGURE 5-16. Operational Integrator Circuit.

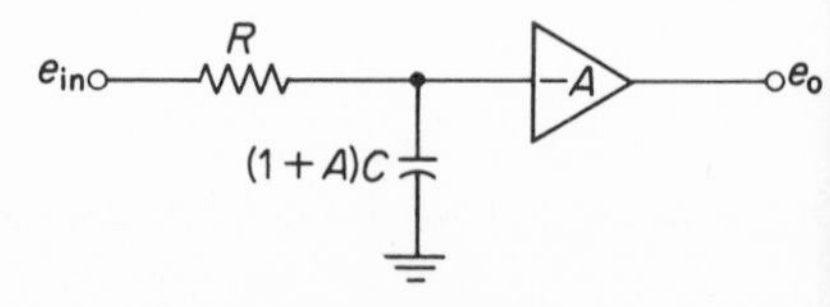

FIGURE 5-17. Equivalent Circuit of Operational Integrator.

or

$$\frac{E_o(s)}{E_{in}(s)} = \frac{-1}{1/A + (1 + 1/A)RCs} \tag{5-33}$$

In Eq. (5-33), if $|A| \to \infty$,

$$\frac{E_o(s)}{E_{in}(s)} \to \frac{-1}{RCs} \tag{5-34}$$

which indicates that for A very large, the circuit approaches an ideal integrator with negative polarity. This relationship holds in the linear operating region of the amplifier. Further insight into the operational integrator of Fig. 5-16 may be obtained by considering the equivalent circuit of Fig. 5-17.

The transfer function of the circuit in Fig. 5-17 is

$$\frac{E_0(s)}{E_{in}(s)} = \frac{(1/(1 + A)Cs)}{R + (1/(1 + A)Cs)}(-A) = \frac{-A}{1 + R(1 + A)Cs}$$

which is identical to Eq. (5-32). By use of the feedback arrangement of Fig. 5-16, the effective RC time constant has been magnified by a factor of $(1 + A)$. Also, for relatively large values of A (in most applications, $A > 5{,}000$

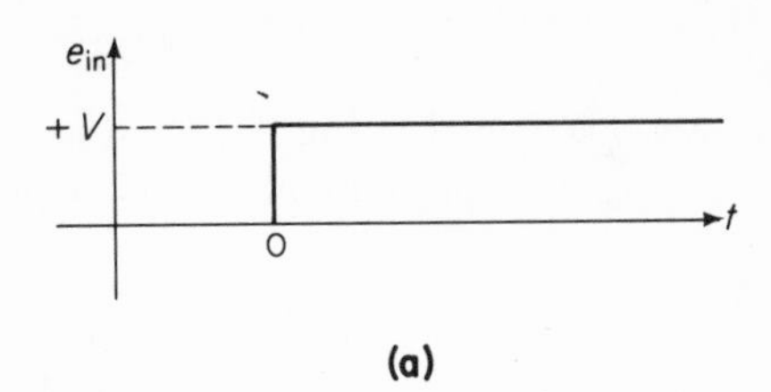

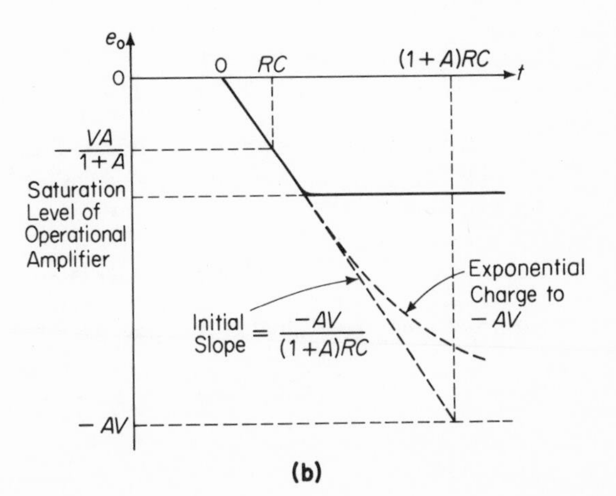

FIGURE 5-18. (a) Applied Step and (b) Resultant Output of Operational Integrator.

will suffice) variations in the gain A will have negligible effect on the gain c time constant of the feedback circuit. It should be noted that this effectiv magnification of the RC time constant is similar to the Miller effect i vacuum-tube triodes where the grid-to-plate capacitance is effectively $(1 + A)C_{gp}$. Therefore, the operational integrator is also referred to as a *Mille integrator.*

If a step voltage input, $+V\,u(t)$, is now applied to the integrator at tim $t = 0$ (where $|V| \leq$ the output saturation level of the operational amplifier and the capacitor C is assumed to have zero initial charge (at $t = 0$), then th waveform of Fig. 5-18(b) results.

The relationship for the output $e_o(t)$ in the linear operating region ma be found as follows:

$$E_{\text{in}}(s) = \frac{V}{s} \qquad \text{(5-35}$$

$$E_o(s) = \frac{-AV}{s[1 + (1 + A)RCs]} \qquad \text{(5-36}$$

The inverse transform of Eq. (5-36) may be obtained by use of Laplace trans form tables, which yield

$$e_o = -AV(1 - \epsilon^{-t/(1+A)RC}) \qquad \text{(5-37}$$

Eq. (5-37) is plotted in Fig. 5-18 and is a familiar exponential charge from 0 V to $-AV$ with time constant $\tau = (1 + A)RC$. The initial slope is found a follows:

$$\frac{de_o}{dt} = \frac{-AV}{(1 + A)RC}\epsilon^{-t/(1+A)RC} \qquad \text{(5-38}$$

At $t = 0^+$,

$$\frac{de_o}{dt} = \frac{-AV}{(1 + A)RC} \qquad \text{(5-39}$$

If the sweep were absolutely linear, then in RC sec, e_o would be equa to $-(VA/1 + A)$ V. Actually, the value of e_o at RC sec is found from Eq. (5-37):

$$e_o(RC) = -AV(1 - \epsilon^{-1/(1+A)}) \qquad \text{(5-40}$$

Eq. (5-40) may be evaluated by using the series expansion of ϵ^x,

$$\epsilon^x = 1 + x + \frac{x^2}{2!} + \frac{x^3}{3!} + \cdots \qquad \text{(5-41}$$

which yields

$$e_o(RC) = -AV\left(1 - 1 + \frac{1}{1+A} - \frac{1}{(1+A)^2 2!} + \cdots\right)$$

$$e_o(RC) = -\frac{AV}{1+A} + \frac{AV}{(1+A)^2 2!} - \cdots \qquad (5\text{-}42)$$

Deviations from an ideal sweep may be expressed either in terms of errors in sweep speed (i.e., variations in the slope of the waveform from the required linear slope) or in terms of absolute differences between the sweep waveform and the required output. The latter definition is further affected when some type of gain adjustment is used to set the end of the sweep waveform to its final value, as illustrated in Fig. 5-19(a). Assume that a waveform is

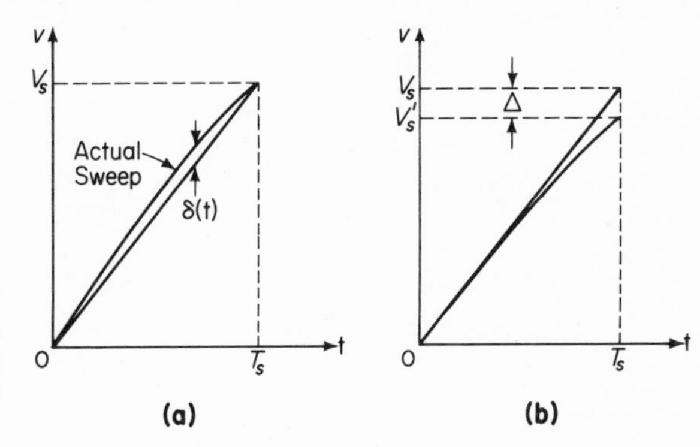

FIGURE 5-19. Deviation of Exponential Sweeps from Linear Waveform with (a) End Points Adjusted to Be Identical and (b) Initial Slope and Start Point Identical.

required to rise linearly from 0 to V_s in T_s sec. The *nonlinearity* (*NL*) is then defined by Eq. (5-43).

$$NL = \frac{\delta(\max)}{V_s} \times 100\% \qquad (5\text{-}43)$$

An alternate defintiton of error, illustrated by Fig. 5-19(b), relates to the maximum deviation of an exponential rise relative to a linear sweep with the same initial slope. This error Δ is approximately four times as large as $\delta(\max)$ for exponential sweeps where T_s is small compared to the RC time constant. An alternate definition of nonlinearity (NL') is given by Eq. (5-44).

$$NL' = \frac{\Delta}{V_s} \times 100\% \qquad (5\text{-}44)$$

In Eq. (5-42), then, the first term, $|-(AV/1+A)|$, represents the desired value of the sweep, and the sum of the remaining terms represents the total deviation (Δ). For A large, only the second term is of significance; therefore, the NL' is

$$NL' \approx \frac{[AV/(1+A)^2 2!]}{(AV/1+A)} \times 100\% = \frac{1}{2(1+A)} \times 100\%$$

For $A \geq 5{,}000$, the total nonlinearily NL' is less than 0.01%. When very precise sweeps are required, such as in radar and sonar displays, changes in initial slope due to variations in R and C are more significant than errors due to nonlinearity. An adjustment control (usually on the value of R) is normally provided to compensate for initial variations and changes due to aging of components.

When long term stability and accuracies of better than $\pm 0.1\%$ are required, a digitally controlled sweep is the logical choice. This approach is discussed in Chap. 7.

Transistor Miller Integrator Sweep Circuit. The integrator circuit of Fig. 5-16 can be synthesized using a direct-coupled discrete component amplifier. A single transistor, for example, will yield adequate gain and impedance matching for many sweep generator applications. Historically, the Miller integrator has been designed in this manner. The approach provides a simple, low cost design, with the added advantage of providing larger voltage swings than are available from operational amplifiers.

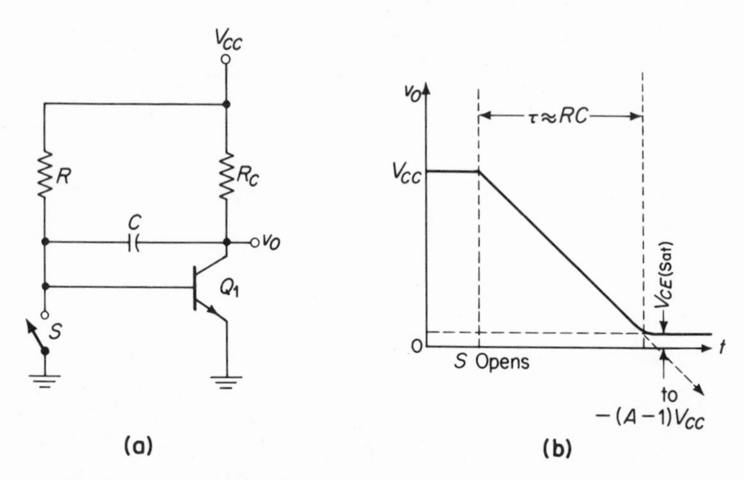

FIGURE 5-20. (a) Transistor Miller Integrator Sweep Circuit; (b) Resultant Idealized Output Voltage Waveform.

Operation of the sweep generator is simialr to that of the operational integrator of Fig. 5-16. The lower input impedance, lower gain, and higher output impedance of a single transistor amplifier, however, does affect overall performance. Considerable improvement in input impedance and gain may be achieved by use of a pair of transistors in the compound configuration (referred to as the Darlington connection) of Fig. 5-21. Further improvement in input impedance (at the expense of reduced voltage gain) may be obtained by using a small external resistor in the emitter circuit.

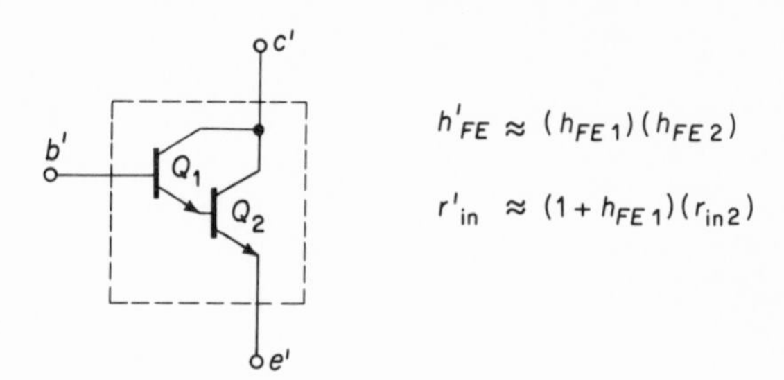

FIGURE 5-21. Compound Transistor Configuration.

The output sweep, shown in Fig. 5-20(b), is initiated by the opening of switch S. In a practical circuit the switch may be a suitable solid-state device such as a transistor driven from saturation to cutoff. Prior to opening S, transistor Q_1 is in cutoff. Assuming the leakage current is negligible, $v_o = V_{CC}$. When S is opened, Q_1 is biased ON and the circuit enters its linear operating region. If S is left open, then after approximately RC sec, Q_1 will be in saturation and the output will be near 0 V. In many applications the opening of S is controlled by a monostable multivibrator whose period is set to the sweep duration desired.

Recovery of the Miller integrator takes place after S is closed. The capacitor should be fully charged to V_{CC} before initiating a new sweep. Charging of C is through resistor R_C; therefore, a minimum recovery period of approximately $5R_CC$ should be allowed between the end of a sweep and the start of the next one. This places conflicting requirements on R_C since a large value of R_C is desirable to obtain higher voltage gain.

5-7 ASTABLE (FREE-RUNNING) MULTIVIBRATOR

The *astable multivibrator* is a closed loop regenerative feedback device, similar to the monostable multivibrator, except that it is free-running and does not require external triggering. The output is a continuous, rectangular wave. The output may be differentiated and rectified and the resultant train of pulses used for timing of clocked digital circuits, when precision timing is not required.

The discrete component design of symmetrical astable multivibrators has been analyzed in detail in the litetature.[3,4] The very limited usage of the astable multivibrator in present digital system design does not warrant repetition of this material.

A more current design approach to the astable multivibrator utilizes linear integrated circuit amplifiers,[5] and is illustrated in Fig. 5-22. The operational amplifier is assumed to have the characteristics outlined in Fig. 5-5. R_3 and R_4 provide a fixed level of positive feedback and R_1, R_2, and C provide

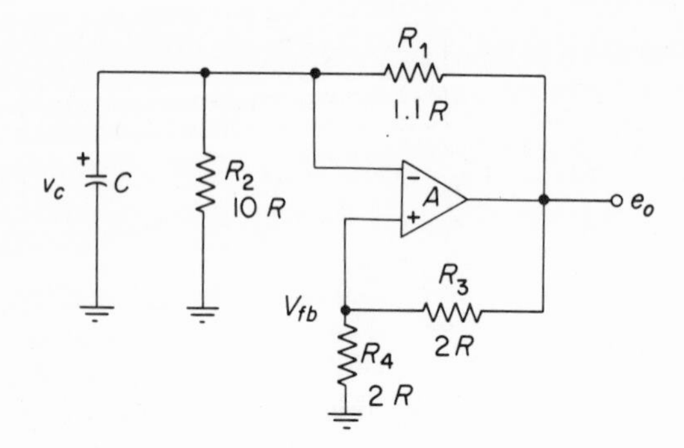

FIGURE 5-22. Operational Amplifier Design of Astable Multivibrator.

a frequency dependent level of negative feedback. At high frequencies the negative feedback is reduced and the circuit becames unstable. The circuit cannot "hang up" in either output state and is self-starting.

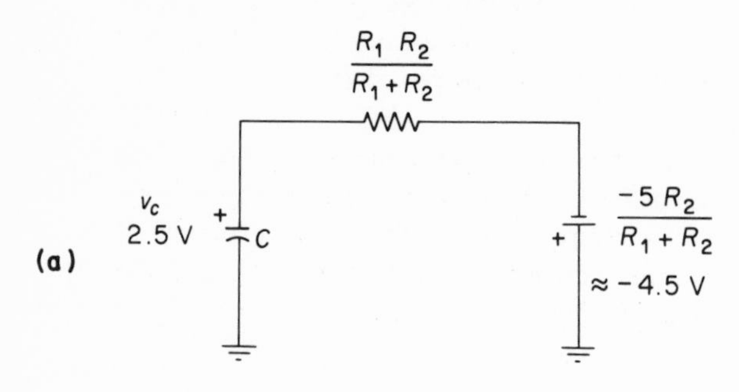

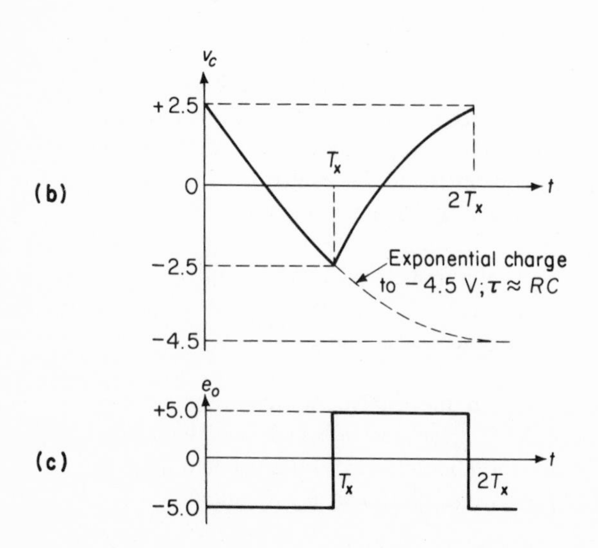

FIGURE 5-23. (a) Thévenin Equivalent of Charging Circuit and (b) Waveform of v_c, and (c) Output Waveform e_o.

When the output is in saturation at $+5$ V, $v_{fb} = +2.5$ V, and when at -5 V, $v_{fb} = -2.5$ V. Assuming that the output were at $+5$ V, then when C charges to $v_C \geq +2.5$ V, the output will switch rapidly to -5 V. This action is identical to that of a Schmitt trigger. The output is then at -5 V, the $v_{fb} = -2.5$ V, and the voltage waveform of v_C is as shown in Fig. 5-23(b) during the period $0 < t < T_x$.

The time T_x for v_C to charge from $+2.5$ V to -2.5 V is found as follows:

$$+2.5 - 7(1 - \epsilon^{-(T_x/RC)}) = -2.5 \qquad 7\epsilon^{-(T_x/RC)} = 2.0 \qquad T_x = 1.25RC$$

The charging time to $+2.5$ V is solved in a similar manner; therefore, the total period of the astable multivibrator is

$$\tau = 2T_x = 2.5RC \tag{5-45}$$

The range of R is limited at the lower end by the loading effect on the operational amplifier and at the upper end by drift resulting from input offset current. A reasonable range for R is 500 Ω to 20 K. R_3 and R_4 are chosen to present the same equivalent resistance as R_1 and R_2 at the (+) and (−) inputs to the amplifier. If R_4 is made smaller than R_3, the period τ is reduced, and if R_3 is made smaller than R_4, τ is increased. This provides a convenient method for making small adjustments in the period.

The maximum frequency that the multivibrator can be operated at is limited by the slew rate of the operational amplifier. Using a high-speed differential comparator, operation at frequencies in the 5-10 MHz range is feasible. Differential comparators, however, do not generally have symmetrical saturation levels at the output. The μA710, for example, saturates at approximately

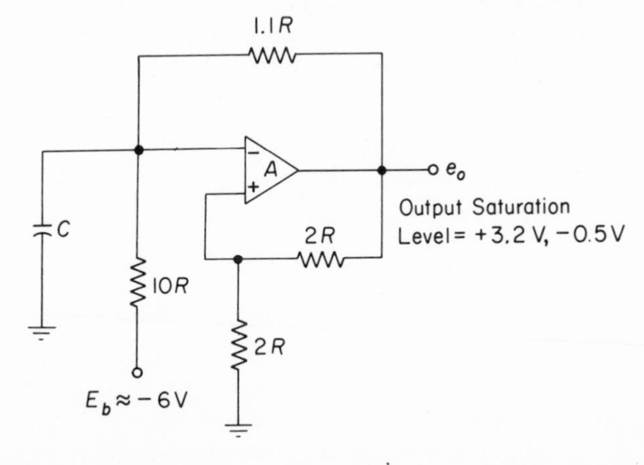

FIGURE 5-24. Differential Comparator Circuit of Astable Multivibrator.

−0.5 and +3.2 V. In order to obtain equal charge and discharge periods, th input must be biased as shown in Fig. 5-24.

It is left as an exercise to the reader (Prob. 11) to show that the timin relationship for the circuit of Fig. 5-24 is different than that given i Eq. (5-45).

REFERENCES

1. Wisseman, L., Robertson, J. J., "High Performance Integrated Operationa Amplifiers," *Appl. Note AN-204*, Motorola Semiconductor Products, 196
2. *Integrated Operational Amplifiers*, Microelectron. Div., Radiation Inc Melbourne, Florida, 1966.
3. Giles, J. N., ed., *Linear Integrated Circuits Applications Handbook*, Fai child Semiconductor Inc., 1967.
4. Millman, J. and Taub, H., *Pulse, Digital, and Switching Waveform* McGraw-Hill Book Company, New York, 1965.
5. Strauss, L., *Wave Generation and Shaping*, McGraw-Hill Book Compan New York, 1960.

PROBLEMS

1. In the summing circuit shown below, e_1, e_2, and e_3 are connected to th Q outputs of three flip-flops. The input voltages are +5 in the **1** state an 0 V in the **0** state. Assume the output resistance of the flip-flops is negli gible. If the e_1 input is from the least significant bit etc., what values o R_2 and R_3 will make e_0 proportional to the binary number represented b the flip-flop states?

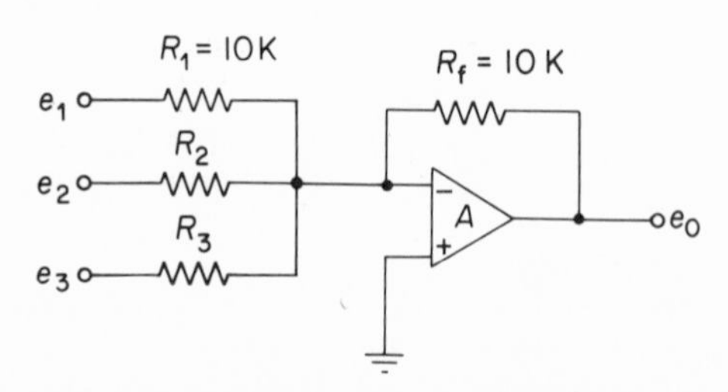

FIGURE P5-

2. In the circuit of Prob. 1, assuming A is very large, what is e_0 when th input is in the 101 state? What is the maximum error in e_0 if ±1% tolerance resistors are used?

3. A Schmitt trigger is designed using the circuit of Fig. 5-6 with $R_1 = 5$ K, $R_2 = 100\ \Omega$, the amplifier gain $A = 1{,}000$, and the amplifier saturation level at $V_0 = +10$ V and -3 V. Plot V_o vs. V_{in} for $V_{\text{TH}} = +2$ V.

4. The Schmitt trigger shown in Fig. 5-8 has the emitter circuit of Q_2 modified as in Fig. P5-4. Find the new trigger thresholds V_{T1} and V_{T2}. (Note that R_x does not affect the operating characteristics when Q_2 is in cutoff.)

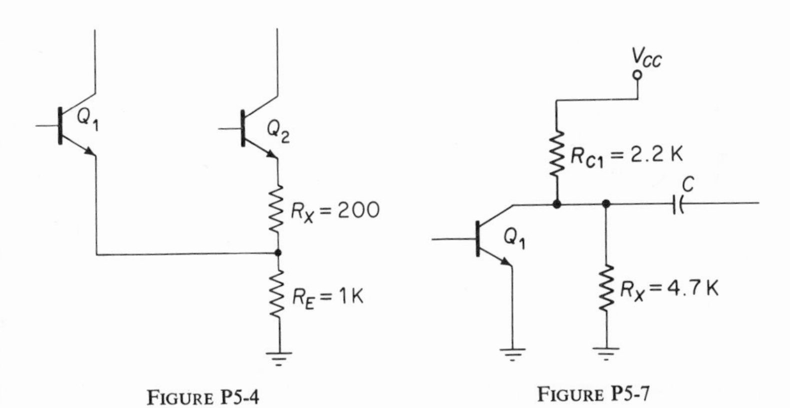

FIGURE P5-4

FIGURE P5-7

5. The transistors Q_1 and Q_2 in Fig. 5-8 are replaced by a low gain pair ($h_{FE} = 10$). What effect if any does this have on V_{T1} and V_{T2}? Would making $R_{C1} = 33$ K, $R_1 = 47$ K, and $R_2 = 33$ K improve the operation? Explain.

6. Resistor R_{C2} in Fig. 5-8 is changed to 4 K. Will the circuit function properly? What effect would a further increase—e.g., to 22 K—have on the circuit's operation?

7. The collector circuit of Q_1 in the monostable multivibrator (shown in Fig. 5-11) is changed as shown in Fig. P5-7. What effect does R_x have on the width of the output pulse?

8. In the circuit of Prob. 7 silicon transistors are used for Q_1 and Q_2. If the base-emitter reverse breakdown voltage is 5 V, what value of R_x would permit proper operation?

9. Design a monostable multivibrator (similar to the circuit shown in Fig. 5-14) using two NOR gates. Indicate the input reset and trigger levels and sketch the output waveform.

10. The sweep circuit shown in the figure below is controlled by switching transistor Q_1 ON and OFF. When e_{in} is positive, Q_1 is fully saturated ON ($V_{CE} = +0.01$ V) and, when negative, Q_1 is in cutoff. The amplifier gain $A \geq 5{,}000$ and its saturation levels are ± 15 V.

    (a) Assuming the output was initially in steady state, sketch $e_o(t)$ for the input pulse shown.

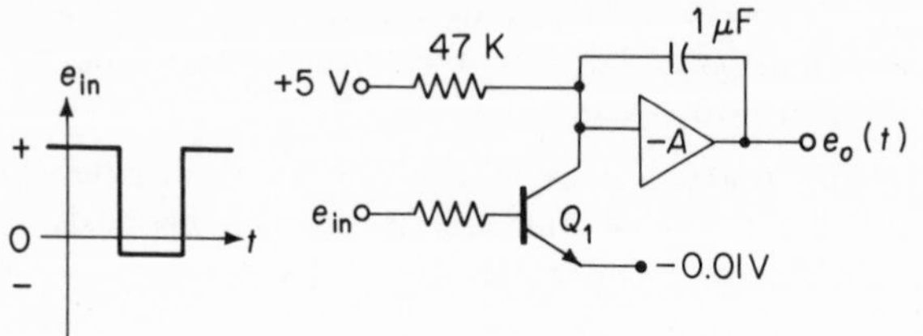

FIGURE P5-1

(b) If the emitter of Q_1 is returned to ground and the saturation voltage is +0.01 V, would the circuit function properly? Explain.

11. (a) In the astable multivibrator circuit of Fig. 5-24, calculate the exact value of bias voltage E_b that makes the output e_o a symmetrical square wave (equal charge and discharge periods).
(b) What is the period of the output waveform?

CHAPTER 6

COUNTING CIRCUIT DESIGN

Counting circuits utilize binary memory elements, or flip-flops, and fall into the general category of *sequential circuits*. Sequential circuits and the concept of state diagrams were introduced in Chap. 3. It was noted that the output of a sequential circuit depends not only on the input but also on the previous state of the circuit. This is in comparison with a combinational circuit whose output is a function only of the input variables. In general a sequential circuit contains both memory elements and combinational circuits.

Many digital counters are relatively simple and can be designed by intuition in a straightforward manner. However, after the common characteristics of the counters are discussed, the general approach of counter design as a sequential circuit using *state diagrams* and *state transition tables* is presented. The latter method provides a powerful design tool for general purpose digital systems and is conveniently illustrated in the design of counters.

Digital pulse counters have many applications in industrial processes and communications, and they are fundamental building blocks of digital computers. These counters change condition or state with each externally applied input pulse. Counters of different capacities, or scales, are constructed by connecting different numbers of flip-flops in a series. Some of the many applications of counters include the measurement of distance, time, frequency, and events; synchronization; frequency division; and waveform generation.

6-1 BINARY STORAGE

There are variations in the design of the basic flip-flop. Two very useful types are the *RST* flip-flop, with set, reset, and toggle inputs; and the *JK* flip-flop, with a clocked input and an auxiliary set and reset inputs, P_j and P_k.

The symbols for these circuits are shown in Fig. 6-1. Variations of these flip-flops include positive and negative logic, leading and trailing edge triggering, positive and negative triggering pulses, ac and dc coupling, clocked inputs, master-slave operation, and various combinations of these. To maintain consistency in the treatment of counting circuits the flip-flops used will be the same as those discussed in Chap. 3.

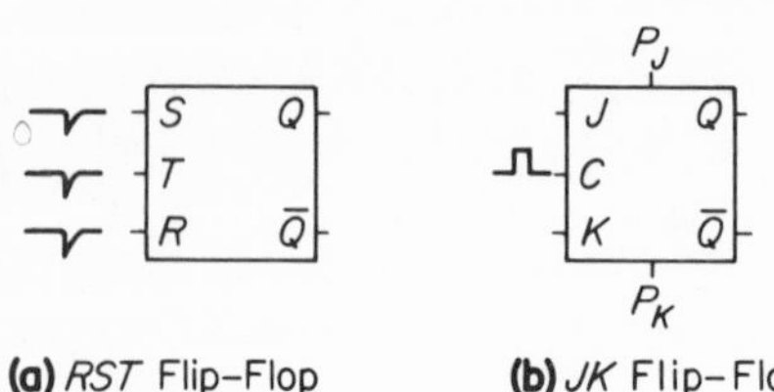

FIGURE 6-1. Basic Flip-Flop Symbols.

The *RST* flip-flop of Fig. 6-1(a) will use positive TRUE logic and be triggered by either positive pulses or the use of negative trailing edge logic. The set and reset operation of this flip-flop is consistent with the design illustrated in the toggle input shown in Sec. 3-8.

The *JK* flip-flop used is the integrated circuit design shown in Sec. 3-7. Positive TRUE logic also is used, and the flip-flop is dual rank with a clocked input. The flip-flop is dc triggered and therefore the clock pulse is virtually free of waveform restrictions. The truth tables for the *JK* flip-flop are shown in Fig. 6-2.

J	K	Q^{n+1}
0	0	Q^n
0	1	0
1	0	1
1	1	$\overline{Q^n}$

Synchronous Inputs at Clock Time

P_j	P_k	Q^{n+1}
0	0	Q^n
0	1	0
1	0	1
1	1	*

Asynchronous Inputs
*See Text

FIGURE 6-2. Truth Tables for the *JK* Flip-Flop.

The synchronous inputs strictly follow the definition of a *JK* flip-flop. For the preset inputs, however, the case of P_j and P_k both being positive will have no immediate effect on the outputs. The input that falls last will control the final state of the flip-flop. This is denoted by the asterisk in Fig. 6-2.

Presetting should be performed with the clock line LOW. However, in applications such as ripple counting, in which the state of the clock line cannot be predicted, presetting can be accomplished by lowering the J input while raising P_j or lowering the K input while raising P_k. If the preset inputs are not used, they should be tied down to **0**.

The *JK* flip-flop can be toggled by tying the J and K inputs together to a **1** and clocking at the C input. If simultaneous clocking of all stages is not required, toggling of individual flip-flops can be accomplished at the C input with the J and K inputs tied to a **1**.

Of course, it is very important that the logic circuit designer be thoroughly familiar with the operation of the type of basic circuits to be used in his design. The operational data are generally available from the integrated circuit manufacturers. The counting circuits in this chapter are described using the flip-flops as defined above. However, of primary importance are the digital principles and techniques used; whereas the circuit configuration will vary with the use of different types of basic circuits.

6-2 CASCADED FLIP-FLOPS AS A COUNTER

A chain of integrated circuit JK flip-flops used as a counter is shown in Fig. 6.3. The J and K terminals, with the connections shown, are at the **1** level when counting; and therefore each circuit with the input at terminal C is a toggle flip-flop. This is the simplest method of counting that employs an asynchronous technique known as *ripple carry*, or simply called a *ripple-through* counter.

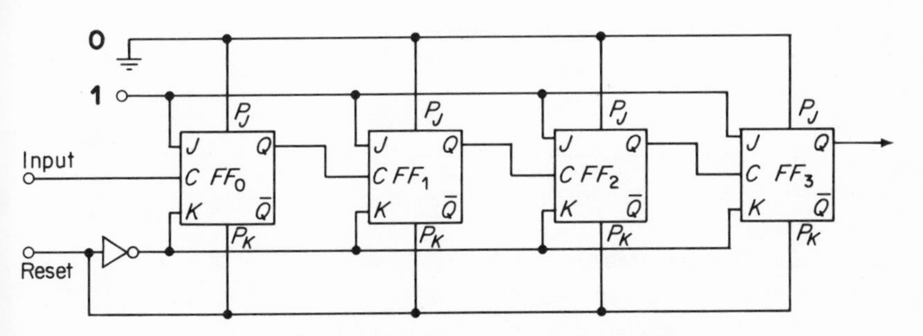

FIGURE 6-3. Binary Ripple-Through Counter.

The basic flip-flop circuit changes state with each input pulse and returns to the same state after every second pulse. Each output waveform, therefore, is a square wave at one-half the frequency of its input. By using the output of one stage as the input to the next, the frequency is successively halved. The pertinent waveforms are shown in Fig. 6-4, illustrating the successive frequency division.

When the Q output of a flip-flop is HIGH the flip-flop is said to be in the *set* state or it is said to be storing a binary **1** (positive true logic). The waveforms are consistent with the dual rank flip-flops, i.e., the master flip-flop stores the input information when the clock voltage is high and transfers it to the slave when the clock voltage is low.

A reference starting point is chosen when all the flip-flops are in the **0** state, i.e., all the Q output voltages are LOW. The waveforms that appear at the Q outputs of the individual flip-flops, as the result of 16 successive input pulses, are shown in Fig. 6-4. The first input pulse applied to FF_0 causes this flip-flop to change from state **0** to state **1**. As a result of this transition a positive voltage step is applied to the C input of FF_1. The first or master rank of FF_1 is set, but the second rank remains in the **0** state as long as C remains high. After the first trigger pulse, FF_0 is in state **1**; the remaining flip-flops are in state **0**.

The second input pulse causes FF_0 to return from state **1** to state **0**. Flip-flop FF_1 now receives a negative voltage step, as indicated in Fig. 6-4, which

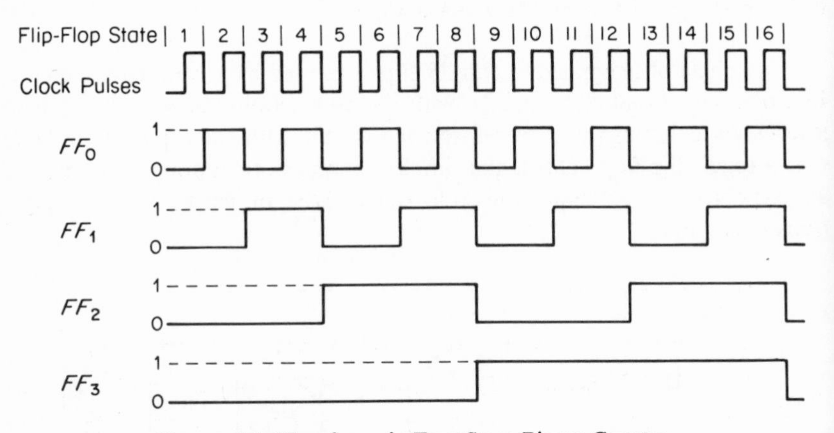

FIGURE 6-4. Waveforms in Four-Stage Binary Counter.

changes its state; FF_1 makes a transition from state **0** to state **1**. Therefore, after receiving two input pulses, FF_0 is in state **0**; FF_1, in state **1**; and FF_2 and FF_3, in state **0**. Additional input pulses cause similar transitions in the flip-flops as indicated in the waveforms. These waveforms may be verified by applying the following rules:

1. Flip-flop FF_0 changes state with each external input pulse.
2. Each of the other flip-flops changes state when and only when the preceding flip-flop makes a transition from state **1** to state **0**.

If discrete component flip-flops were used containing pnp transistors, then the Q output in the **1** state would be a lower voltage than in the **0** state. This is termed *negative logic*, and the waveforms of Fig. 6-4 would become inverted.

Figure 6-5 lists the states of all the flip-flops in the counter chain as a function of the number of the externally applied trigger pulses. This table may be compared with the waveforms in Fig. 6-4 for verification. Note that in Fig. 6-5 the flip-flops are ordered from right to left; whereas in the counter chain of Fig. 6-3 they are ordered from left to right. It also is important to note that the state of flip-flops in Fig. 6-5 indicates the count of the number of input pulses in the binary number system. The chain of flip-flops, of course, counts in the binary system. A chain of n flip-flops counts 2^n pulses before it resets itself to its original state.

No. of Input Pulses	State of Flip-Flop			
	FF_3	FF_2	FF_1	FF_0
0	0	0	0	0
1	0	0	0	1
2	0	0	1	0
3	0	0	1	1
4	0	1	0	0
5	0	1	0	1
6	0	1	1	0
7	0	1	1	1
8	1	0	0	0
9	1	0	0	1
10	1	0	1	0
11	1	0	1	1
12	1	1	0	0
13	1	1	0	1
14	1	1	1	0
15	1	1	1	1
16	0	0	0	0
17	0	0	0	1
18	0	0	1	0
19	0	0	1	1

FIGURE 6-5. States of the Four-Stage Flip-Flop Counter.

It is clear from the previous discussion that the first flip-flop FF_0 in a binary counter changes state with each input pulse; FF_1, with every second input pulse; FF_2, with every fourth input pulse; FF_3, with every eighth input pulse; etc. Therefore, the decimal count stored in a four-stage binary counter is

$$C_D = 8Q_3 + 4Q_2 + 2Q_1 + 1Q_0 \tag{6-1}$$

where the coefficients are the *weights* of the binary digits. For example, if the count 1011 is stored in a counter; i.e., FF_0 in state **1**, FF_1 in state **1**, FF_2 in state **0**, and FF_3 in state **1**, the decimal count is

$$C_D = 8 + 2 + 1 = 11 \tag{6-2}$$

In gereral, for an n-stage counter, the decimal count is

$$C_D = 2^nQ_n + 2^{n-1}Q_{n-1} + 2^{n-2}Q_{n-2} + \cdots + 2^2Q_2 + 2Q_1 + 1Q_0 \tag{6-3}$$

Note that when any stage of a ripple-through counter is to change state, all preceding stages also must be changing from **1**'s to **0**'s. This condition is analogous to adding one to a binary number containing all **1**'s. A **1** added to the least significant bit causes that bit to become a **0** with a carry to the next bit. Each bit in turn becomes a **0** with a carry to the next bit. The carry propagates bit by bit until it finally encounters a **0** and ends there. In a similar manner, in the ripple-through counter, an input causes the least significant stage to change to **0**, which in turn causes the next stage to change to **0**, and so on until it terminates at a stage that changes from **0** to **1**. A finite amount of time elapses as each stage changes sequentially, and the carry appears to ripple through the counter.

When a ripple-through counter is used for frequency division, the important thing is the frequency of the output at the final stage. However, when it is necessary to use the specific binary coded outputs of two or more stages as inputs to logic gates, the time difference between changing outputs can cause errors. Since the individual delays are cumulative, total delay is aggravated when the outputs are from stages widely separated in the counter.

Special gating of the outputs will be required. However, this problem of delays between stages is avoided with the use of synchronous counters (see Sec. 6-5).

6-3 UP-DOWN BINARY COUNTER

A counter that can be made to count in either a forward or reverse direction is called an *up-down* or *reversible* counter. Forward counting is accomplished, as discussed above, when the toggle input of a succeeding flip-flop is coupled to the Q output of a preceding flip-flop. The count will proceed in the reverse direction if the coupling is made instead to the $\bar{Q}$ output.

If a flip-flop changes from state **0** to **1**, then the $\bar{Q}$ output will change from **1** to **0**. The negative-going transition at $\bar{Q}$ will cause the succeeding flip-flop to change state. For reverse binary counting the following rules will apply:

1. Flip-flop FF_0 changes state with each external input pulse.
2. Each of the other flip-flops changes state when and only when the preceding flip-flop makes a transition from state **0** to state **1**.

For example, these rules are applied to the numbers in Fig. 6-5. Assume that the counter had been counting in the forward direction and received ten input pulses. The counter would register 1010; i.e., FF_3 in state **1**, FF_2 in state **0**, FF_1 in state **1**, and FF_0 in state **0**. The counter is now transferred to count in the reverse direction. The next input pulse changes the state of FF_0 from state **0** to **1**. This change in FF_0, state **0** to **1**, causes a change in FF_1 from state **1** to

0, which in turn does not cause any further transitions and FF_2 remains in state **0** and FF_3 in state **1**. The result is that the counter registers 1001, which represents the number 9. Since we started with 10 and ended with 9, a reverse count has taken place.

A block diagram of an up-down counter with controls to either add or subtract counts is shown in Fig. 6-6. A limitation of this circuit as shown is that a change in the up-down control signal can also cause a change in the state of the counter. For example, if an input to one of the coupling AND gates is a **1** from the output of its preceding flip-flop, and the other input from the up-down control changes state, then the voltage change is applied to the next flip-flop. This may cause a change in the state of the counter without the application of an input pulse. This false counting can be avoided by the use of clocked *JK* flip-flops in a synchronous counter application (see Sec. 6-5 and Prob. 6-3).

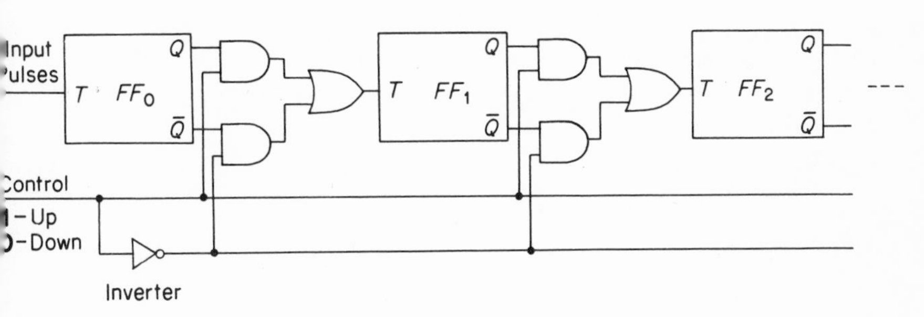

FIGURE 6-6. Up-Down Binary Counter.

6-4 APPLICATION OF COUNTERS

Counters find many applications in the control of industrial processes. The events or items to be counted first have to generate an electrical signal or impulse. For counting objects this is usually accomplished by passing them single file on a conveyor belt through a photocell-light source combination.

An application of direct counting is the measurement of the length of strip metal going through a rolling mill or processing line. A pulse tachometer is geared to a processing roll. When the haed end of the strip is detected by a metal detector, the tachometer pulses are gated to an up-counter. The count can be continuously displayed for the operation, or it can be printed out when the tail end of the strip leaves the measuring area.

This measuring system can be easily expanded to initiate a cut fater a predetermined length of strip has passed the measuring roll. The operator uses switches to set the desired length between cuts. The switch setting is then compared to the length measured by the counter. When the two are equal, a contact closure initiates the shearing operation.

Counters also find application in digital computers. A problem is solved in a digital computer by subjecting the data to a sequence of operations in accordance with the program of instructions introduced into the computer. Counters may be used to count the operations as they are performed and to call forth the next operation from the memory when the preceding one has been completed.

Measurement of Time Intervals

The time interval between two pulses, or events, may be measured with a high degree of precision using the circuit shown in Fig. 6-7. Initially the

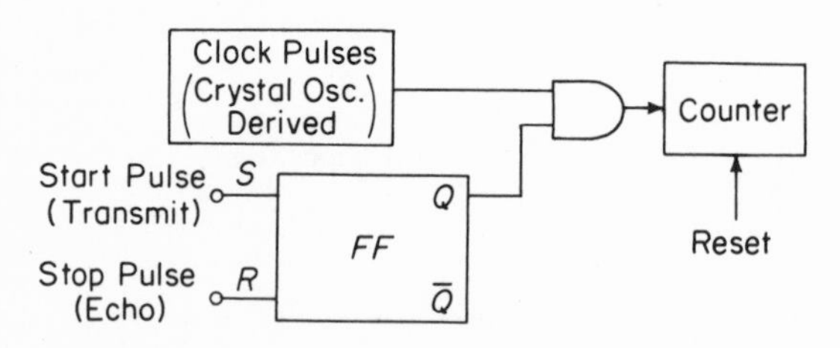

FIGURE 6-7. Time Interval Measurement.

counter is reset to **0**. The first input pulse sets the flip-flop and allows the clock pulses from the crystal oscillator to pass through the AND gate and be counted in the counter. The second pulse closes the gate, and the counting is stopped. The count registered in the counter is proportional to the length of time the gate is open and therefore gives the desired time interval. For example, if the crystal oscillator frequency is 1 MHz, each count represents 1 μsec. The resolution is $\pm T$, the oscillator period, which in this example is ± 1 μsec.

Distance Measurement. In sonar applications the time interval between the transmitted pulse and return echo is proportional to the two-way propagation path. Knowing the propagation velocity through the medium (approximately 4800 ft/sec in seawater), we can determine the target distance.

Speed Measurements. A speed determination may be converted into a time measurement. For example, if two photocell-light source combinations are set a fixed distance apart, the average speed of an object passing between the photocells is proportional to the time intervals between the generated pulses. Velocities of projectiles have been measured in this manner.

Measurement of Frequency

The application of counters for the precise determination of frequency is illustrated in Fig. 6-8. The principle of this circuit is to count the cycles of the

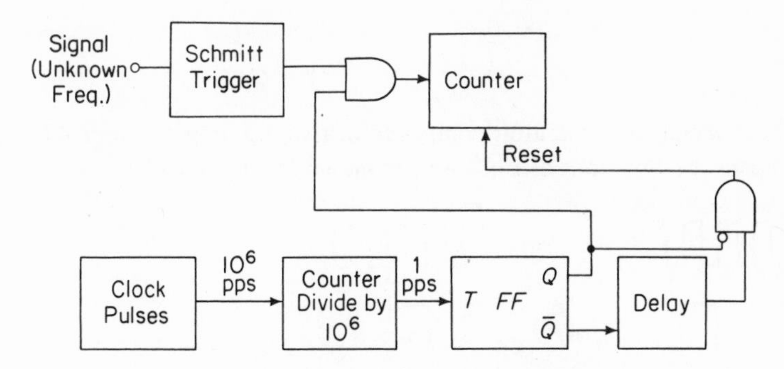

FIGURE 6-8. Digital Frequency Measurement.

unknown frequency for a precisely known time interval. The time interval is generated by a crystal oscillator and a flip-flop counter divider chain. If the oscillator frequency is 1 MHz and the counter chain divides by one million, pulses of one pulse per second (1 PPS) are obtained for the timing interval, with an accuracy determined by the crystal oscillator frequency tolerance. The 1-PPS pulses toggle a flip-flop that gates the pulses of unknown frequency to the counter. If the counter registers 10,000 pulses at the end of the one-second interval, then the unknown frequency is 10 kHz. A delay is shown to reset the counter prior to the next counting interval. If additional time is required to read the content of the counter, the additional control logic can be added.

Waveform Generation and Synchronization

The waveforms that occur at the outputs of the stages in binary counters can be combined directly or with other logic circuits to generate complex pulse-type waveforms. A typical example of waveform generation is the voltage waveforms shown in Fig. 6-9, which may be used to synchronize a sequence of operations. Such waveforms are used frequently in communications when multiple data channels are multiplexed, sampled, coded, and transmitted over a single communications line.

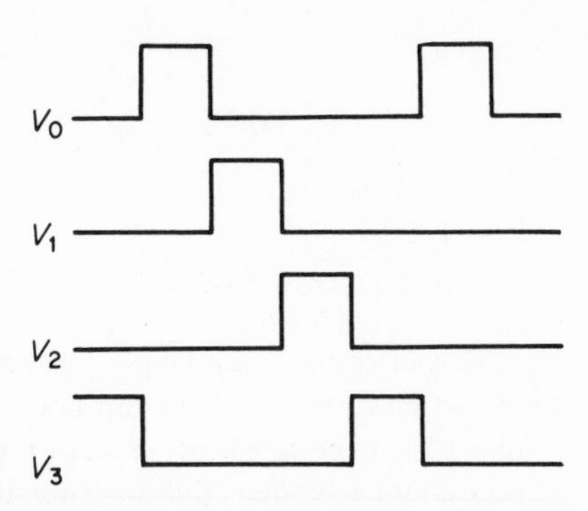

FIGURE 6-9. Waveform Generation.

Since there are four different waveforms in Fig. 6-9, and two-stage counters are capable of four different states, it is assumed that the waveforms can be generated with two flip-flops. The waveforms of the Q and $\bar{Q}$ outputs of a two-stage counter are shown in Fig. 6-10(a). By comparing the waveforms with the desired waveforms in Fig. 6-9 the required logic functions follow:

$$V_0 = \bar{Q}_0\bar{Q}_1 \qquad V_2 = \bar{Q}_0 Q_1$$
$$V_1 = Q_0\bar{Q}_1 \qquad V_3 = Q_0 Q_1 \tag{6-4}$$

The final waveform generator, comprised of two flip-flops and four AND gates to perform the functions of Eq. (6-4), is shown in Fig. 6-10(b).

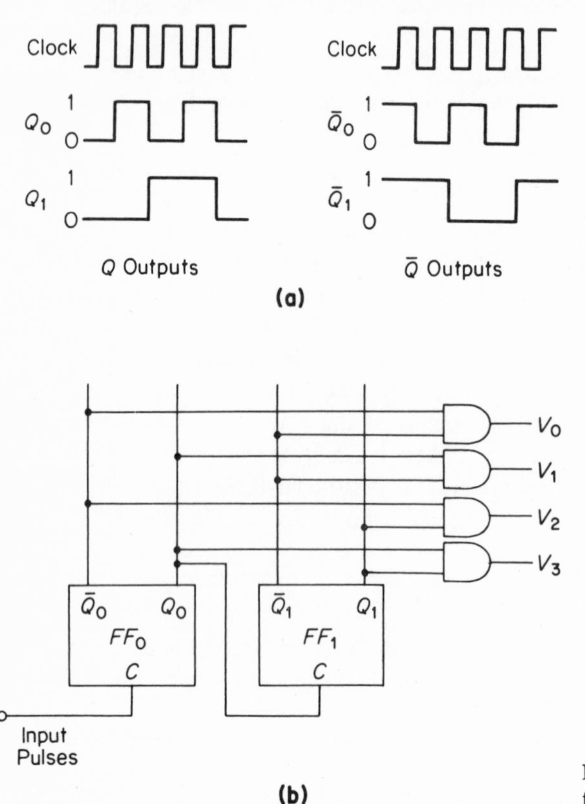

FIGURE 6-10. (a) Flip-Flop Waveforms. (b) Waveform Generation.

6-5 SYNCHRONOUS COUNTERS

The carry time in a ripple-through counter is in the order of the sum of the transition times of all the flip-flops. If the chain is long the carry time can be longer than the intervals between input pulses and it will not be possible to read the counter between pulses. This situation can be reduced through the use of *synchronous* techniques. Synchronous counters employ a clock input signal that is common to all stages, and therefore all outputs that are scheduled to change do so simultaneously. The synchronous counter falls into a general class of switching circuits known as *synchronous sequential circuits*. Sequential circuits are introduced in Chap. 3.

The output of a sequential circuit is a function of both the input and the previous history or *state* of the circuit. *State diagrams* are used to show diagrammatically the sequential progressions of the circuit from one state to the next.

A synchronous sequential circuit contains a combinational switching circuit with feedback from memory elements. A source of clocked input pulses

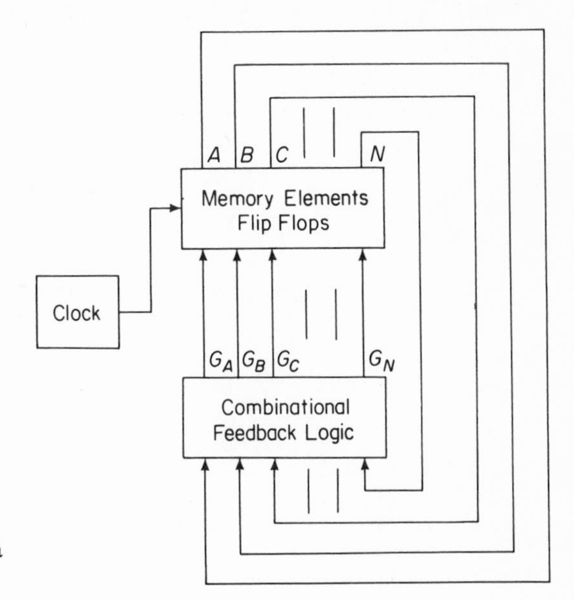

FIGURE 6-11. Block Diagram of a Synchronous Sequential Circuit.

is gated into the circuit. A general block diagram of this type circuit is shown in Fig. 6-11. The memory elements in this case are the flip-flops, and the combinational switching circuit provides the feedback logic for the counter to progress in the desired sequence.

As an example, we will design a synchronous counter that will count 16 pulses and then recycle to its original state. The state diagram for this counter is shown in Fig. 6-12. To facilitate the design of the required combinational circuit

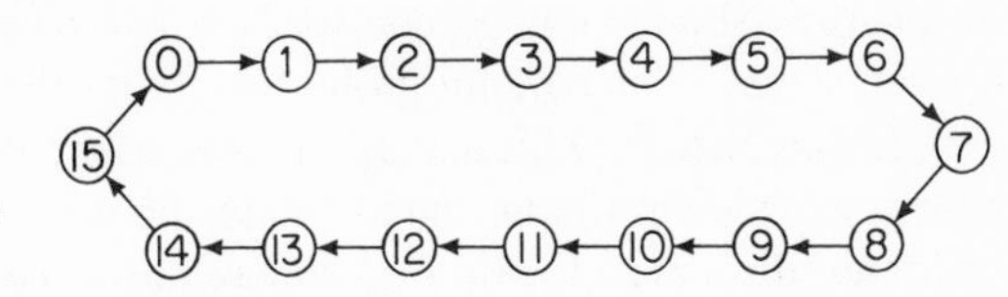

FIGURE 6-12. State Diagram for Binary Counter.

a *state transition table* is constructed. The state transition table shows the *present state* of each memory element and the *next state*, i.e., the states of the memory elements after the next clock pulse. The required number of binary memory elements *n* is equal to the number of binary digits required to represent the total number of states *m* in the binary number system. Therefore,

$$2^{n-1} < m \leq 2^n \qquad (6\text{-}5)$$

The 16-state counter requires four flip-flops. The state transition table is shown in Fig. 6-13. The column at the extreme left is a numerical listing of the sequences of events or states for the circuit. The next column group lists the

State Sequence	Present State D^n	C^n	B^n	A^n	Next State D^{n+1}	C^{n+1}	B^{n+1}	A^{n+1}	Toggle Function T_D	T_C	T_B	T_A
0	0	0	0	0	0	0	0	1	0	0	0	1
1	0	0	0	1	0	0	1	0	0	0	1	1
2	0	0	1	0	0	0	1	1	0	0	0	1
3	0	0	1	1	0	1	0	0	0	1	1	1
4	0	1	0	0	0	1	0	1	0	0	0	1
5	0	1	0	1	0	1	1	0	0	0	1	1
6	0	1	1	0	0	1	1	1	0	0	0	1
7	0	1	1	1	1	0	0	0	1	1	1	1
8	1	0	0	0	1	0	0	1	0	0	0	1
9	1	0	0	1	1	0	1	0	0	0	1	1
10	1	0	1	0	1	0	1	1	0	0	0	1
11	1	0	1	1	1	1	0	0	0	1	1	1
12	1	1	0	0	1	1	0	1	0	0	0	1
13	1	1	0	1	1	1	1	0	0	0	1	1
14	1	1	1	0	1	1	1	1	0	0	0	1
15	1	1	1	1	0	0	0	0	1	1	1	1

FIGURE 6-13. State Transition Table for Synchronous Binary Counter Using Toggle Flip-Flops.

present states of the memory elements. The four flip-flops are designated *A*, *B*, *C*, and *D*, where *A* indicates the least significant digit. The superscript, also designated by the letter n, indicates the present state, and the superscript $n + 1$ indicates the state after the next clock pulse. The superscript letters indicating the bit-time should not be confused with n's used to indicate the number of flip-flop stages. The column group indicating the next state is listed beside the present state group. This actually is redundant because the states are listed in sequential order in the left-hand group; however, for clarity this double listing facilitates the determination of the required combinational gating function. Additional columns can be listed at the right for any desired purpose.

At this point it is necessary to decide on the type flip-flop that is to be used. The counter can be designed to use either toggle or *RS* flip-flops. We shall use *JK* flip-flops as synchronous toggles by tying the *J* and *K* terminals together. These combined terminals are designated by T, e.g., T_A is the input to J and K for the A flip-flop. Whenever the T input is a **1**, the flip-flop will change state at the next clock pulse. The required T inputs are listed in the right-hand column in Fig. 6-13 by inspection, by comparing the present state and next state columns. The Boolean expressions for the gating functions can now be written from the T columns; however, since we would want to use the final expressions in reduced form, it is best to map these functions first. The four maps for the T gates and the simplified functions are shown in Fig. 6-14.

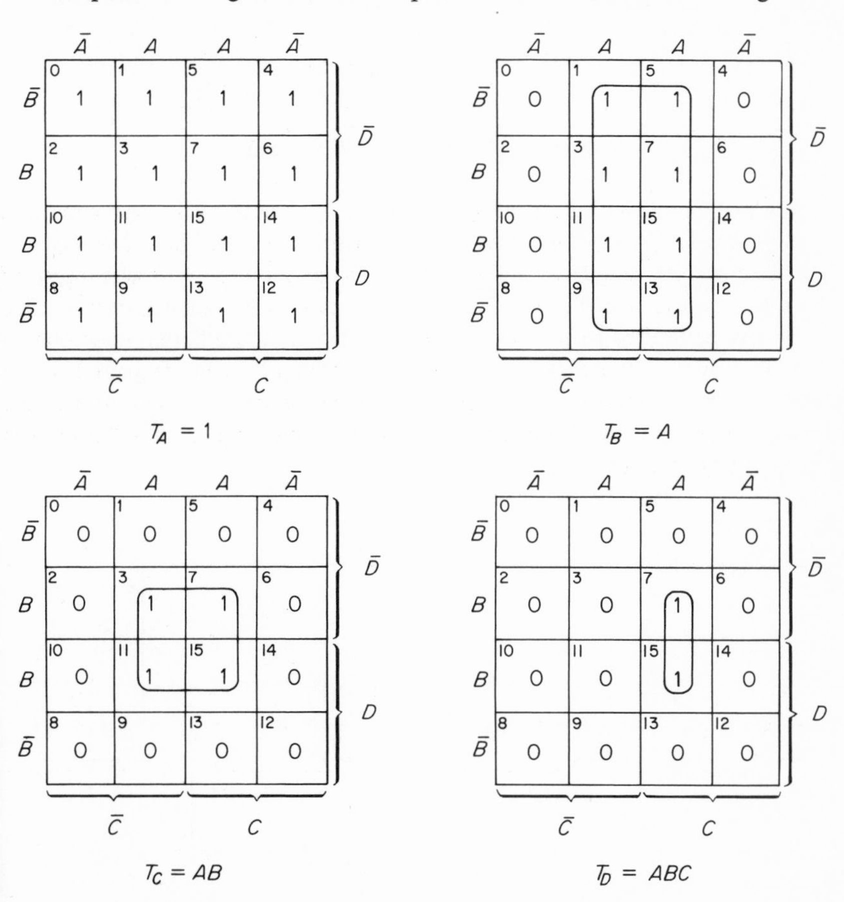

FIGURE 6-14. Maps for Gating Functions of Fig. 6-13.

The first flip-flop receives every clock pulse. The second flip-flop is gated by the TRUE output of the first, and therefore it receives every second clock pulse. The third flip-flop is gated by the TRUE outputs of the first and second flip-flops, and therefore receives every fourth clock pulse; etc. The circuit for this synchronous binary counter is shown in Fig. 6-15.

All flip-flops in the circuit of Fig. 6-15 that change state do so together.

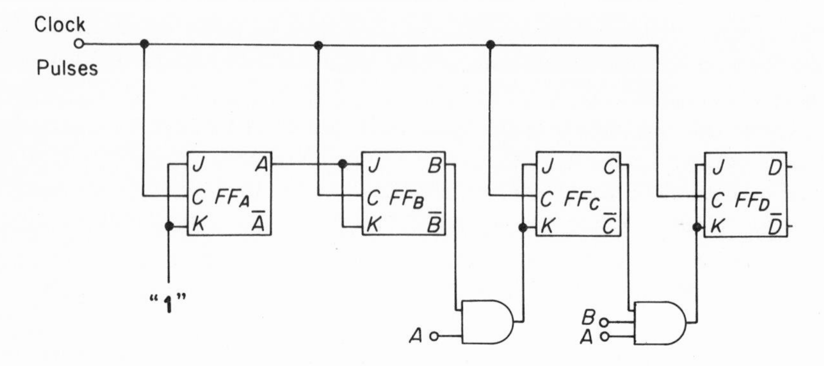

FIGURE 6-15. Synchronous Binary Counter.

The only differences result from small, inherent variations of the response times of the individual stages that in practice are negligible. This has advantages in decoding specific counts over the ripple-through counter. A disadvantage of this circuit is the large fan-in of the AND gates in long counter chains. The fan-in can be limited to two for each gate with a slightly different interconnection, as shown in the circuit of Fig. 6-16. The flip-flops are triggered simultaneously, but the maximum counting rate of the counter is somewhat reduced because the gating waveforms have to propagate down the chain of AND gates before the next input pulse arrives.

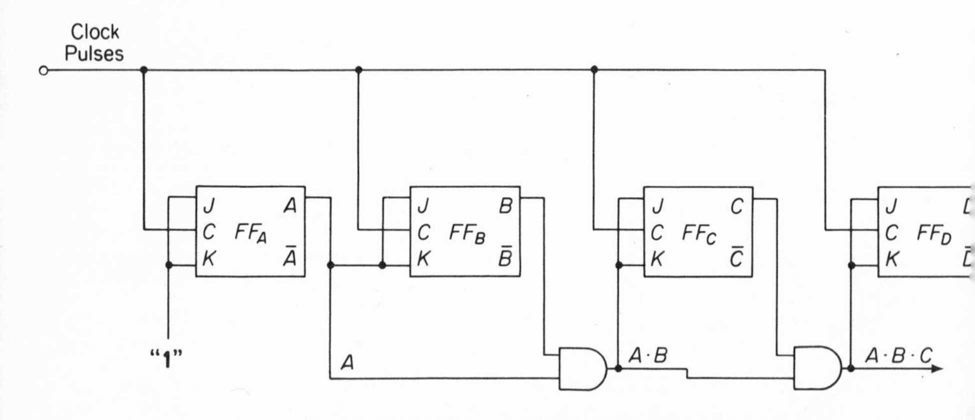

FIGURE 6-16. Alternate Synchronous Binary Counter.

The transition time of the flip-flop relative to the duration of the clock pulse can become critical in the circuits of Figs. 6-15 and 6-16. The triggering pulse must be completed before a binary output voltage changes, otherwise, it could cause the output of an AND gate to change and result in an incorrect triggering of a flip-flop. This can result in a *race* condition. In some cases it even may be necessary to add an artificial or additional delay to prevent incorrect operation. This situation is avoided when dual rank *JK* flip-flops are used for this application in a clocked counter system. This is one of many advantages of using dual rank, integrated circuit *JK* flip-flops in synchronous systems.

As a further illustration of the steps taken in the design of a general synchronous switching circuit, we will design the same synchronous binary counter using only the *set* and *reset* terminals of the flip-flops. The *J* and *K* terminals of the flip-flop will be gated separately. The state transition table is the same as in Fig. 6-13, but the *J* and *K* gating functions will be listed in the right-hand columns. The new state transition table is shown in Fig. 6-17.

State	Present State				Next State				Set Function				Reset Function			
Sequence	D^n	C^n	B^n	A^n	D^{n+1}	C^{n+1}	B^{n+1}	A^{n+1}	J_D	J_C	J_B	J_A	K_D	K_C	K_B	K_A
0	0	0	0	0	0	0	0	1	0	0	0	1	X	X	X	0
1	0	0	0	1	0	0	1	0	0	0	1	0	X	X	0	1
2	0	0	1	0	0	0	1	1	0	0	X	1	X	X	0	0
3	0	0	1	1	0	1	0	0	0	1	0	0	X	0	1	1
4	0	1	0	0	0	1	0	1	0	X	0	1	X	0	X	0
5	0	1	0	1	0	1	1	0	0	X	1	0	X	0	0	1
6	0	1	1	0	0	1	1	1	0	X	X	1	X	0	0	0
7	0	1	1	1	1	0	0	0	1	0	0	0	0	1	1	1
8	1	0	0	0	1	0	0	1	X	0	0	1	0	X	X	0
9	1	0	0	1	1	0	1	0	X	0	1	0	0	X	0	1
10	1	0	1	0	1	0	1	1	X	0	X	1	0	X	0	0
11	1	0	1	1	1	1	0	0	X	1	0	0	0	0	1	1
12	1	1	0	0	1	1	0	1	X	X	0	1	0	0	X	0
13	1	1	0	1	1	1	1	0	X	X	1	0	0	0	0	1
14	1	1	1	0	1	1	1	1	X	X	X	1	0	0	0	0
15	1	1	1	1	0	0	0	0	0	0	0	0	1	1	1	1

FIGURE 6-17. State Transition Table for Synchronous Binary Counter *JK* Flips-Flops in the *RS* Mode.

When the *J* input is **1**, the flip-flop will set to the **1** state with the next clock pulse. Likewise, when the *K* input is **1**, the flip-flop will reset to the **0** state. The *X* indicated in the table means that at that time the input can be either a **0** or a **1**, i.e., *don't care* condition. The maps for these eight *J* and *K* functions are shown in Fig. 6-15. The required switching equations are indicated under each map. The implementation of these switching functions is straightforward.

The above synchronous binary counters could have been designed by inspection; however, the detailed orderly steps illustrate the procedure in the design of a general synchronous switching circuit. These steps are

1. Develop a state diagram.
2. Determine the number of memory elements required.
3. Construct a state transition table.
4. Determine and simplify the required switching functions.

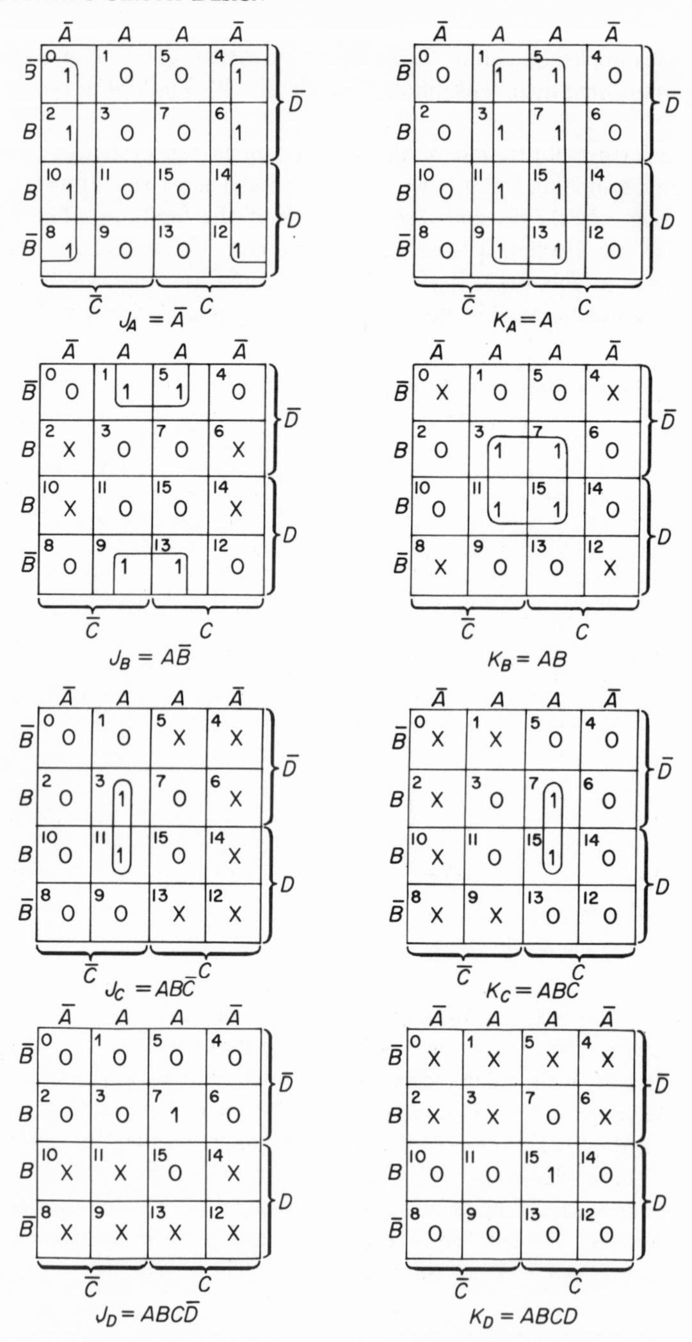

FIGURE 6-18. Maps for Gating Functions of Fig. 6-17.

This procedure is necessary when designing complex counters, and yields the simplified form of the required feedback logic.

6-6 COUNTING TO A BASE OTHER THAN A POWER OF TWO

A chain of n flip-flops can count 2^n input pulses before returning to its initial state; for some applications it may be desirable to use a number system other than the binary system. Since we are most familiar with the decimal system, frequent applications require fabricating counters using the base 10. Constructing a decade counter using flip-flops requires the use of external gating circuits. By use of Eq. (6-5), four flip-flops must be used for a decade counter. In using four flip-flops, only 10 of the possible 16 states are used; and therefore many possible designs can be implemented.

The design approach for a synchronous decade counter uses the same procedure as for the general synchronous sequential circuit used in the previous section for the binary counter. The state diagram for a decade counter is shown in Fig. 6-19. There are 10 states indicated, zero through nine; and it also is desired that the counter sequence is in binary form from zero through nine. This direct binary-coded-decimal representation (BCD) is used frequently. The counting sequence is illustrated in the state transition table of Fig. 6-20 using toggle-type flip-flops for the counter.

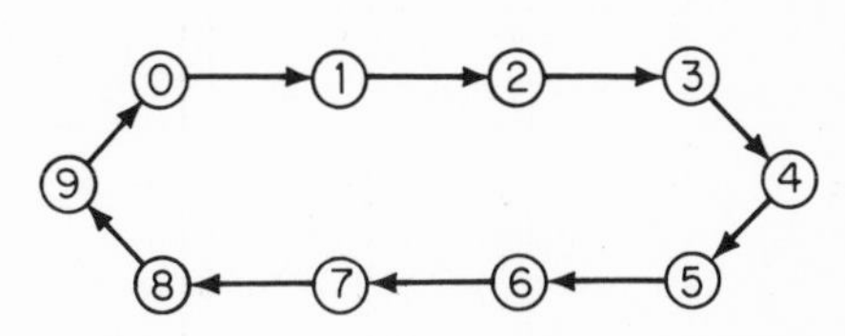

FIGURE 6-19. State Diagram for Decade Counter.

State Sequence	Present State D^n	C^n	B^n	A^n	Next State D^{n+1}	C^{n+1}	B^{n+1}	A^{n+1}	Toggle Function T_D	T_C	T_B	T_A
0	0	0	0	0	0	0	0	1	0	0	0	1
1	0	0	0	1	0	0	1	0	0	0	1	1
2	0	0	1	0	0	0	1	1	0	0	0	1
3	0	0	1	1	0	1	0	0	0	1	1	1
4	0	1	0	0	0	1	0	1	0	0	0	1
5	0	1	0	1	0	1	1	0	0	0	1	1
6	0	1	1	0	0	1	1	1	0	0	0	1
7	0	1	1	1	1	0	0	0	1	1	1	1
8	1	0	0	0	1	0	0	1	0	0	0	1
9	1	0	0	1	0	0	0	0	1	0	0	1

FIGURE 6-20. State Transition Table for a Synchronous BCD Counter.

Decade Counter Using Toggle Flip-Flops

The Boolean switching functions for gating the flip-flops are determined by constructing maps for the toggle functions from the table. The maps and resultant switching expressions are shown in Fig. 6-21. The six binary states that are not used in the decade counter are indicated as *don't care* conditions.

The implementation of the switching functions in Fig. 6-21 is routine. The one remaining consideration is to ensure that the circuit will not *hang-up* if initially, or because of a noise transient, it finds itself in the 10, 11, 12, 13,

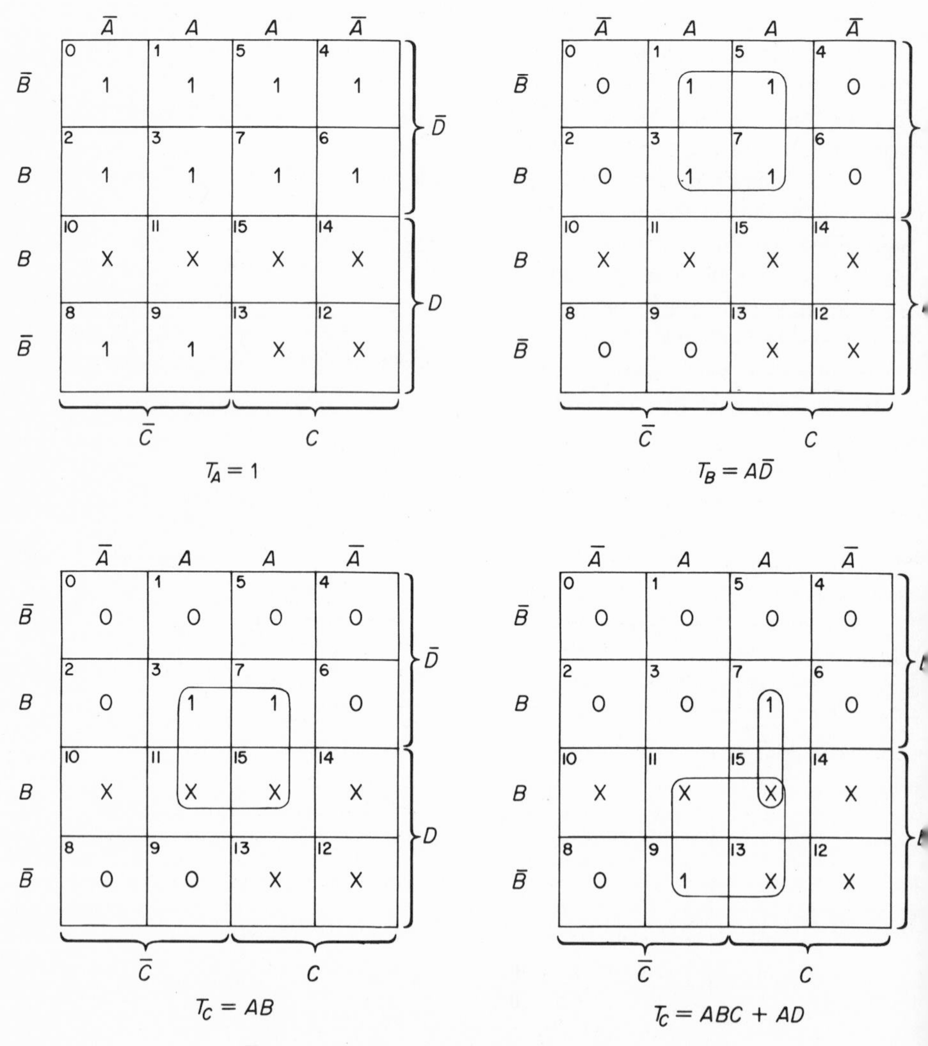

FIGURE 6-21. Maps for Gating Functions of Fig. 6-20.

14, or 15 state. The situation that will be encountered is determined by constructing a state transition table for the *don't care* states using the conditions for the gating functions chosen on the maps in Fig. 6-21. The table is shown in Fig. 6-22, and the gating values are used to determine the next state.

State	Toggle Function				Present State				Next State				
	T_D	T_C	T_B	T_A	D^n	C^n	B^n	A^n	D^{n+1}	C^{n+1}	B^{n+1}	A^{n+1}	
10	0	0	0	1	1	0	1	0	1	0	1	1	(11)
11	1	1	0	1	1	0	1	1	0	1	1	0	(6)
12	0	0	0	1	1	1	0	0	1	1	0	1	(13)
13	1	0	0	1	1	1	0	1	0	1	0	0	(4)
14	0	0	0	1	1	1	1	0	1	1	1	1	(15)
15	1	1	0	1	1	1	1	1	0	0	1	0	(2)

FIGURE 6-22. State Transition Table for the Don't Care States.

Assuming that the counter starts in state 10, its next clock pulse changes flip-flop A and the circuit changes to state 11. On the following clock pulse flip-flops A, C, and D change state, causing the counter to go into state 6, which does not result in a hang-up condition. Similarly, if the circuit is in state 12, we see that it will advance to state 13 and then to state 4. Upon starting in state 14, the counter will advance to 15 and then to state 2. This completes the design procedure. The complete state diagram for this counter is shown in Fig. 6-23. A state diagram, such as this, with no hang-ups is referred to as a *bush*.

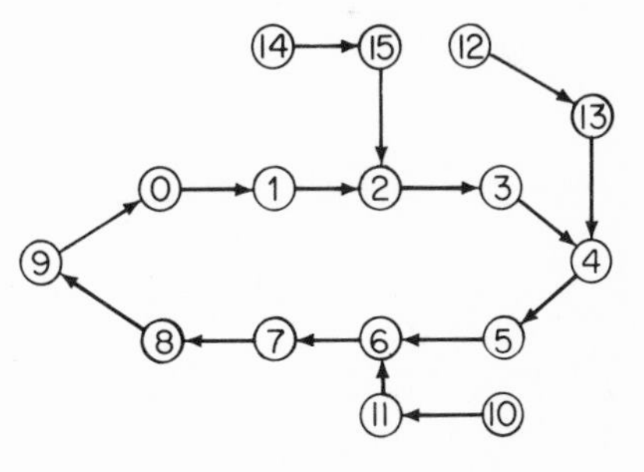

FIGURE 6-23. Complete State Diagram for Decade Counter Design.

If, for the example, the don't care conditions had been chosen such that the counter sequenced 12-13-15-12, then the circuit could hang up. For counter applications this may be objectionable, and the don't care choices would have to be altered.

Nonsynchronous Counters. A nonsynchronous decimal counter using a feedback gating technique with *RST* flip-flops is shown in Fig. 6-24. The operation of this counter is easily seen by referring to the waveform chart in Fig. 6-24(b). It is convenient to choose a starting point when all flip-flops are in the

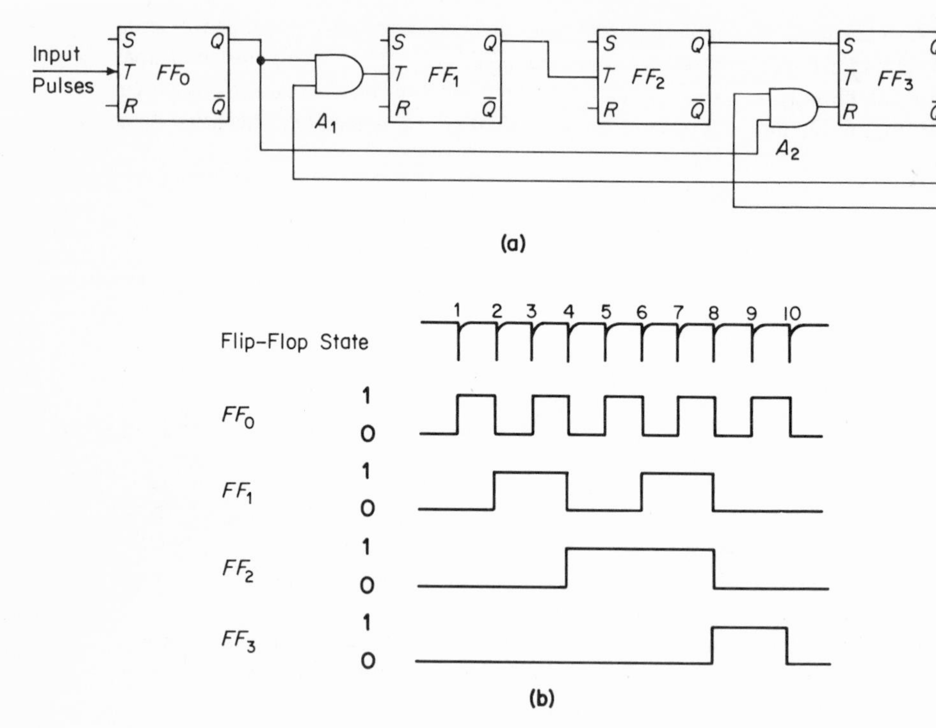

FIGURE 6-24. (a) Decade Counter Using Gates. (b) Waveform Chart.

0 state. It is observed that FF_0 is triggered by each input pulse and therefore operated as a normal toggle flip-flop. The output of FF_0 goes to FF_1 through the AND gate A_1, which initially has a **1** input from $\bar{Q}$ of FF_3. Therefore, the pulses go through gate A_1 and toggle FF_1 and FF_2 in a normal manner until the eighth input pulse. At the count of eight FF_1 resets to **0** and FF_2 also resets to **0**, which in turn sets FF_3 to state **1** (note the output of FF_2 is coupled to the set input of FF_3). FF_3 now applies a **0** to gate A_1 and a **1** to A_2. The count of nine sets FF_0 to a **1**. At the count of 10 FF_0 is reset to **0**, and its output is inhibited by A_1 but is passed through A_2, and FF_3 is reset to **0**. After 10 input pulses all flip-flops have returned to their **0** states and a division by 10 has been achieved. This counter counts in a binary sequence from zero through nine and therefore has a 1-2-4-8 binary code. The state diagram is the same as Fig. 6-19.

Direct-Feedback Decade Counter. Another method of counting to some base R, which is not a power of two, is to introduce direct feedback from later flip-flop stages to earlier ones. A decade counter employing one of many possible feedback arrangements is shown in Fig. 6-25. The counters are coupled

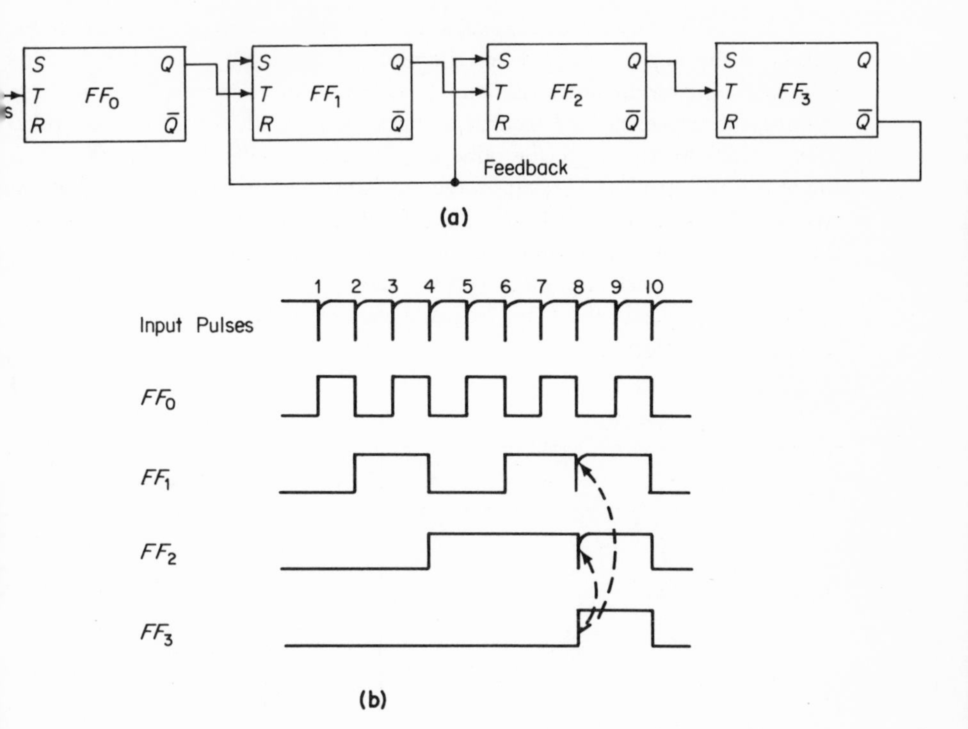

FIGURE 6-25. (a) Decade Counter Employing Feedback. (b) Waveform Chart.

from the Q output to the toggle input of the next stage in a normal manner. The counting proceeds in a binary sequence until the count of eight is reached. The eighth pulse causes FF_0, FF_1, and FF_2 to reset to **0** and FF_3 to set to state **1**. The Q output of FF_3 is fed back to set FF_1 and FF_2 back to state **1**. Therefore, after the counter has settled, the eighth input pulse has caused the counter to advance from 0111, seven, to 1110, fourteen. The feedback has caused the skipping of six binary states, and the counter recycles after the count of 10. The state diagram for this decade counter is shown in Fig. 6-26. The weighting

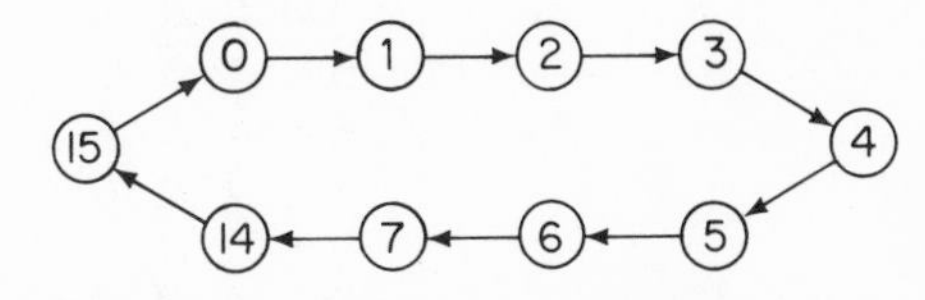

FIGURE 6-26. State Diagram for a 1-2-4-2 Binary-Coded-Decimal Counter.

code for the binary-coded decimal counter shown is no longer 1-2-4-8, which is obvious from the last two states. However, the decimal requirement is met with a weighted code of 1-2-4-2.

It is likely for *race* conditions to exist in feedback counters. If the input pulse to a flip-flop still exists when the feedback pulse is applied, then the flip-flop can settle in an indeterminate state. This condition is avoided by pulse shaping the triggering and feedback pulses with *RC* differentiating networks. The network for the triggering pulse should have a smaller time constant than the feedback network. This causes the feedback pulse to be delayed so that the trigger pulse is returned to **0** before the feedback appears. Feedback counters also are limited in the maximum counting rate because of the large number of transitions that must take place between clock pulses. If there is no feedback to the first stage, as in the 1-2-4-2 decade counter, the counting speed is determined by the first stage.

SEQUENCE

Decimal Count	D	C	B	A
0	0	0	0	0
1	0	0	0	1
2	0	0	1	0
3	0	0	1	1
4	0	1	0	0
5	0	1	0	1
6	0	1	1	0
7	0	1	1	1
8	1	0	0	0
9	1	0	0	1
0	0	0	0	0
		etc.		

(a)

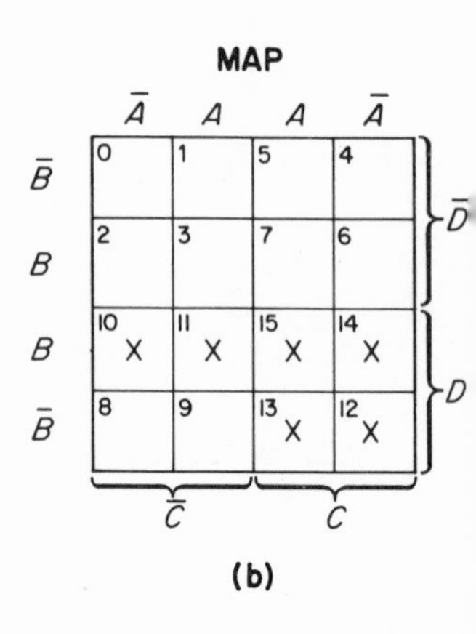

(b)

DECODING

0 = $\bar{A}\ \bar{B}\ \bar{C}\ \bar{D}$
1 = $A\ \bar{B}\ \bar{C}\ \bar{D}$
2 = $\bar{A}\ B\ \bar{C}$
3 = $A\ B\ \bar{C}$
4 = $\bar{A}\ \bar{B}\ C$
5 = $A\ \bar{B}\ C$
6 = $\bar{A}\ B\ C$
7 = $A\ B\ C$
8 = $\bar{A}\ D$
9 = $A\ D$

(c)

FIGURE 6-27. Decoding Analysis for 1-2-4-8 BCD Counter.

6-7 DECIMAL DECODING

Frequently it is required to decode the decimal counter for an output presentation, e.g., to display the range of a target or to indicate a frequency measurement. The decoding network has the binary-coded-decimal inputs and a 10-wire output. The output lines can be used to provide control for drivers of decimal digit display tubes, projection readout devices, etc.

The decoding matrix, which is a group of AND gates, can be built with diodes, transistors, or integrated circuit gates. Minimization of the AND gate functions is readily accomplished by use of mapping. The decoding logic for a 1-2-4-8 BCD counter is determined with the diagrams in Fig. 6-27. The map shows the count sequence, and the omitted counts are plotted as don't care conditions for simplifying the decoding circuit. The decoding function for each state is read directly from the map. The don't care conditions allow some savings, as seen with fewer inputs to the AND gates for states 2 through 9. A matrix design using diode AND gates is shown in Fig. 6-28 for decoding this decimal counter.

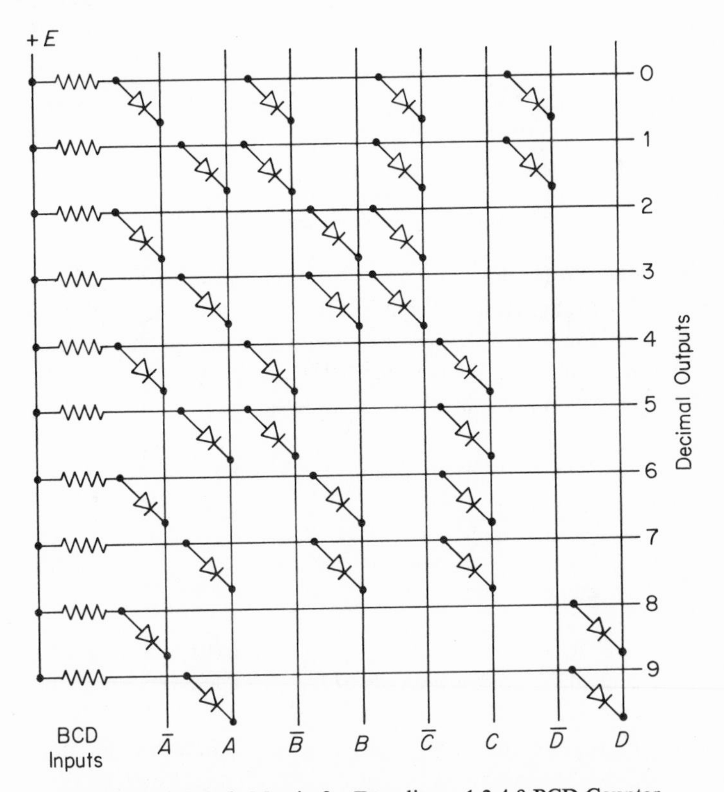

FIGURE 6-28. Diode Matrix for Decoding a 1-2-4-8 BCD Counter.

6-8 THE SHIFT REGISTER

The shift register is a chain of binary storage elements interconnected in such a manner that the state of each flip-flop is shifted to the right (next stage) upon the application of a clock or shift pulse. The input **0**'s and **1**'s are shifted down the register with each clock pulse, with a spillover occurring from the output of the last stage. Applications of shift registers in digital systems include counters, multiplexers, shifting numbers for arithmetic operations, digital signal processing, input-output modules for serial-to-parallel conversion, etc.

A shift register circuit using *JK* flip-flops is shown in Fig. 6-29. The input

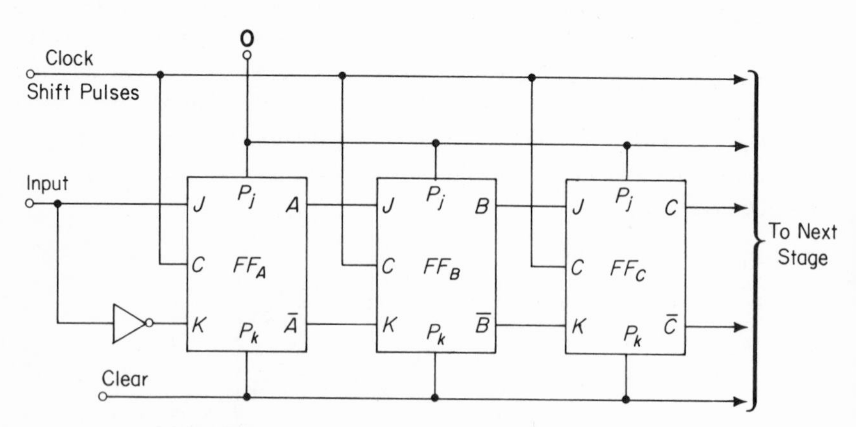

FIGURE 6-29. Shift Register Designed with *JK* Flip-Flops.

to the first stage can be either a **0** or **1**, and it is shifted to the right with each clock pulse. This is achieved by connecting the set and reset outputs of a flip-flop directly to the *J* and *K* inputs of the next stage, as shown in Fig. 6-29. With each clock pulse the content of each stage is shifted to the right. The states of each flip-flop are represented as follows:

$$A^{n+1} = \text{input at time } (n + 1), \text{ either } \mathbf{0} \text{ or } \mathbf{1}$$

$$B^{n+1} = A^n$$

$$C^{n+1} = B^n$$

etc.

If the serial input to a six-stage shift register is **110101**, then the states of the register after each clock pulse are shown in Fig. 6-30, assuming that the register is cleared to all **0**'s initially.

The shift register may be cleared by the simultaneous input of a clear pulse to all P_k (reset) terminals or by setting a **0** input and successively clocking the **0**'s through the length of the shift register.

Time	Input	A	B	C	D	E	F
0	0	0	0	0	0	0	0
1	1	1	0	0	0	0	0
2	0	0	1	0	0	0	0
3	1	1	0	1	0	0	0
4	0	0	1	0	1	0	0
5	1	1	0	1	0	1	0
6	1	1	1	0	1	0	1

FIGURE 6-30. States of Shift Register with **110101** Input.

Shift Register Counter

A shift register can be used as a counter when a single **1** is entered at the input and is shifted down the entire length of the register. When the **1** leaves the final stage it is reinserted into the input. If the register contains n stages, then n shift pulses are required to shift the **1** to the last stage and a divide-by-n is achieved.

The shift register can be set to any predetermined value by gating control signals to the desired P_j and P_k terminals. In using the shift register as a counter, the **1** to be written into the input after n shift pulses can be generated with a logic gate using the output from the first $(n-1)$ outputs. For example, a divide-by-five counter is shown in Fig. 6-31. Regardless of the initial state of each

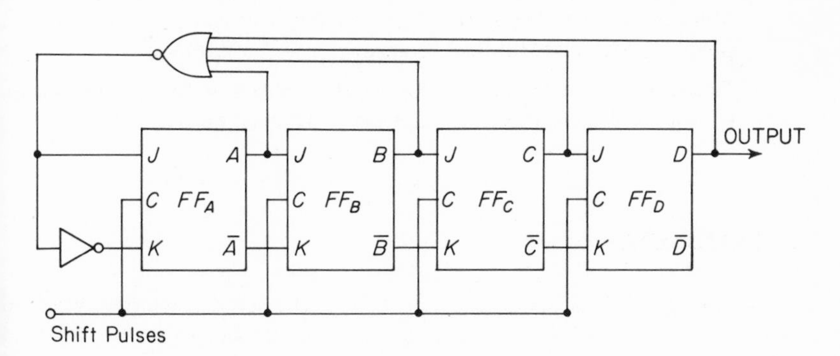

FIGURE 6-31. Divide-by-Five Shift Register.

stage when the power is first applied, the register will assume correct operation within the time of four shift pulses. Also, regardless of the initial state, a **1** will not be written into the input until there are all **0**'s in the first four stages. Then, from the output of the NOR gate a **1** will be inserted into FF_A. The **1** will be shifted down the register and will be automatically entered into the input at every fifth shift pulse.

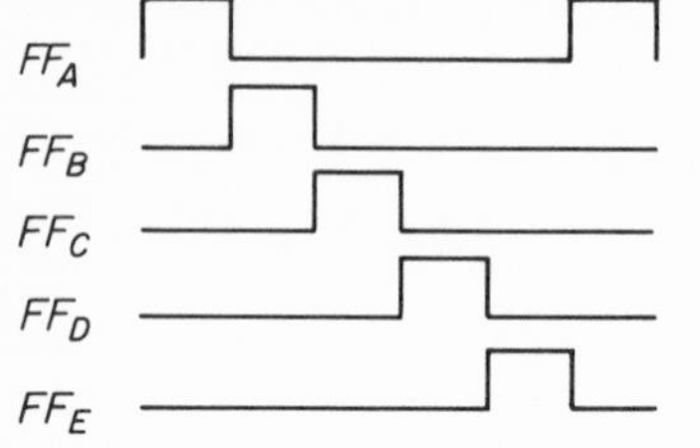

FIGURE 6-32. Output Waveforms of Divide-by-Five Shift Register.

The shift register shown does not efficiently use the flip-flops for counter applications. A divide-by-five counter, of course, can be built with three flip-flops. A general procedure for designing counters and sequence generators with shift registers is given in Chap. 8. One useful application of the shift register is for the generation of sequential switching waveforms for use in multiplexing. The waveforms at the output of each flip-flop in the shift register in Fig. 6-31 are shown in Fig. 6-32. These waveforms can be used to gate sequentially signals or other logic voltages in performing a multiplexing function. Sequential waveforms also can be generated from a counter output using AND gates to form a multiposition switch, as shown in Figs. 6-10 and 6-28. The designer should choose the best method based on the circuit available, the size of the switch, and the operations to be performed.

REFERENCES

1. Maley, G. A. and Earle, J., *The Logical Design of Transistor Digital Computers*, Prentice-Hall, Englewood Cliffs, New Jersey, 1963.
2. Millman, J. and Taub, H., *Pulse, Digital, and Switching Waveforms*, Chap. 18, McGraw-Hill Book Company, New York, 1965.
3. Torng, H.C., *Introduction to the Logical Design of Switching Systems*, Chap. II, Addison-Wesley Publishing Company, Reading, Massachusetts, 1964.

PROBLEMS

1. Develop a logic block diagram for a traffic speed measuring system to determine the speed of each vehicle. The vehicles pass over two road switches that are a short fixed distance apart and extend across the road perpendicular to the traffic flow.
2. Design a synchronous decade counter, 1-2-4-8 code, using the set and reset features of *JK* flip-flops.
3. Design a synchronous divide-by-five counter using *JK* flip-flops in either the toggle or *RS* modes.

4. Design a waveform generator, using the counter of Prob. 3, to produce the voltage waveforms in Fig. P6-4.

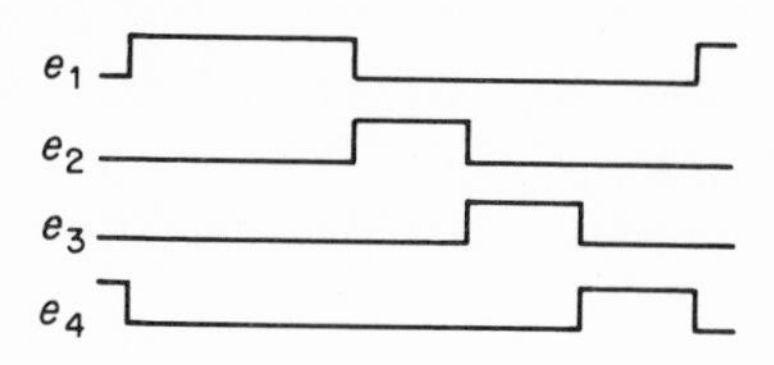

FIGURE P6-4

5. (a) Design a synchronous decade counter for the 1-2-4-2 BCD code using the *JK* flip-flop as a toggle element.
   (b) Repeat the design using the set-reset features of the *JK* flip-flop.
6. Sketch the output waveforms of a decade counter for a 1-2-2-4 BCD weighting code.
7. Design a synchronous decade counter for the 1-2-2-4 BCD code using the *JK* flip-flop as a toggle element.
8. Design a synchronous divide-by-seven counter using *JK* flip-flops.
9. Design a synchronous counter, using *JK* flip-flops as toggle elements, that counts in the following sequence: 2, 1, 0, 5, 4, 3, 9, 8, 7, for every nine pulses received.
10. Design a synchronous counter, using *JK* flip-flops in the *RS* mode, that counts in the following sequence: 1, 3, 5, 7, 9, 11, 13, 15, for every eight pulses received.
11. Design a ripple-through (nonsynchronous) decade counter of the 1-2-2-4 BCD code using feedback and gating techniques.
12. Design a ripple-through 1-2-4-8 BCD decade counter using the *JK* flip-flop with no additional external logic gates. Use the input gates in the *JK* flip-flop to the fullest advantage.
13. Design a decoding network for the 1-2-2-4 BCD decade counter using a minimum number of gates.
14. Design a divide-by-13 counter. Synchronous triggering is not a requirement in its application.
15. Draw a logic schematic for a shift register that contains the feature for parallel entry of data.
16. Draw a logic schematic for a shift register that can shift data either to the right or left. Shift-right and shift-left control signals are available as inputs to the circuit.

CHAPTER 7

COMMUNICATIONS AND DATA CONVERSION CIRCUITS

The information to be acted upon by digital processing and communication equipment is very often in the form of analog data and must be converted to digital form. These analog data may be voltages such as the outputs of radar and sonar receivers, thermocouples, and electromechanical transducers or may be a mechanical shaft position such as the elevation and azimuthal angles of a radar antenna. The device that produces the conversion is the *analog-to-digital converter*, also called a *quantizer* or *encoder*. The complexity of these devices increases with high precision and fast conversion times.

The analog signal is sampled at a predetermined rate and temporarily stored in a *sample-and-hold* circuit. The stored voltage is measured in an analog-to-digital converter circuit and encoded into a binary number. The binary numbers representing the sampled data can be recorded on magnetic tape or punched paper tape. These digital data also can be stored in magnetic or semiconductor memories for further use in data processing. If the digital data are to be transmitted over a communications channel, the binary words are *pulse code modulated* prior to transmission.

In digital data processing it often is necessary to convert the output or final processed data back to analog form. *Digital-to-analog converters* are used for this purpose. These converters also are used to perform intermediate control functions within processing circuits and in output equipment to drive analog recording instruments.

-1 PULSE CODE MODULATION (PCM)

In pulse code modulation for signal transmission the analog or continuous ime function is sampled at periodic intervals and the magnitude of each sample measured and converted to form a binary word. In this measurement proedure the final value is rounded off to the nearest increment of measurement. 'his rounding off process to an integral number of increments is called *quantiation*. The number of quantizing levels determines the resolution of the meauring system. If we should want to transmit the sampled values in 1-V steps, ay from 0 to 15 V, then a four-digit binary code is required to represent these 6 quantizing levels. This is illustrated in Fig. 7-1 with an analog signal and he resultant code sequence after quantization. An actual signal would most ikely have both positive and negative excursions, as in a bipolar ac signal. 'requently ac signals are made unipolar by adding a dc bias to the signal so t assumes only positive values, as shown in Fig. 7-1. It also may be convenient ɔ quantize the magnitude of an ac signal and to use a separate bit to indicate he sign.

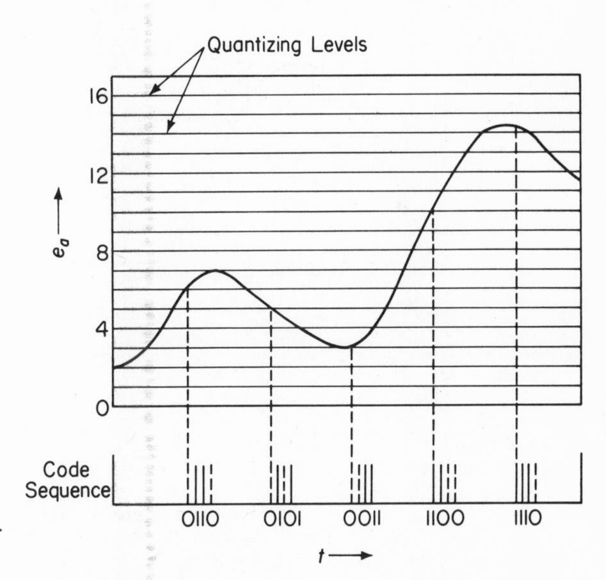

'IGURE 7-1. Pulse Code Modulaion.

It is useful to represent the actual thresholds of quantization as halfway between the increments shown in Fig. 7-1. That is, the value *four* encompasses from level $3\frac{1}{2}$ to $4\frac{1}{2}$; similarly, the value 10 encompasses from level $9\frac{1}{2}$ to $10\frac{1}{2}$, tc. The zero value is from $-\frac{1}{2}$ to $+\frac{1}{2}$, and if the signal is positive only the ero value is from level 0 to $+\frac{1}{2}$. The important point here is that the maximum rror because of quantization is $\pm\frac{1}{2}$ the quantizing increment, or $\pm\frac{1}{2}$ of the

least significant bit. Therefore, the resolution of the measuring system is equa
to the smallest quantization increment. This rounding off of the measuremen
introduces an error in reconstructing the signal waveform. The error is calle
quantization noise. If a signal is quantized to 128 levels, or seven bits, then th
maximum error is $\pm\frac{1}{2}$ part in 128, or less than $\pm 0.5\%$.

7-2 QUANTIZATION

In reconstructing the signal waveform from a PCM encoded message each word is converted to a corresponding voltage level, and the resultan staircase-type waveform is applied to a low-pass filter for smoothing. This i illustrated in Fig. 7-2, where k denotes the increment of quantization. Th quantized signal is an approximation of the actual signal. After passage throug a low-pass smoothing filter, we can think of the output as being comprised o the actual signal plus quantizing noise, as illustrated by the voltage waveform

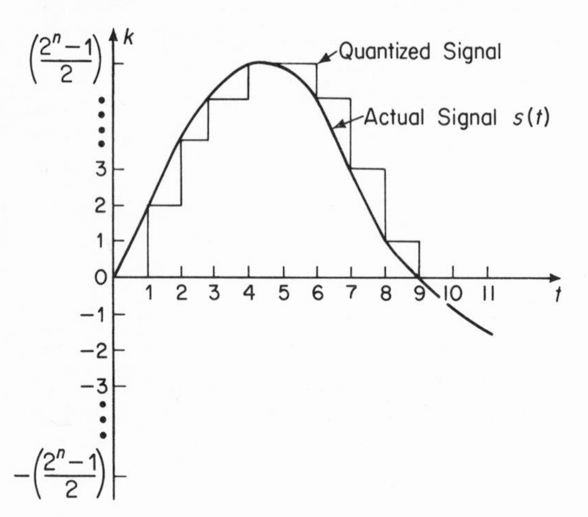

FIGURE 7-2. Quantized Approximation of Signal.

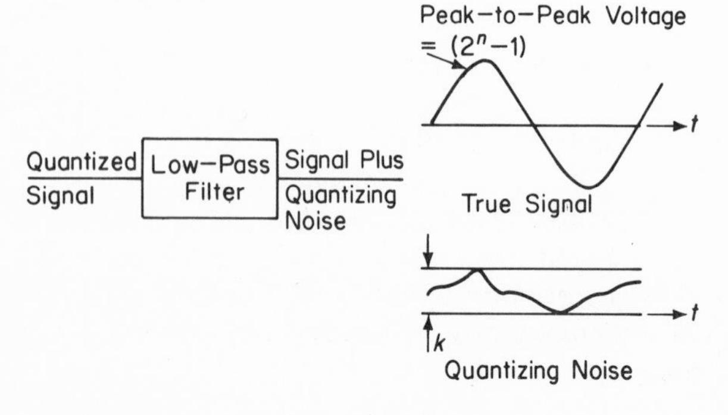

FIGURE 7-3. Signal Plus Quantizing Noise.

n Fig. 7-3. The maximum or *peak* value of the signal is $\frac{1}{2}(2^n - 1)k$ and of the oise is $k/2$. In calculating power we shall use the conventional approach f assuming the voltage waveform is applied across a standard 1-Ω resistor. herefore, the *peak instantaneous signal power* is

$$S_p = (2^n - 1)^2 \frac{k^2}{4} \tag{7-1}$$

nd for the noise,

$$N_p = \frac{k^2}{4} \tag{7-2}$$

he signal-to-noise ratio of the peak instantaneous power is given by Eq. (7-3),

$$\frac{S_p}{N_p} = (2^n - 1)^2 \approx 2^{2n} \tag{7-3}$$

or large signal-to-noise ratios and is 10 dB or more when $2^n \gg 1$. The signal-o-noise ratio increases exponentially with the number of bits used to encode he signal.

In general the signal-to-noise ratio of the peak instantaneous powers is not he same as the *average-power* signal-to-noise ratio, which of course is our major oncern. However, since both the signal and quantizing noise are well bounded ve can expect these two ratios to have similar values. Calculation of the aver-ge power requires a knowledge of the signal and noise time waveforms or tatistics.

If the quantizing noise after filtering, as shown in Fig. 7-3, has a constant mplitude distribution between $\pm k/2$ (i.e., any amplitude within the interval is equally likely), it can be described with the probability density function (v_n) for the amplitude of the noise voltage, as shown in Fig. 7-4.[1] For this

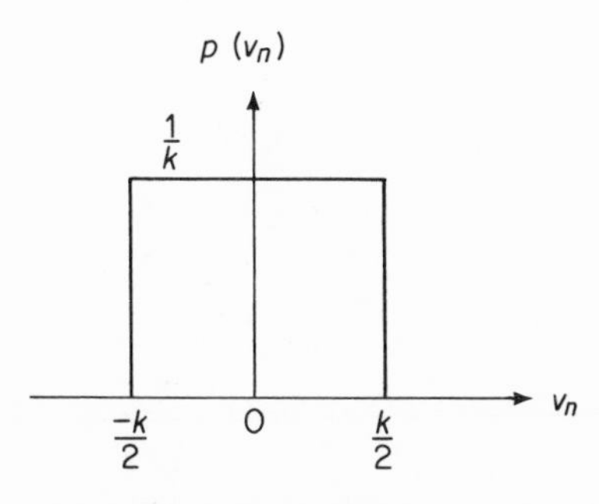

IGURE 7-4. Probability Density unction of Quantizing Noise mplitude.

listribution the average power of the noise, which is the mean square noise oltage across a 1-Ω resistor, is given by

$$N_o = \bar{v}_n^2 = \int_{-\infty}^{\infty} v_n^2 p(v_n)\, dv_n \tag{7-4}$$

$$= \int_{-k/2}^{k/2} v_n^2 \frac{1}{k}\, dv_n = \frac{v_n^3}{3k}\bigg|_{-k/2}^{k/2} = \frac{k^2}{12}$$

This result states that for a uniform amplitude distribution of the quantizing noise, the average noise power is one-twelfth of the square of the quantizing voltage increment. The larger the number of bits used to encode the samples, the smaller the quantizing increment in a given voltage range, and therefore the noise power is lower.

If the signal is a sine wave of peak amplitude $(2^n - 1)k/2$, then the average signal power is

$$S_s = \frac{[(2^n - 1)/2]^2 k^2}{2} = \frac{(2^n - 1)^2}{8} k^2 \tag{7-5}$$

$$\approx \frac{2^{2n} k^2}{8} \qquad \text{for} \quad n > 3 \tag{7-6}$$

The signal-to-noise ratio for a PCM sine wave resulting from the quantizing error is, from equations (7-4) and (7-5),

$$\frac{S_s}{N_o} = \frac{3}{2}(2^n - 1)^2 \tag{7-7}$$

If we let the total number of discrete levels equal m, i.e.,

$$2^n - 1 = m \tag{7-8}$$

then

$$\frac{S_s}{N_o} = \frac{3m^2}{2} \tag{7-9}$$

The total harmonic distortion D_h can be expressed as a ratio of the rms voltage of the noise to the rms voltage of the fundamental sinusoid:

$$D_h = \sqrt{\frac{N_o}{S_s}} = \frac{\sqrt{6}}{3m} \tag{7-10}$$

Figure 7-5 lists values of Eq. (7-10) for various values of m.

The above analysis is based upon a uniform amplitude distribution of the quantizing noise, Fig. (7-4). For just a few levels of quantization we might question this assumption since the approximate sine wave is made up of just a few levels of the staircase structure. Actual analysis of a staircase-type sine wave of 10 levels—i.e., five levels in a half-cycle—yields a total harmonic distortion of

m Total Number of Quantization Levels	n Number of Binary Digits Required for Encoding	D_h % of Distortion $\frac{\sqrt{6}}{3m} \times 100\%$
7	3	11.68
10	4	8.16
15	4	5.44
20	5	4.08
31	5	2.64
50	6	1.63
63	6	1.30
100	7	0.82
127	7	0.64

FIGURE 7-5. Distortion of a Sine Wave Resulting from Different Levels of Quantization.

7.55%, compared with 8.16% determined from Eq. (7-10). The total harmonic distortion for a 100-level staircase-type sine wave is 0.81%; and this is compared with 0.82% obtained from Eq. (7-10). This analysis yields very useful results, especially for 15 or more levels of quantization, $n \geq 4$. Therefore, the assumption of a uniform amplitude distribution of the quantizing noise is valid.

If the signal waveform is not sinusoidal, then its average power will have to be calculated from either a waveform analysis or from the statistics of the signal. As another example a complex waveform will be considered that traverses all the levels of quantization with equal probability; i.e., we will assume that the signal also has a constant amplitude distribution. For example, a triangular waveform has this property. The probability density function is similar to Fig. (7-4), only the single increment k is now replaced by the entire span of mk levels. Therefore, by analogy with Eq. (7-4), the average signal power is

$$S_o = \frac{m^2 k^2}{12} \tag{7-11}$$

Use of Eq. (7-8) yields

$$S_o = \frac{(2^n - 1)^2 k^2}{12} \tag{7-12}$$

The signal-to-noise ratio of the average power is, from Eqs. (7-4) and (7-12),

$$\frac{S_o}{N_o} = (2^n - 1)^2 = m^2 \tag{7-13}$$

This is the same result as the signal-to-noise ratio of the peak instantaneous power obtained in Eq. (7-3), which of course is expected since both amplitude-bounded waveforms have the same statistics.

Expressing Eq. (7-13) in decibels gives Eq. (7-14).

$$D = 20 \log_{10} m \tag{7-14}$$

This relationship showing the improvement in signal-to-noise ratio with the number of quantizing levels is tabulated in Fig. 7-6. When the exact signal

Quantization Levels	m	4	8	16	32	64	128
$10 \log_{10} \frac{S_O}{N_O}$	D	12	18	24	30	36	42

FIGURE 7-6. Quantization Signal-to-Noise Ratio in Decibels for a Complex Waveform.

waveform is unknown, Eq. (7-14) can be used as a guide to determine the number of bits to encode for PCM for a given reconstruction signal-to-noise ratio. Of course, if the signal statistics are known, then the actual signal power and signal-to-noise ratio can be calculated for different numbers of bits of encoding.

Another form of error introduced into the conversion process is due to the accuracy of the converter itself. Noise introduced due to converter errors is called *converter noise*. The error or inaccuracies in the various parts of the converter circuits must be held to less than one quantization level. These errors are discussed in the various circuits that follow. However, converter noise and quantization noise are independent and add on an rms basis. Therefore, the actual signal-to-noise ratio due to the conversion equipment will be approximately 3 dB less than that of the quantization noise alone.

The dynamic range of an input signal has a significant effect on determining the number of bits for encoding. The maximum signal level must span all the levels of quantization. If the quantization voltages and system gain are not set properly, the quantizer will either saturate, or all the levels will not be used. If the quantizer saturates the effect is clipping the signal and introducing excessive distortion. Similarly, if all levels of quantization are not spanned for the maximum signal, the quantizer is not being used efficiently and higher quantization noise will result.

When the input signal is at the low end of the dynamic range, the number of quantizing levels spanned is small and the resulting quantization noise increases. Therefore, to maintain an allowable level of distortion the number of quantizing levels should be determined for the low level signals. If these levels are increased uniformly to span the waveforms for the largest signal level, then there are many more levels used than required to stay within a fixed percentage of allowable distortion. In order to make full use of the dynamic range it appears that the quantizing increments should be spaced closer together for the lower level signals and further apart for the higehr level signals. This leads to nonuniform quantization of the signals.[1] Nonuniform quantization can be achieved directly by nonlinear spacing of the quantizing levels in the encoder or by first logarithmically compressing the input signal and then applying uniform quantization.

7-3 SAMPLING THEOREM

In the previous section we discussed the resolution to which a sample of an analog signal should be quantized to describe it adequately. If we now take a continuous analog waveform, we must consider the following question: "How *often* must a continuous waveform be sampled and quantized to describe it completely, i.e., to be able to reconstruct the original waveform without a loss of information?" This section shows that an analog waveform can be sampled at discrete time intervals and be completely specified, if it is continuous, single valued, and frequency band limited.

The process of sampling is best understood by considering the frequency content of signals. Information signals can be considered band limited when all the information is contained in the frequencies within the band. For example, the information contained in the human voice can be transmitted within a 3-kHz bandwidth with reasonable fidelity. The telephone companies take advantage of this by band limiting and frequency multiplexing many 3-kHz bands on a single line. The information is translated in frequency by the familiar heterodyning or mixing process.

If an audio signal $A \cos \omega_a t$ is multiplied by another signal $C \cos \omega_c t$, it is well known that two sideband signals are generated.

$$f(t) = (A \cos \omega_a t)(C \cos \omega_c t) \tag{7-15}$$

$$= AC \cos \omega_a t \cos \omega_c t \tag{7-16}$$

Using trigonometric identities we get

$$f(t) = \tfrac{1}{2} AC \cos (\omega_c - \omega_a)t + \tfrac{1}{2} AC \cos (\omega_c + \omega_a)t \tag{7-17}$$

$$= \tfrac{1}{2} AC[\cos 2\pi(f_c - f_a)t + \cos 2\pi(f_c + f_a)t] \tag{7-18}$$

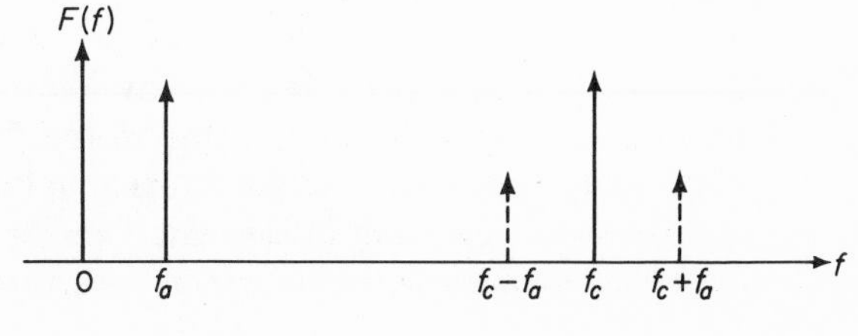

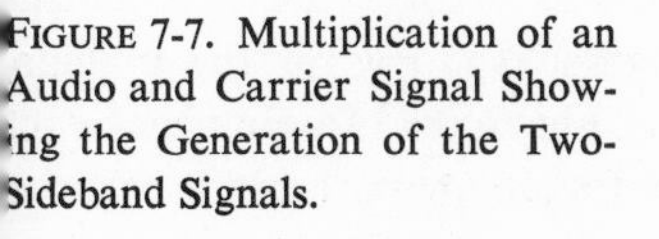

FIGURE 7-7. Multiplication of an Audio and Carrier Signal Showing the Generation of the Two-Sideband Signals.

This is illustrated in a frequency plot in Fig. 7-7, where the audio and carrier frequencies are indicated by solid arrows and the two sideband signals generated by dotted arrows. Assume that the magnitudes of A and C are both unity.

Information, however, is not contained in a single frequency but in a band of frequencies. Complex signals are comprised of a sum of single frequencies within the band, as is well known from Fourier analysis. Therefore, in a similar manner, band-limited signals multiplied by a single frequency f_c generate upper and lower sidebands. This is illustrated in Fig. 7-8(a).

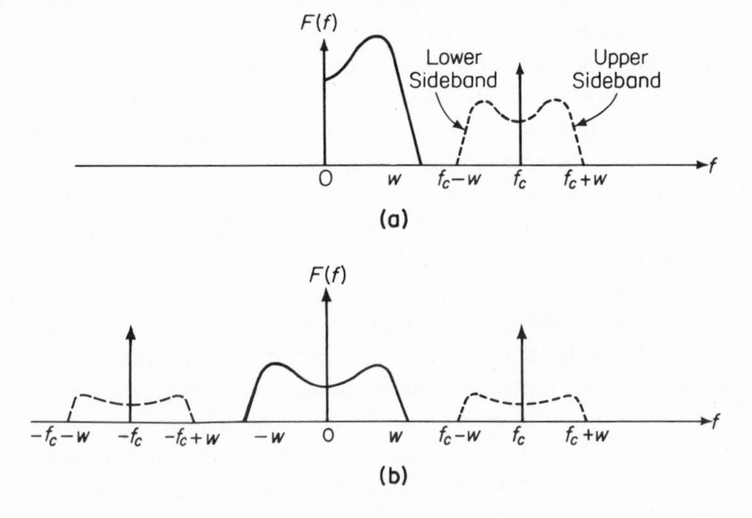

FIGURE 7-8. (a) Spectrum and Double Sidebands. (b) Equivalent Plot Using Both Positive and Negative Frequencies.

As a mathematical and graphical convenience it is helpful to think of a single frequency or spectrum as containing both positive and negative frequencies. The concept of negative frequencies appears in Fourier analysis when the series are written in the following closed exponential form:

$$f(t) = \frac{1}{T} \sum_{n=-\infty}^{\infty} c_n e^{jn\omega_0 t} \tag{7-19}$$

where

$$c_n = \int_{-T/2}^{T/2} f(t) e^{-jn\omega_0 t}\, dt \tag{7-20}$$

The summation for both positive and negative values of n provides mathematical symmetry about the vertical axis, as illustrated in Fig. 7-8(b), and also provides simplification in mathematical analysis. This model implies that each frequency component contains both a positive frequency part and a negative requency part, each one-half of the voltage. Note that if the spectrum in Fig. '-8(b) is folded at the vertical axis and the components are added, the spectrum of Fig. 7-8(a) is obtained.

The Sampling Function

The sampling function is a narrow, rectangular pulse train of unity amplitude. It is multiplied by the input signal to obtain the sampled data signal. The time waveforms are illustrated in Fig. 7-9. Electronically, the sampler can be thought of as a transmission gate that is normally closed and is opened

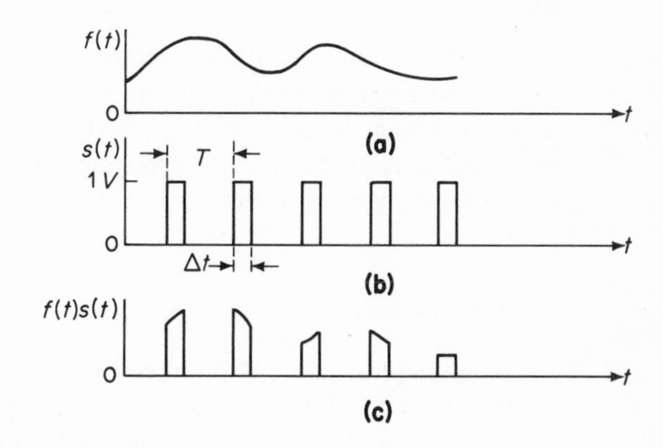

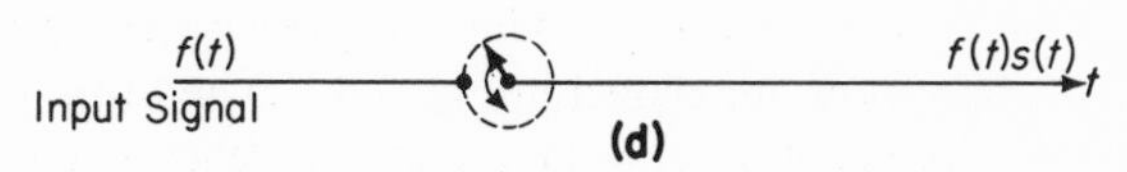

FIGURE 7-9. (a) Analog Signal $f(t)$. (b) Sampling Pulse Train $s(t)$. (c) Sampled Signal. (d) A Mechanical Sampler.

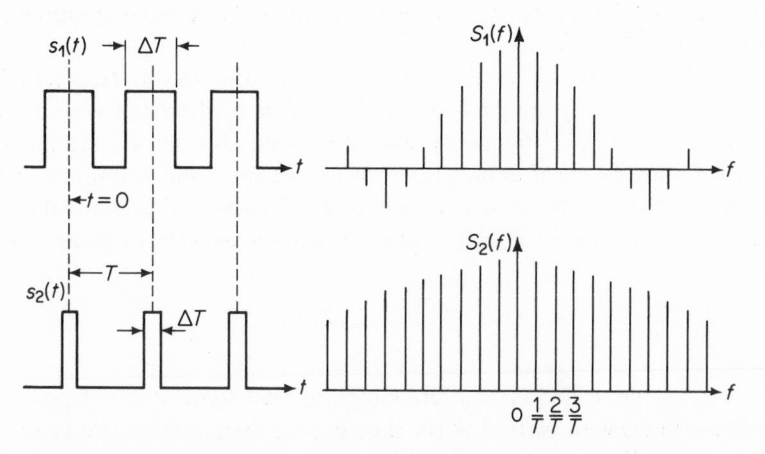

FIGURE 7-10. Sampling Pulse Train and Frequency Spectra for Wide and Narrow Pulses.

during the sampling pulse duration. Mechanically, the sampler can be thought of as a rotating switch, as shown in Fig. 7-9(d). In either event, mathematically the sampler multiplies the signal $f(t)$ by the sampling function $s(t)$, giving the sampled signal $f(t)s(t)$.

The spectrum of $s(t)$ is found by calculating its frequency representation by use of Fourier series. The spectrum depends upon the pulse width ΔT and the period T. Spectra for both a wide and a narrow ΔT with the same period are shown in Fig. 7-10. The fundamental frequency is

$$f_1 = \frac{1}{T} \tag{7-21}$$

and the other frequencies in the spectrum are integral multiples, harmonically related. The envelopes of these spectra are $(\sin x)/x$ functions, and it is noted that the spectrum becomes broader as the pulses becomes narrower. In the limit as ΔT approaches zero, the spectrum becomes infinite with constant amplitude frequency components.

Sampled Data

The sampled data can be represented as the product of two functions $f(t)s(t)$, as shown in Fig. 7-9. We are now going to look at the sampled data function in the frequency domain. Figures 7-11(a) and (b) show a typical function and its corresponding frequency spectrum. The sample function and its spectrum are shown in Fig. 7-11(c) and (d). In multiplying the two functions together, $f(t)s(t) = x(t)$, each frequency component in the sampling function generates a double sideband, producing the multiple sideband spectrum $X(f)$ shown in Fig, 7-11(f).

In Fig. 7-11(f) we see that the original spectrum can be recovered from the sampled signal by using a low-pass filter. This is illustrated with a dotted line around the lowest frequency spectrum. Note also that the sampling frequency f_s must be equal to or greater than twice the signal bandwidth $2W$ or the adjacent sidebands will overlap and proper filtering will be impossible. This leads to the relationship of Eq. (7-22) and is referred to as the *sampling theorem*.

$$f_s = \frac{1}{T} \geq 2W \tag{7-22}$$

Eq. (7-22) is also called the Nyquist sampling rate. Since it is not possible to construct filters with vertical skirts, the sampling frequency should be selected larger than $2W$ to separate the multiple sidebands. In practice, values of 2.5 to $3.5W$ are used to allow the use of a more economical filter. Actually, a trade off of sampling rate and filter slope requirements should be made for a specified allowable distortion.

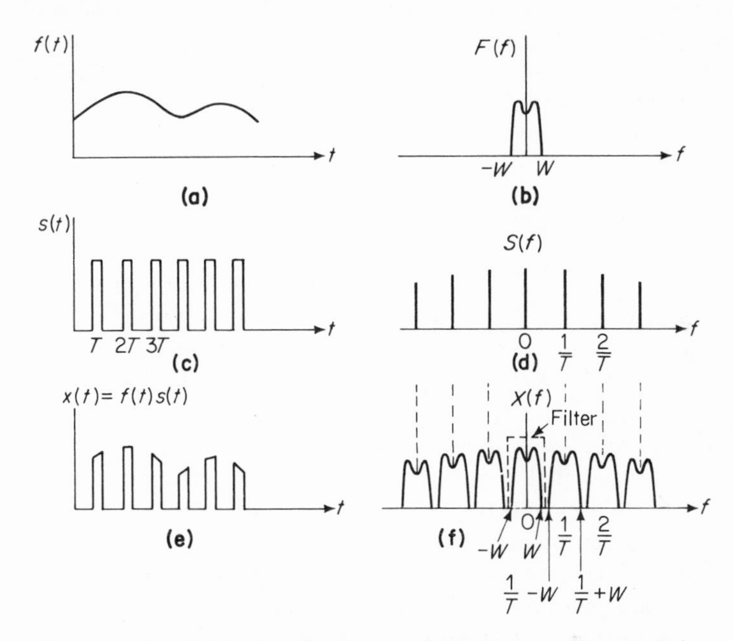

FIGURE 7-11. Signal: (a) Waveform and (b) Spectrum. Sample Pulses: (c) Waveform and (d) Spectrum. Sampled Data: (e) Waveform and (f) Spectrum.

In the above discussion the spectrum of the analog signal extends down to 0 Hz. If a time function $f(t)$ is composed of a band of frequencies displaced from 0 Hz with a bandwidth of W Hz, with a highest frequency of f_h, then the minimum sampling rate is given by $2 f_h/m$, where m is the largest integer not to exceed f_h/W. All higher sampling rates are not necessarily usable because of possible overlapping of the many sidebands generated.[2] The proof of this is similar to the analysis above but becomes somewhat more complex. This proof is not stated here, but the results are indicated in Fig. 7-12.

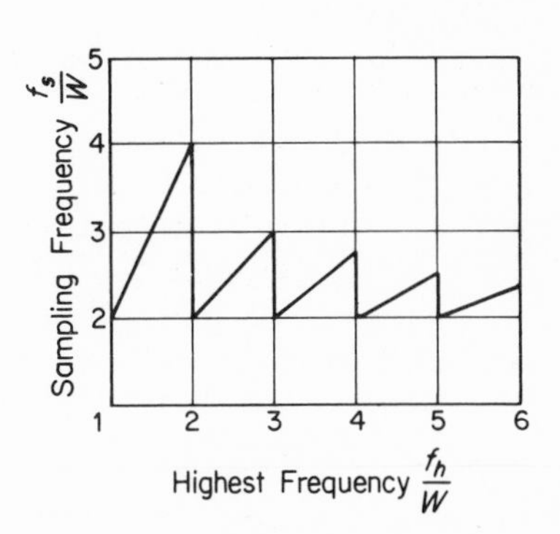

FIGURE 7-12. Minimum Sampling Frequency vs. Highest Signal Frequency Normalized with Bandwidth.

As an example, consider a signal with the highest frequency $f_h = 10$ kHz and the lowest frequency $f_l =$

7 kHz. The signal bandwidth $W = 3$ kHz. To use Fig. 7-12 the ratio of f_h to W is required: $f_h/W = 3.33$. Then, from the graph, $f_s/W = 2.25$. This relationship states that a minimum sampling rate of $f_s = 6.75$ kHz is required. It is good engineering practice to plot out the multiple spectra to assure that foldover and overlapping do not occur. This will aid in the selection of the proper filter characteristics.

7-4 DIGITAL-TO-ANALOG CONVERTERS

The preceding sections discussed the number of bits to which an analog signal should be encoded for an acceptable signal-to-noise ratio due to quantization and the rate at which samples should be taken to preserve the information content of the signal. We now will discuss the methods and circuits used to perform the conversion between analog and digital data. Digital-to-analog converters are discussed first because they also may be used when converting analog signals to digital form in analog-to-digital conversion circuits. The complexity of these circuits is generally a function of the degree of precision and the speed of response required.

To determine the decimal equivalent of a binary number, recall that the various digits are multiplied by their corresponding weights and that the results are added. For example, 100101 in decimal notation is

$$1(32) + 0(16) + 0(8) + 1(4) + 0(2) + 1(1) = 37 \tag{7-23}$$

In a similar manner, to convert the binary number to an analog voltage, voltages with the proper binary weighting factors are added together. A simplified digital-to-analog converter illustrating this principle is shown in Fig. 7-13. The switches may be relay contacts or semiconductor switches, depending upon the equipment requirements and the speed of operation. One disadvantage of this

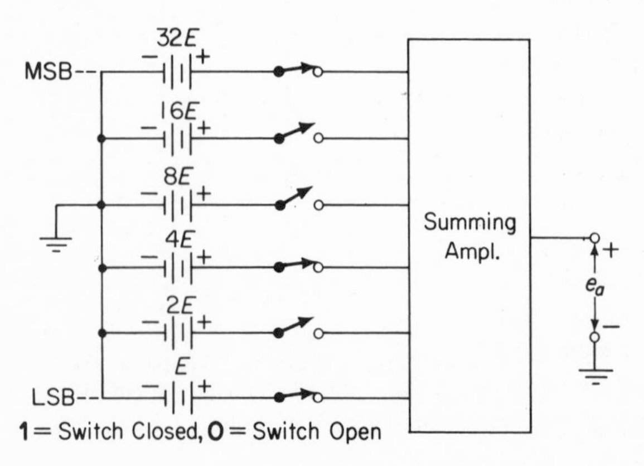

FIGURE 7-13. Simplified Digital-to-Analog Converter.

circuit is that many standard voltage references are required. A similar circuit will be discussed later where one reference voltage is used and the weighting is accomplished with precision resistors. In this type of digital-to-analog converter the voltages are summed in a summing amplifier.

The Summing Amplifier

The summing amplifier is used frequently in computer circuits. The basic part of this circuit is the operational amplifier, which has been discussed in Chap. 5 and is shown in block diagram form in Fig. 7-14. In the figure $-A$ represents the gain of the amplifier, which provides signal inversion and negative feedback, and R_f and R_i are the feedback and input impedances, respectively.

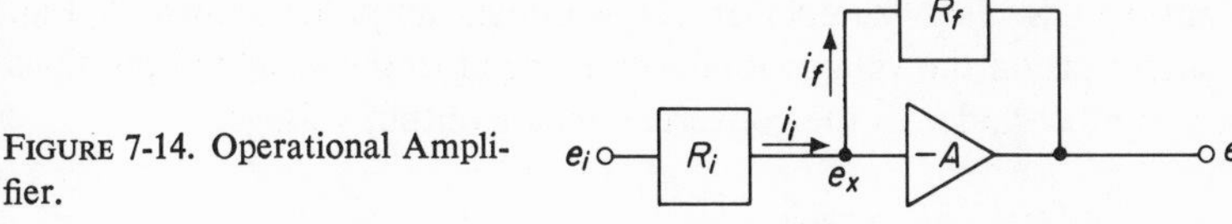

FIGURE 7-14. Operational Amplifier.

In our analysis we shall assume that the amplifier has a very high input resistance—i.e., much larger than the other circuit resistances R_i and R_f—and therefore draws negligible current. With this assumption,

$$i_i = i_f \tag{7-24}$$

The analog output voltage e_a is

$$e_a = -Ae_x \tag{7-25}$$

The amplifier gain $-A$ in this application is very high, usually 10^3–10^6; therefore,

$$e_x = -\frac{e_a}{A} \approx 0 \tag{7-26}$$

For example, if $A = 10{,}000$ and e_a is restricted to a maximum voltage of 20 V, then

$$e_{x\max} = \frac{e_{a\max}}{A} = \frac{20}{10{,}000} = 0.002 \text{ V} \tag{7-27}$$

Since e_x is always very close to 0 V, this point is referred to as a *virtual* ground. Using these relations we find that the circuit currents reduce to

$$i_i = \frac{e_i - e_x}{R_i} \approx \frac{e_i}{R_i} \tag{7-28}$$

and

$$i_f = \frac{e_x - e_a}{R_f} \approx \frac{-e_a}{R_f} \tag{7-29}$$

Combining Eqs. (7-24) and (7-29) with the condition that $e_x \approx 0$ gives

$$-e_a = R_f i_i \tag{7-30}$$

where the minus sign indicates signal inversion.

The operational amplifier is used as a summing amplifier by having multiple inputs applied through resistors. This was treated in detail in Sec. 5-2. Some of the material is repeated here with notations that correspond to the digital input signals. For example, in the summing amplifier shown in Fig. 7-15, the subscripts on the voltage inputs correspond to the weighted positions of a binary number and e_a is the converted analog output voltage.

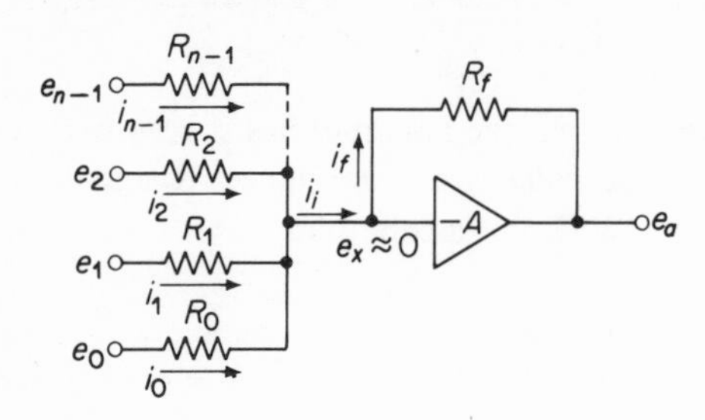

FIGURE 7-15. Operational Summing Amplifier.

The input voltages e_0 through e_{n-1} cause the corresponding input currents to flow; therefore,

$$i_i = i_0 + i_1 + i_2 + \cdots + i_{n-1} \tag{7-31}$$

$$i_i = \frac{e_0}{R_0} + \frac{e_1}{R_2} + \cdots + \frac{e_{n-1}}{R_{n-1}} \tag{7-32}$$

Substituting Eq. (7-32) into Eq. (7-30) gives

$$-e_a = R_f\left(\frac{e_0}{R_0} + \frac{e_1}{R_1} + \frac{e_2}{R_2} + \cdots + \frac{e_{n-1}}{R_{n-1}}\right) \tag{7-33}$$

If all the above resistors are equal, i.e.,

$$R_f = R_0 = R_1 = R_2 = \cdots = R_{n-1} = R \tag{7-34}$$

Then

$$-e_a = e_0 + e_1 + e_2 + \cdots + e_{n-1} \tag{7-35}$$

which indicates that the output is the true sum of the inputs. This is the type of circuit used in the summing amplifier block in Fig. 7-13.

Note that Eq. (7-33) indicates that by choosing the appropriate values for R_f, R_0, R_1, etc. weighted sums of the input voltages can be obtained. For example, if

$$R_f = R_0 = R, \qquad R_1 = \frac{R}{2}, \qquad R_2 = \frac{R}{4}, \tag{7-36}$$

$$R_3 = \frac{R}{8}, \ldots, R_{n-1} = \frac{R}{2^{n-1}}$$

then Eq. (7-33) becomes

$$-e_a = R\left(\frac{e_0}{R} + \frac{2e_1}{R} + \frac{4e_2}{R} + \frac{8e_3}{R} + \cdots + \frac{2^{n-1}e_{n-1}}{R}\right) \tag{7-37}$$

$$-e_a = e_0 + 2e_1 + 4e_2 + 8e_3 + \cdots + 2^{n-1}e_{n-1} \tag{7-38}$$

where e_0, e_1, $e_2, \cdots$, e_{n-1} are either E or 0 V, depending upon whether the input is a binary **1** or **0**. This choice of resistor values eliminates the need for multiple reference voltage sources.

The above analysis was based upon the input to the operational amplifier being at virtual ground, i.e., e_x is at approximately 0 V. We illustrated the small value of e_x for a typical amplifier gain by Eq. (7-27); ideally, this voltage is zero. However, this would require an infinite gain amplifer. The actual amplifier gain required in a digital-to-analog converter is related to the desired number of levels of quantization. Encoding to n bits gives $2^n - 1$ levels of quantization. The required amplifier gain can be calculated, referring to Fig. 7-14. Using Eq. (7-24), and substituting in the expressions for the currents i_i and i_f gives

$$\frac{e_i - e_x}{R_i} = \frac{e_x - e_a}{R_f} \tag{7-39}$$

If $R_i = R_f$,

$$-e_a = e_i - 2e_x \tag{7-40}$$

The analog signal differs from the input by $2e_x$, and we would desire to keep this value less than one level of quantization voltage. Therefore,

$$2e_{x\,\max} \ll \frac{E_{\max}}{(2^n - 1)} \tag{7-41}$$

Note that the total input signal excursion is $E_{\max}$, i.e., peak-to-peak voltage. Then,

$$e_{x\,\max} \ll \frac{E_{\max}}{(2^n - 1)2} \tag{7-42}$$

Using Eq. (7-25) with the maximum values for the variables, i.e.,

$$E_{\max} = -Ae_{x\,\max} \tag{7-43}$$

and substituting in the inequality, Eq. (7-42), gives

$$|A| \gg (2^n - 1)2$$

or

$$|A| \gg 2^{n+1} \tag{7-44}$$

An amplifier gain of 10×2^n is usually satisfactory in this expression, Eq. (7-44). This will give an error voltage that is one-tenth the quantization voltage interval to obtain the required accuracy for the number of bits being used. The analysis shows that the finite amplifier gain affects the accuracy of digital systems and never should be overlooked when operational amplifiers are used in digital circuits.

The maximum error voltage in the previous example, Eq. (7-27), is 2 mV. With an input range of 0 to 20 V the maximum error voltage is 1/10,000 of the input range, and quantizing to more than 13 bits, 1 part in 8,192, would be meaningless. It is important to remember that there may be other, similar sources of error in the system, and the overall accuracy may limit the quantization to less than 13 bits. Using the factor of 10 mentioned above, the converter would be limited to 10 bits of quantization.

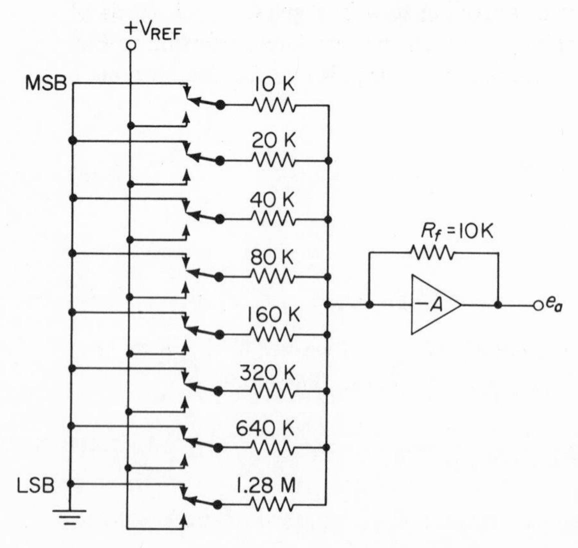

FIGURE 7-16. Eight-Bit Digital-to-Analog Converter with a Single Reference Voltage.

A digital-to-analog converter for 8 bits is shown in Fig. 7-16. For a large number of bits the resistors in the network cover a very large range of values, since each successive resistor is double that of the preceding one. Practically, this range has to be limited since the largest resistor must be much smaller than the input resistance of the operational amplifier to prevent errors in summing. Also, in most applications the binary switches would be semiconductor switches rather than mechanical and therefore would present a small, nonzero resistance. Any variation in switch resistance should be swamped out by a much larger $R/2^{n-1}$ resistor; otherwise the variation would affect the most significant bit and contribute an appreciable error. This problem of a very wide range of resistor values can be avoided by using the ladder network described below.

Resistive Ladder Network

A digital-to-analog converter using a resistive ladder network, employing just two values of resistors, R and $2R$, is shown in Fig. 7-17. For simplicity

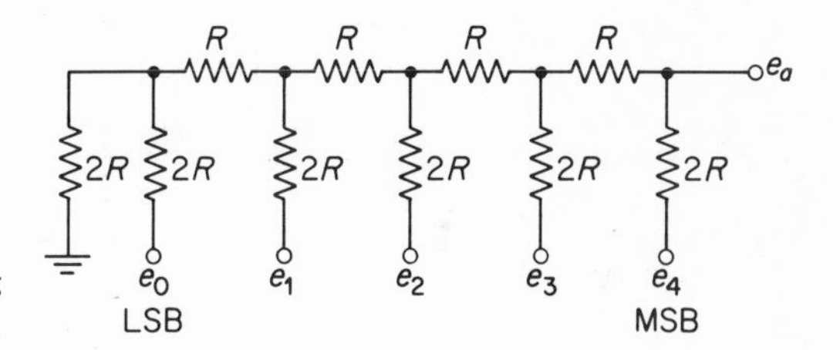

FIGURE 7-17. Digital-to-Analog Ladder Network.

this circuit shows just a 5 bit input; however, it can be extended to n bits. The open circuit analog output voltage is one-half the voltage of the most significant bit e_4, plus $\frac{1}{4}e_3$, plus $\frac{1}{8}e_2$, etc. For 5 bits,

$$e_a = \tfrac{1}{2}e_4 + \tfrac{1}{4}e_3 + \tfrac{1}{8}e_2 + \tfrac{1}{16}e_1 + \tfrac{1}{32}e_0 \tag{7-45}$$

Therefore, the analog output voltage is a properly weighted sum of the input binary digits.

The ladder network of Fig. 7-17 is easily analyzed by using the superposition principle. By superposition, the response of a linear network is identical to that found by considering each voltage source alone, with all other sources removed and replaced by short circuits, and then summing the individual responses. First, considering the most significant bit e_4 alone, the network is redrawn in Fig. 7-18(a). This is a conventional ladder network, and the resistance looking to the left, from the left side of any R resistor, is equal to R. The equivalent circuit reduces to that shown in Fig. 7-18(b) and, clearly,

$$e'_a = \tfrac{1}{2}e_4 \tag{7-46a}$$

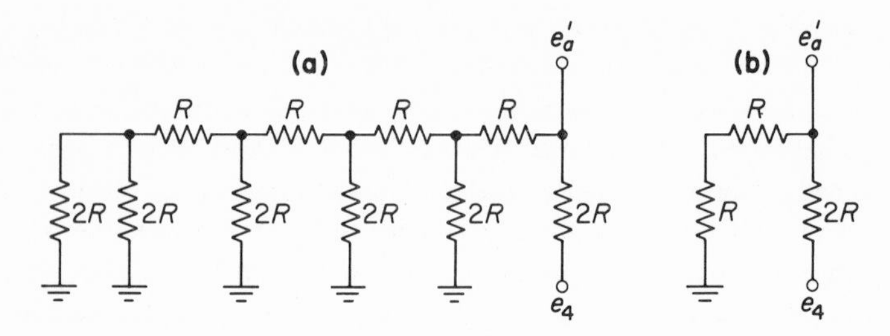

FIGURE 7-18. (a) Ladder Network with Only e_4 Present. (b) Equivalent Circuit.

Next the analysis is performed with only input e_3 present, and the resultant equivalent circuit is shown in Fig. 7-19(a). Breaking the circuit at point x and replacing the left side with the Thévenin equivalent results in the network shown in Fig. 7-19(b). Clearly,

$$e_a'' = \tfrac{1}{4}e_3 \tag{7-46b}$$

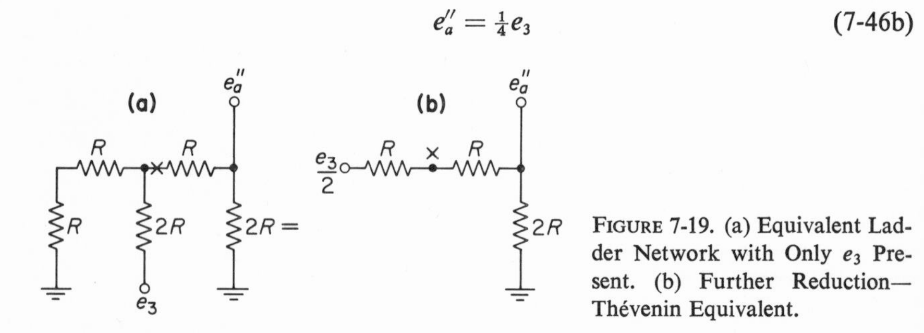

FIGURE 7-19. (a) Equivalent Ladder Network with Only e_3 Present. (b) Further Reduction—Thévenin Equivalent.

Upon continuing this process for each input bit, and summing the results, Eq. (7-45) is obtained. This completes the proof and shows that the analog voltage output from the ladder network is the proper weighted sum of the binary input signals. The calculations are based upon having no load at the output terminal. Loading the output with a resistance will cause attenuation of the analog voltage. Of course, this can be calculated and taken into account in the overall system design.

Error Considerations. Conversion errors in the ladder network are attributed to resistor tolerances and the voltage reference source, including the switch resistance. The total source resistance should be small compared to the value of the resistors in the ladder network to swamp out any changes in the reference supply and switch resistances. The voltage drops across the switches also should be small compared to the reference voltage used. However, a constant voltage drop across each switch would just reduce its effective reference potential and is not considered an error. This can be calibrated out with a voltage adjustment if the absolute value is a critical parameter. Perhaps the largest source of error is in the resistor ladder network itself, and precision resistors must be used. The worst-case error must be such that the output voltage error is much less than the voltage increment attributed to the least significant bit.

The error analysis can be performed on the network using the Thévenin equivalent circuits. This is a tedious task, considering both plus and minus errors in the initial values of the resistors. Computer circuit analysis programs can aid in an exact analysis. Since the voltages corresponding to the most significant digits contribute the most weight, it is seen from the equivalent circuits that the accuracy of the higher significant bit resistors should be of the order of the number of levels of quantization, i.e., less than 0.1% for 10 bits. It is the differences in the resistor values that are of concern. If the resistor values increase or decrease together with changes in temperature etc., no errors are introduced because the output voltage is a function of the ratio of the resistor values. Therefore, all precision resistors of the same type should be used. Typical values of R in these networks are 20 K.

Bipolar Signal. In the digital-to-analog converters discussed above the reference voltages are $+E$ V and ground for a **1** and **0**, respectively, producing a single polarity signal. If the ground reference in Fig. (7-16) is replaced by a source of $-E$ V, then the analog output will be an ac or bipolar signal.

Semiconductor Switches

In many digital system applications semiconductor switches are used to obtain high speed operation. A digital-to-analog converter using a ladder network driven from transistor switches is shown in Fig. 7-20. The reference voltages, $+V_{REF}$ and ground, are switched by the transistors, Q_2 and Q_3, to the resistor ladder network. The transistor pairs function as single-pole, double-throw switches. The input voltages for the **0**'s and **1**'s cut off or saturate the transistor switches. This complementary arrangement of transistors provides low offset voltage and is driven from a single input line. In some applications the npn and pnp transistors are operated in the inverse direction to obtain very low offset voltages.[3]

A positive input signal turns on the input transistor Q_1, which in turn presents a ground-level voltage V_{CE}(sat) to the input of the switches. This causes transistor Q_2 to be forward biased and connects the reference voltage $+V_{REF}$ to the output. A ground-level input turns Q_1 off and presents $+V_{CC}$ to the inputs of the transistor switches. This causes Q_3 to go into saturation and connects the output to the ground reference. The bias resistors and supply voltage are selected such that the pnp transistor is reverse biased when the npn transistor is conducting.

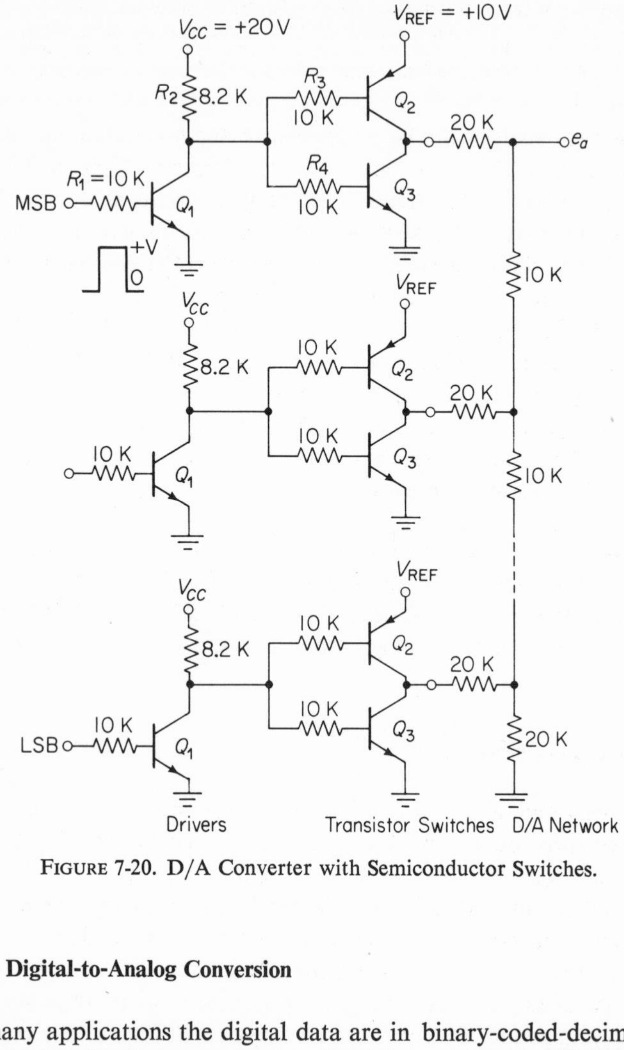

FIGURE 7-20. D/A Converter with Semiconductor Switches.

BCD Digital-to-Analog Conversion

In many applications the digital data are in binary-coded-decimal form, and in converting the data to an analog signal it would be desirable to use the BCD data directly. The digital-to-analog conversion ladder network produces an output analog voltage in equal steps due to the contribution of each section from the binary input signals. The BCD conversion network should provide similar voltage steps only from a decimal input control, i.e., a division between decades of 10 to 1. If a binary ladder network were partitioned in groups of four, and the input controls from 1-2-4-8 BCD signals were applied to the network input switches, an incorrect analog output would result. Although the BCD signals are in binary form from zero through nine, the voltage increases in each section are in steps of one-sixteenth increments instead of the required one-tenth increments. This can be compensated for by using different reference voltages in each BCD section; but the use of many different reference voltages results in a cumbersome design. An alternate approach that we will consider is to design a resistor ladder network using a single reference voltage, where each decimal section provides the correct voltage contribution of steps in one-tenth increments.

A diagram of the desired form of a BCD ladder network is shown in Fig. 7-21 for three decimal sections. The form of the ladder network is such that additional sections can be added. The network can be broken at point a and

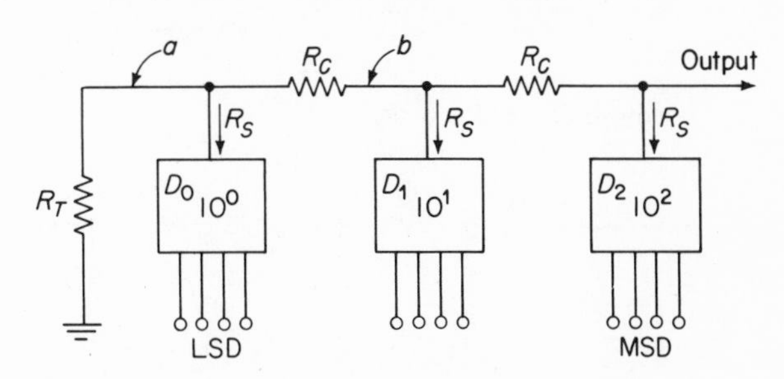

FIGURE 7-21. BCD D/A Converter Diagram.

another section, as contained between points a and b, can be inserted. Therefore, the resistances of this ladder network have the relationship that R_S in parallel with R_T when added to R_C equals R_T. That is,

$$\frac{R_T R_S}{R_T + R_S} + R_C = R_T \tag{7-47}$$

Let the parallel combination of R_T and R_S equal R_P;

$$R_P = \frac{R_T R_S}{R_T + R_S} \tag{7-48}$$

Then Eq. (7-47) becomes

$$R_P + R_C = R_T \tag{7-49}$$

This network will be analyzed in a similar manner to the one in Fig. 7-17. Frequent use will be made of the superposition and Thévenin theorems.

The circuit in Fig. 7-21 is redrawn with the decade networks D_0, D_1, and D_2 replaced with their Thévenin equivalent circuits. The new circuit is shown in Fig. 7-22. For each D network the Thévenin equivalent

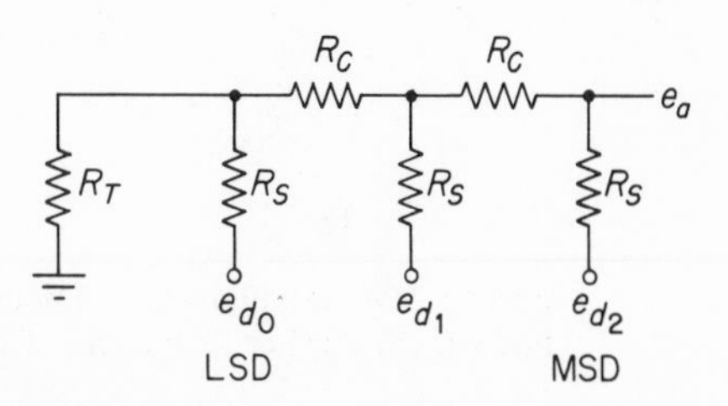

FIGURE 7-22. Equivalent Decimal Ladder Network.

circuit is a voltage source e_d, corresponding to the BCD voltage for the decade, and a source resistance that is R_S by definition. Using superposition, the output voltage will be calculated due to the MSD voltage e_{d_2} alone, with the other inputs returned to ground. The equivalent circuits used for this calculation are shown in Fig. 7-23.

The output due to e_{d_2} is

$$e'_a = e_{d_2}\frac{R_P + R_C}{R_P + R_C + R_S}$$

$$= e_{d_2}\frac{R_T}{R_T + R_S} \tag{7-50}$$

where $R_T = R_P + R_C$.

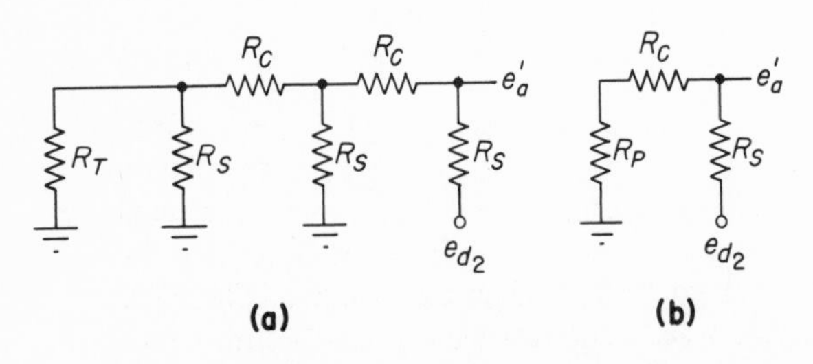

FIGURE 7-23. (a) Decimal Ladder Network with Only e_{d_2} Present. (b) Equivalent Circuit.

Next we consider the output response because of the next most significant digit, e_{d_1}. Fig. 7-24(a) is used for this purpose. Breaking the circuit at point x and replacing the left side with the Thévenin equivalent gives the circuit in Fig. 7-24(b). The output due to only e_{d_1} present is

$$e''_a = e_{d_1}\left(\frac{R_T}{R_T + R_S}\right)\left(\frac{R_S}{R_T + R_S}\right) \tag{7-51}$$

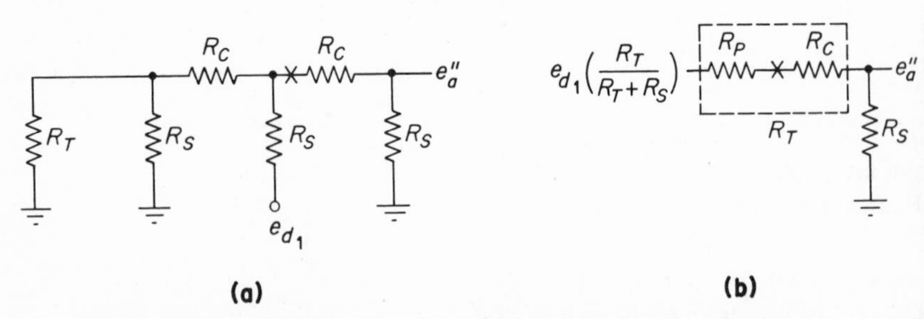

FIGURE 7-24. (a) Equivalent Decimal Ladder Network with Only e_{d_1} Present. (b) Further Reduction – Thévenin Equivalent.

If each decade has a ten-to-one contribution to the output voltage, then

$$e''_a = \tfrac{1}{10}e'_a \tag{7-52}$$

for the condition where $e_{d_1} = e_{d_2}$. This is equivalent to saying that in the decimal number 77 (both digits the same, corresponding to $e_{d_1} = e_{d_2}$), the MSD has a contribution of 70 and the next most significant digit has a contribution of one-tenth that, or seven. Let

$$e_{d_1} = e_{d_2} = e_d \tag{7-53}$$

and substituting Eqs. (7-50) and (7-51) into Eq. (7-52) gives

$$e_d\left(\frac{R_T}{R_T + R_S}\right)\left(\frac{R_S}{R_T + R_S}\right) = e_d\left(\frac{1}{10}\right)\left(\frac{R_T}{R_T + R_S}\right) \tag{7-54}$$

Solving for R_S,

$$R_S = \tfrac{1}{9}R_T \tag{7-55}$$

which is the requirement for a decimal ladder network. The basic design relationships are Eqs. (7-47) and (7-55). In these two equations there are three variables, R_S, R_C, and R_T; therefore, the value of one of them has to be selected.

The remaining step in the design of the BCD D/A converter is to determine the decimal networks, D's in Fig. 7-21. These networks have an equivalent source resistance equal to R_S, and the output voltage corresponds to the BCD code. For a 1-2-4-8 code, the network shown in Fig. 7-25 fulfills the requirements, where

$$\frac{1}{R_S} = \frac{1}{R} + \frac{1}{2R} + \frac{1}{4R} + \frac{1}{8R} \tag{7-56}$$

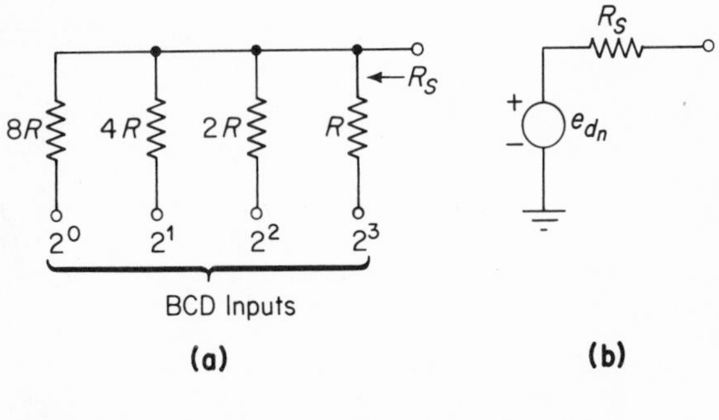

FIGURE 7-25. (a) Single-Decade BCD Network. (b) Thévenin Equivalent Circuit.

The complete BCD digital-to-analog conversion network for three decades is shown in Fig. 7-26. We shall choose $R = 10$ K and calculate the values of the other resistors. From Eq. (7-56),

$$\frac{1}{R_S} = \frac{1}{10} + \frac{1}{20} + \frac{1}{40} + \frac{1}{80} \tag{7-57}$$

$$R_S = 5.33 \text{ K}$$

Using Eq. (7-55) we get

$$R_T = (9)(5.33) = 48 \text{ K} \tag{7-58}$$

Also, from Eq. (7-47),

$$R_C = R_T - \frac{R_T R_S}{R_T + R_S} = 48 - 4.8 = 43.2 \text{ K} \tag{7-59}$$

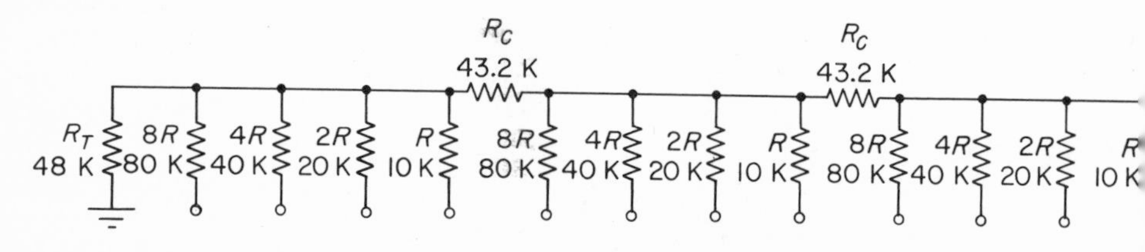

FIGURE 7-26. BCD D/A Converter (Three Decades, 1-2-4-8 Code).

This concludes the design approach for BCD digital-to-analog converters, and the previous comments on errors due to resistor tolerances and nonideal transistor switch source resistances apply. The design approach is general and can be applied to other BCD codes and nondecimal codes for D/A conversion.

7-5 ANALOG-TO-DIGITAL CONVERTERS

The conversion of analog data to digital form can be accomplished by many different methods, from simple mechanical devices to completely solid state circuits. The type of analog-to-digital converter to be used in a given application usually depends upon the conversion time and precision required. We will discuss the basic methods of electronic conversion.

The analog-to-digital conversion technique cannot be more accurate than the analog equipment employed. The total errors of all the analog conditioning circuits must be less than one quantization interval for the total number of bits to be meaningful. There are errors in the voltage comparators, operational summing amplifiers, and digital-to-analog resistive ladder networks. It cannot be overemphasized that to digitalize a signal to a given accuracy one must be able to condition and manipulate it to the same or better accuracy by pure analog techniques.

Although a pure analog system is capable of better accuracy than a digital system its accuracy is rarely completely usable because it is presented in a form that cannot be read, recorded, or interpreted to such high accuracy. This is why pure analog, primary standard instrumentation must resort to such techniques as manually adjusted multidecade digital dials, null-balanced, for readout. Digital data, however, are readily presented in numerical form, regardless of the number of digits, and are easily manipulated, processed, stored, and recorded.

Once data are converted to digital form they are less susceptible to noise, may be processed mathematically, analyzed, and used for control much more accurately and rapidly than could analog data. Very little data are in digital form originally, and a very important function in the digital processing equipment is the analog-to-digital conversion.

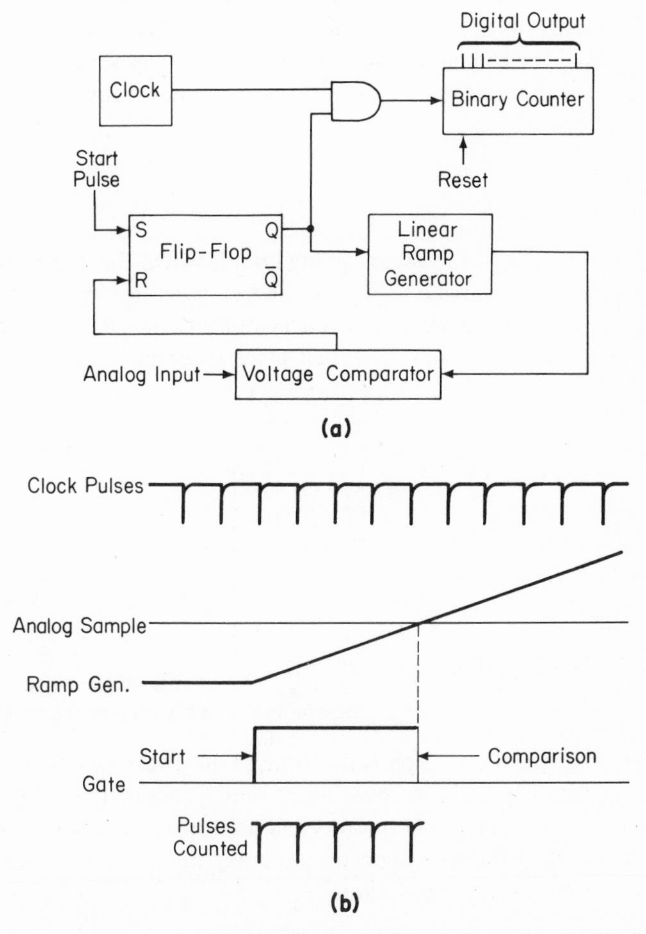

FIGURE 7-27. (a) Sweep Timing Analog-to-Digital Converter. (b) Timing Waveforms.

Sweep-Timing Method

The sweep-timing method of analog-to-digital conversion is illustrated in Fig. 7-27, with the pertinent timing waveforms. At the start of the conversion cycle the flip-flop is set, which simultaneously initiates the ramp generator and opens the gate to pass clock pulses to the counter. The analog input voltage and the output of the linear ramp generator are applied to the voltage comparator circuit. The counting continues until the ramp reaches the level of the analog voltage. At comparison the flip-flop is reset and further clock pulses are inhibited. The digital value of the analog voltage is stored in the counter.

For high precision the ramp must be extremely linear, and the voltage comparator must have a resolution less than 1 bit interval and be fast acting. Conversion accuracies of approximately 1% can be achieved practically. Higher accuracies have been obtained using circuits employing special compensation techniques and adjustments. A slope adjustment is usually required for the ramp generator.

Feedback Method

A modification of the sweep-timing circuit where the linear ramp generator is replaced by a staircase voltage waveform generated by using a digital-to-analog converter at the outputs of the counter is called the feedback method of analog-to-digital conversion. A circuit block diagram is shown in Fig. 7-28.

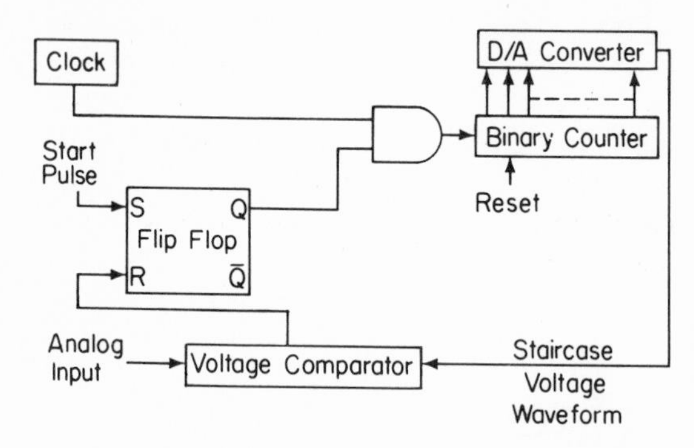

FIGURE 7-28. Feedback Method of Analog-to-Digital Conversion.

The principle of operation is the same as the sweep-timing circuit. This technique is generally preferred over sweep timing because the staircase voltage can be generated with precision resistors and adjustments are not required. The conversion accuracy depends upon the accuracy of the digital-to-analog converter and the comparator resolution.

Tracking Converter

The tracking converter shown in Fig. 7-29 is a variation of the feedback type. The initial conversion period depends upon the magnitude of the analog input signal and can be appreciable for high count converters. However, when the input is continuous—i.e., not in multiplexed operation—the conversion follows the input and the digital output is continuously available. The clock that drives the up-down counter runs continuously, and the up and down controls are energized by the voltage comparator. Whenever the comparator detects that the feedback voltage is less than the analog input, the counter counts up. The counter counts down when the feedback voltage is larger. Since the comparator cannot determine a balance condition, a continuous one-bit variation results around the balance point. If this steady fluctuation of the least significant bit is objectionable, it can be eliminated with additional circuitry that detects changes larger than one-bit and gates the clock pulses.

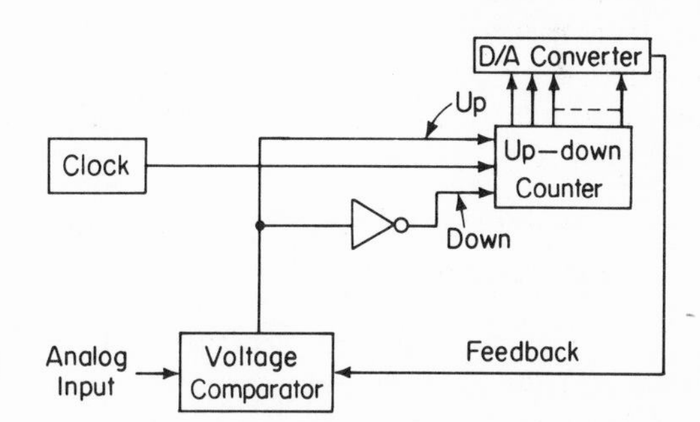

FIGURE 7-29. Tracking-Type Analog-to-Digital Converter.

The clock operates at a very high repetition rate to follow the amplitude variations in the analog signal. The clock frequency is dependent upon the conversion resolution (number of bits) and the frequency content of the analog signal. For example, if the highest frequency component of an analog signal is 1 kHz and the desired quantization is 128 levels or seven bits, then the clock period should be equal to the time it takes the waveform to traverse one bit interval of amplitude, at the point of the fastest rate of change. This is shown in Fig. 7-30 for a sine wave at the maximum signal frequency.

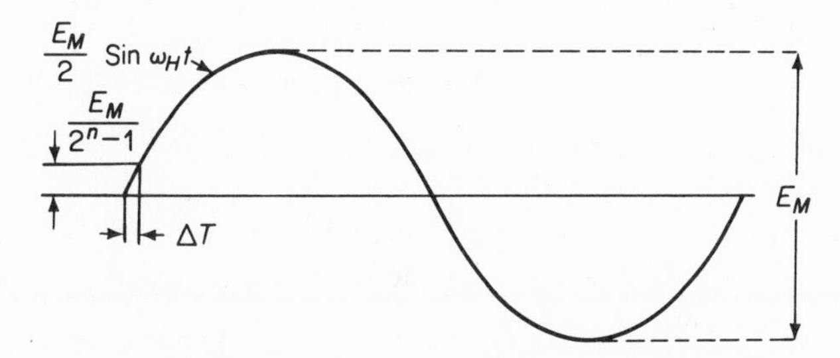

FIGURE 7-30. Determination of Clock Frequency for Tracking Converter.

The fastest rate of change of the sine wave is around the zero value. We will equate one level of quantization to the level of the sine wave near the zero crossing. These increments are shown in Fig. 7-30.

$$\frac{E_M}{2^n - 1} = \frac{E_M}{2} \sin \omega_H \Delta T \tag{7-60}$$

$$\sin \omega_H \Delta T = \frac{2}{2^n - 1} \tag{7-61}$$

For small angles,

$$\omega_H \Delta T = \frac{2}{(2_n - 1)} \text{ rad} \tag{7-62}$$

$$\Delta T = \frac{1}{\pi f_H (2^n - 1)} \tag{7-63}$$

The clock frequency is

$$f_c = \frac{1}{\Delta T} = \pi f_H (2^n - 1) \tag{7-64}$$

Using a seven-bit A/D converter and $f_H = 1$ kHz, e.g.,

$$\Delta T = \frac{1}{\pi(1{,}000)(127)} = 2.5 \ \mu\text{sec} \tag{7-65}$$

$$f_c = 400 \text{ kHz} \tag{7-66}$$

This example indicates that a high clock frequency is required for the converter to track the sine wave. An actual band-limited, analog signal may have a maximum rate of change that is different than that of a sine wave at the highest frequency in the spectrum and that traverses the complete dynamic range. If the actual waveform is known, then the minimum clock frequency can be calculated for the desired number of quantization levels. Otherwise an assumption will have to be made about the waveform, as in the above example, in making the calculations.

It is interesting to compare the bit rate in the tracking converter with what would be required in PCM. We will use the same example for the PCM case and select a sampling rate at twice the Nyquist rate, or at 4 kHz. For encoding to seven bits, the overall bit rate is 28,000 bits/sec. This is compared with 400,000 bits/sec. for the tracking converter. The difference, of course, is that the tracking converter is following the signal in a linear staircase fashion, whereas in PCM the signal is encoded in a logarithmic fashion, i.e., in steps of powers of two. This implies a bit-rate ratio of $2^n/n$. In our example using seven bits of encoding, the two bit rates then would differ by a factor of approximately 18, which is close to our results. If the tracking converter uses a lower clock frequency, then it will not follow the high, signal rates of change. It therefore will lag behind the signal and not follow the waveform, which of course results in distortion.

The binary signal at the output of the voltage comparator in the tracking converter of Fig. 7-29 is used sometimes as a modulating signal in data transmission. This is called *delta modulation.* At the receiver the **0**'s and **1**'s are used to count up and down in a bidirectional counter. The analog information is reconstructed by using a D/A converter at the output of the counter. The required clock or bit rates for delta modulation and the resultant signal-to-noise ratios are discussed in the literature.[4]

In the tracking converter the binary encoded digital data are read from the counter at the desired rate determined by the sampling theorem for PCM storage or transmission. The reading, or gating data from the counter, must be synchronized to take place between clock pulses so the counter does not change during the time the sample is being read.

Successive Approximation Method

The analog-to-digital converters discussed above use direct counting methods, and the conversion time for each sample is appreciable. For example, to convert to a seven bit resolution time has to be allowed for the counter to count up to 128 pulses. If the analog voltage can change during the conversion time, a *sample-and-hold circuit* (see Sec. 7-6) is required to prevent errors in conversion. The conversion time can be significantly reduced by making comparisons on a bit-by-bit basis instead of continuous counting. Converters using this principle fall into the category of *successive approximation.*

The successive approximation technique first compares the analog signal with one-half of the maximum voltage range to determine whether the most significant bit (MSB) is a **0** or **1**. For example, if the analog e_a is greater than one-half the maximum voltage range $\frac{1}{2}E_M$, then the MSB is a **1**; otherwise it is a **0**.

In determining the next most significant bit, the analog voltage is compared with $\frac{3}{4}E_M$ if the MSB is a **1** or $\frac{1}{4}E_M$ if the MSB is a **0**. Similarly, that bit is a **1** if e_a is greater than these values, or it is a **0** otherwise. This process proceeds in a similar fashion until the least significant bit is reached. The process is depicted as a *tree* in Fig. 7-31(a), displaying the many possible paths of successive comparisons. Figure 7-31(b) shows a plot of the same principle where the voltage ranges of the **0**'s and **1**'s for the three most significant bits are indicated.

One method of implementing the successive approximation technique is shown in Fig. 7-32. The input analog voltage is first compared with a voltage equal to one-half the maximum voltage range, which is obtained from a 1-bit digital-to-analog converter (single resistive voltage divider). If the analog signal is greater than $\frac{1}{2}E_M$, the 2^n-bit output (MSB) of the first voltage comparator is a **1**; otherwise it is a **0**. This output leads to all the succeeding digital-to-analog converters. The second most significant bit is determined by comparing

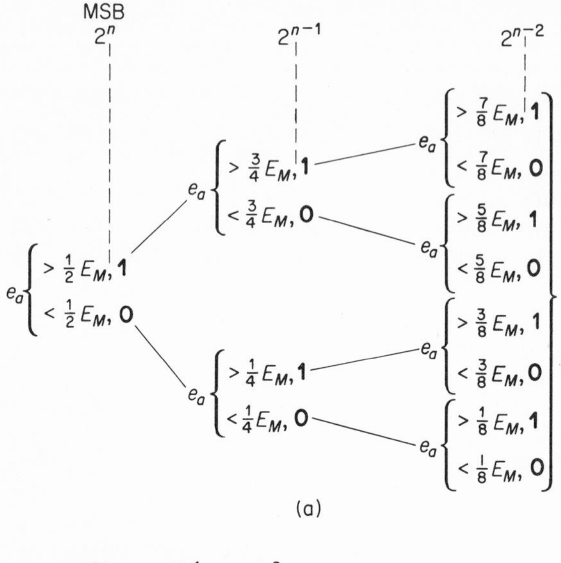

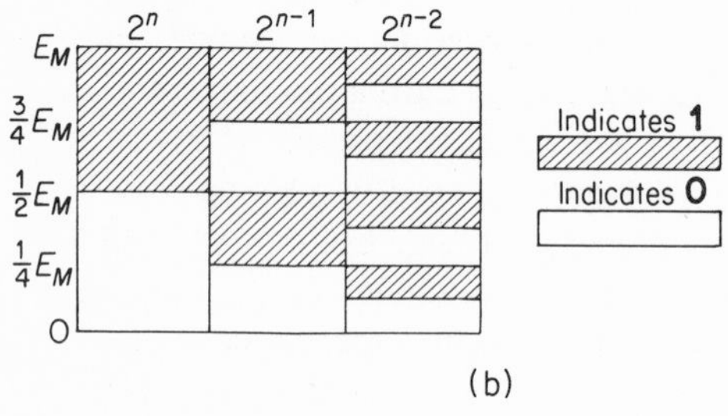

FIGURE 7-31. (a) Successive Approximation *Tree* Depicting Steps of Comparison. (b) Voltage Ranges of Successive Bits.

the analog signal with the output of a two-bit digital-to-analog converter. The second D/A output voltage is either $\frac{3}{4}E_M$ or $\frac{1}{4}E_M$, depending upon whether the MSB is **1** or **0**, respectively. All the D/A converters have a fixed **1** input plus the values of the more significant bits determined by the previous stages to generate the various reference voltage levels, as indicated in Fig. 7-31. The voltage comparators at each stage compare the analog voltage with its reference input to determine the value of its respective binary digit.

The successive approximation circuit of Fig. 7-32 is fast acting; i.e., it has a very short conversion time, and a sample-and-hold circuit is not required at medium and lower frequencies. However, when the digital outputs are being read some means must be provided to prevent the bits from changing during the reading time. This can be accomplished with an auxiliary register (flip-flops) whose inputs are inhibited during the read interval. A sample-and-hold circuit also could be used for this purpose. The accuracy of the converter is limited by the accuracies of the digital-to-analog converters and the resolution of the voltage comparators.

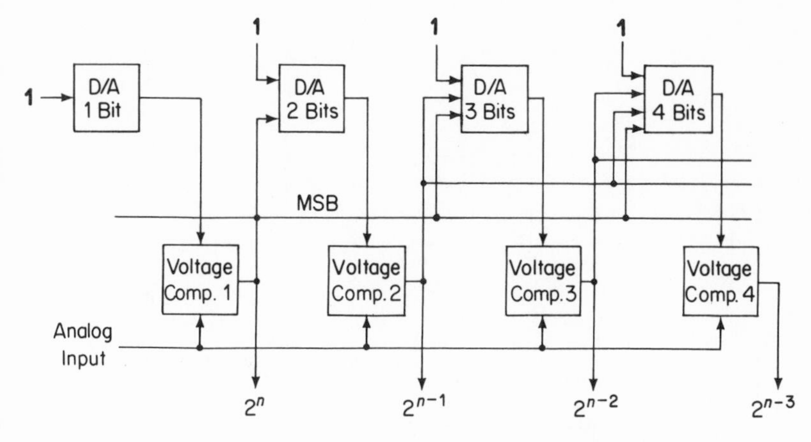

FIGURE 7-32. Successive Approximation Analog-to-Digital Converter.

An alternate method of constructing the successive approximation analog-to-digital converter uses only one voltage comparator. However, additional logic circuits and a counter distributor are required to generate the reference voltages and to make the comparisons in sequence. In this case a sample-and-hold circuit is required to keep the analog voltage constant during the conversion time (see Prob. 17).

There are many ways to handle ac or bipolar signals. For example, the voltage range can be quantized $-E$ to $+E$ V, and any associated digital-to-analog converters will have the same references. Also, the ac signal can be added to a positive bias voltage to produce a single polarity signal. Another technique is to use the most significant bit to represent the sign, **1** for plus and **0** for minus, and the remaining bits to quantize the magnitude of the signal. The most convenient method usually chosen is that which is compatible with the remaining data-processing functions.

7-6 SAMPLE-AND-HOLD CIRCUITS

In many analog-to-digital converters the analog signal may change during the conversion period and errors will result. The amount of uncertainty about the exact time when the analog input signal was at the value represented by the digital output number from the A/D converter is called *aperture time*. In general the aperture time is equal to the conversion time; and it may be reduced by the use of a sample-and-hold circuit at the input to the converter.

A circuit that performs a sample-and-hold function is shown in Fig. 7-33. The sampled voltage is held by use of the storage capacitor C. At the start of a conversion period the storage capacitor has to be discharged. This is accom-

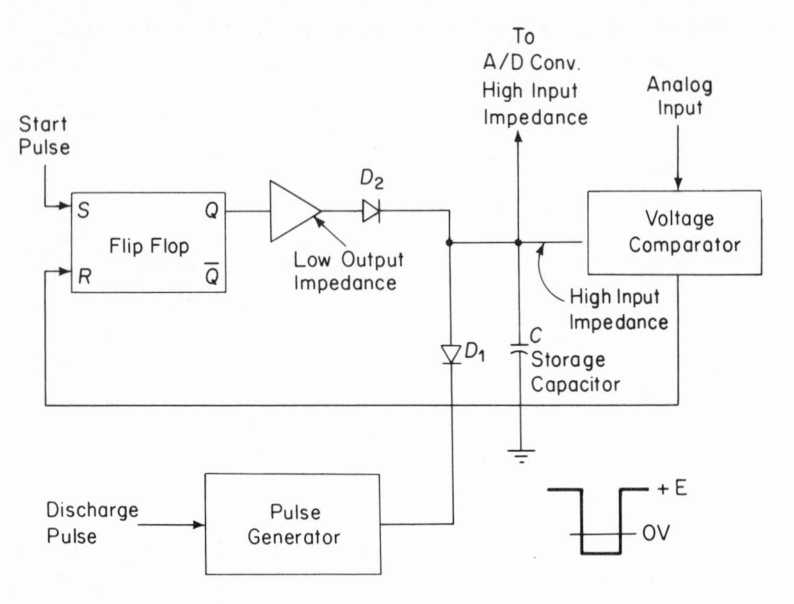

FIGURE 7-33. Sample-and-Hold Circuit.

plished by triggering a pulse generator that delivers a negative pulse to the capacitor through diode D_1. The next step in conversion is to trigger the flip-flop with a start pulse. This places a positive voltage to the low output impedance amplifier that charges the capacitor through diode D_2 with a short time constant. When the voltage across the capacitor equals the analog input voltage, the voltage comparator changes state and resets the flip-flop. Both diodes are now back biased, and the charged capacitor holds the sampled voltage. How well the capacitor voltage holds constant depends upon the size of the capacitor, capacitor leakage, diode leakage, and finite input impedances to the voltage comparator and the analog-to-digital converter circuit. All these factors should be taken into consideration in the design for a given hold interval and conversion tolerance. Once the capacitor is charged to the analog voltage level, any change in the stored voltage ΔV must be less than one interval of quantization, as given by Eq. (7-67).

$$\Delta V \ll \frac{E_M}{2^n - 1} \tag{7-67}$$

The relationship between the capacitor voltage change and the total capacitor current I is given by Eq. (7-68).

$$\Delta V = \frac{\Delta Q}{C} = \frac{\Delta T I}{C} \tag{7-68}$$

ubstituting Eq. (7-68) into Eq. (7-67) gives the following limit on the total apacitor current:

$$I \ll \frac{E_M}{2^n - 1}\left(\frac{C}{\Delta T}\right) \tag{7-69}$$

vhere ΔT is the time required to hold the charge for the A/D conversion.

The above circuit has limitations in speed and accuracy because of the ime response of the voltage comparator and the subsequent resetting of the lip-flop. During the time from voltage comparison to the actual stoppage of harging current to the capacitor, extra charge is building upon the capacitor and causing an error voltage. This problem can be avoided by using a tracking mplifier with feedback as a holding circuit.

Track-and-Hold Circuits

A track-and-hold circuit using an operational amplifier is shown in Fig. 7-34(a). The switches are in the tracking position initially; i.e., the circuit is ampling. In this configuration the amplifier performs as a unity gain inverter.

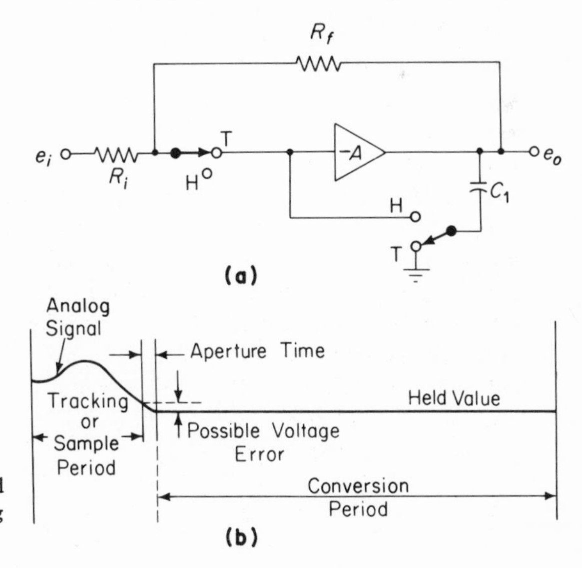

FIGURE 7-34. (a) Track-and-Hold Circuit. (b) Diagram Showing Aperture Time.

The amplifier output drives the capacitor C to follow the signal continuously until a hold command is initiated. In the hold configuration the capacitor is connected as an integrator and the input signal is removed, causing the capacitor voltage (except for the effects of leakage) to remain constant.

In general the switches shown in Fig. 7-34(a) would be solid state to achieve high speed operation. The transition time from track to hold is the aperture time for this circuit. Figure 7-34(b) illustrates the effect of the analog signal changing during the aperture time, resulting in a possible voltage error. Fast switching is required to minimize the error.

It is difficult to switch a capacitor into the feedback path of a high-gain operational amplifier without causing some charge degradation on the capacitor. Fast switching times mimimize the effect, but it is better if capacitor switching can be avoided in this circuit. A modification of the circuit that leaves the capacitor wired in permanently is shown in Fig. 7-35.

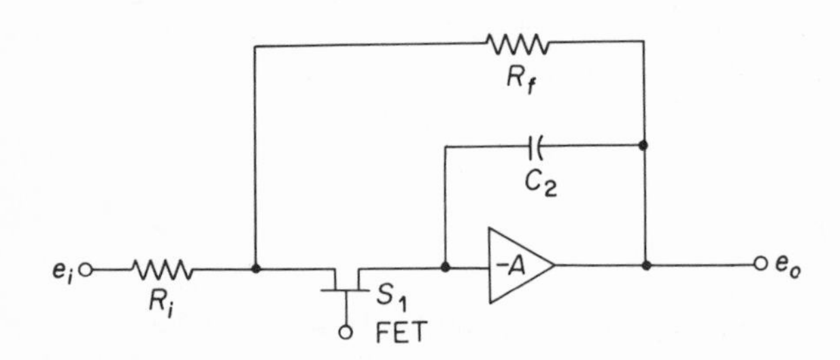

FIGURE 7-35. Modified Track and-Hold Circuit.

This amplifier has a transfer function similar to that of a low-pass RC filter. The transfer function in Laplace transform notation is

$$\frac{E_o(s)}{E_i(s)} \approx \frac{R_f}{R_i}\left(\frac{1}{R_f C_2 s + 1}\right) \tag{7-70}$$

In order for the output to follow the input quickly, the time constant $R_f C_2$ should be small during the tracking time. This implies a small value for both C_2 and R_f. R_f is limited by the current drive capability of the amplifier. C_2, on the other hand, should be large to hold an appreciable charge during the hold portion of the cycle. Therefore, a trade off has to be made between the accuracy requirements during the conversion time and the circuit parameters.

Field effect transistors make excellent switches and can be used to switch between the track and hold positions in the above circuits. Balanced transistor switches, such as the dual transistor chopper, also can be used; however, care has to be taken to provide the proper dc isolation between the transistor switch and the amplifier, e.g., use of a pulse transformer in the base circuit.

7-7 ELECTROMECHANICAL ENCODERS

Encoders or analog-to-digital converters that receive shaft position information are electromechanical, and the voltage input encoders discussed above are electronic. The type of encoding method selected is not limited to the original source of the data. For example, a shaft position can be converted to a voltage by connecting the shaft to the arm of a potentionmeter. Likewise, a voltage can be converted to a shaft position by a servo system.

The accuracy of a *shaft encoder* is determined by the number of conducting and insulating segments that can be placed on the circumference of a disk or, in the case of optical shaft encoders, the number of transparent and opaque sectors around the circumference of a disk. Some shaft encoders use multiple disks with gears of the proper ratios between the disks. In this case the gearing enters into the overall accuracy of the encoder.

An advantage of shaft encoders using coded segments is that the code on the disk can represent special functions corresponding to the shaft position. For example, the digital number produced can represent a trigonometric function of the shaft angle (sine, cosine, tangent, etc.).

In many applications a position input must supply a digital indication of its position at all times. This is accomplished by employing a separate segment track and brush (or photoelectric element) for each binary digit of the number. Figure 7-36 shows a coded segment disk where the length of the segments corresponds to the weights of the binary digits in the coded representation of position. The electrical circuit through the brushes and segments is completed through an extra brush and slip ring to which all segments are connected. The comparison of Fig. 7-36 with the successive approximation voltage chart of Fig. 7-31(b) should be noted.

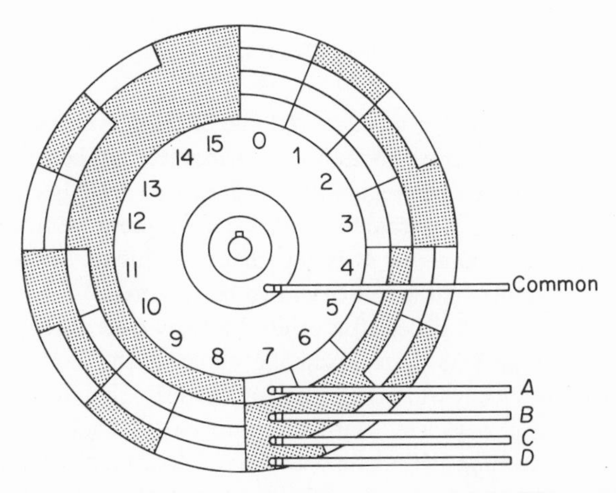

FIGURE 7-36. Binary-Coded Disk.

The encoder in Fig. 7-36 is simple in operation but has the fundamental problem that more than one digit can change simultaneously. The worst case is when the position changes from 1111 to 0000. Since it is not possible to position the brushes perfectly, at the transition there will be positions in which some of the brushes will make contact while some will not, and a gross erro is produced, indicating an erroneous position.

One commonly used method to ensure against conversion ambiguity a sector boundaries is to use a code where successive numbers vary by only on digit. A code with this property is known as a *Gray code*. There are man

Position	Gray Code
0	0000
1	0001
2	0011
3	0010
4	0110
5	0111
6	0101
7	0100
8	1100
9	1101
10	1111
11	1110
12	1010
13	1011
14	1001
15	1000

(a)

0000	0001	0101	0100	
a ← 0010	0011	0111	0110 ← a	
b → 1010	1011	1111	1110 → b	
1000	1001	1101	1100	

(b)

FIGURE 7-37. (a) Gray Code. (b) Map Showing Code Sequence.

possible Gray codes, and one of them is shown in Fig. 7-37(a). Since the prop erty of these codes is that successive numbers change by only one binary digit the number sequence is represented by adjacent squares on a variable map This is illustrated in Fig. 7-37(b), where the line indicates the path of successiv code numbers. It is apparent that this is not the only Gray code, because man similar paths can be drawn on the map.

A coded disk using the Gray code listed in Fig. 7-37(a) is illustrated ir Fig. 7-38. An advantage of using this code in disk fabrication is that th segment width representing the least significant bit may be double the size o that associated with the pure binary code. It has the disadvantage that per forming arithmetic operations is not practical, and the data are usually convertec to pure binary form before undergoing mathematical operations.

The Gray code illustrated in Fig. 7-37(a) is separated by a dotted line a the center. In the top group of binary numbers the MSB's are all **0**'s and i the bottom group all **1**'s. Note that the rest of the binary digits form a mirro image about the dotted line. Gray codes with this property also are callec *reflected codes*. Not all Gray codes are reflected codes.

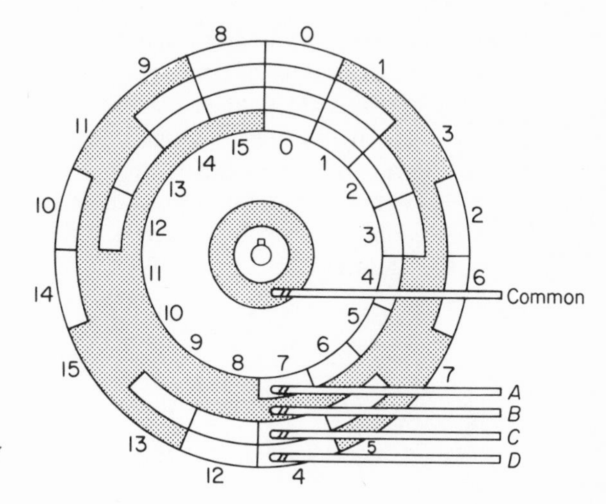

FIGURE 7-38. Disk Using a Gray Code.

Double Brush Binary Encoder

Figure 7-39 shows a method of avoiding errors at the sector boundaries when a pure binary code is used. For ease of illustration this figure shows the sectors as straight lines; however, the same principle applies to the disk. This method is called double brush because two sets of brushes are required. The shaded areas represent the conducting segments, and the black blocks represent the brushes.

In the double brush method the single brush is used to sense the least significant binary digit, 2^0, but there are two brushes for each of the other

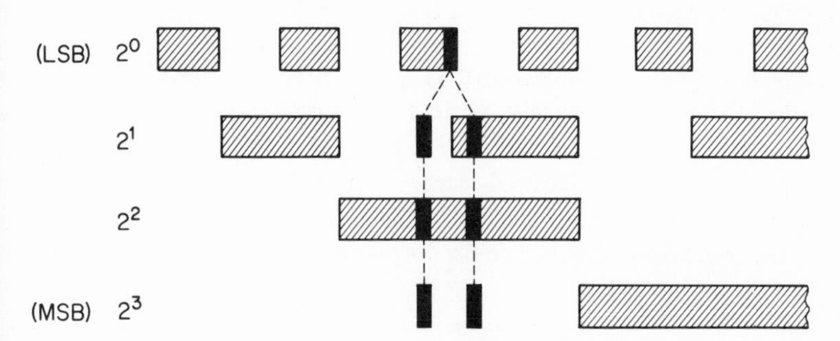

FIGURE 7-39. Double-Brush Method of Conversion.

digit tracks. One set of these brushes is advanced with respect to the position being encoded, and the other set is retarded by an equal distance. The amount of separation between brushes is equal to the width of a segment on the 2^0 (LSB) track.

The principle of operation is to sense the LSB first and then to select either the left or right column of brushes for the higher digits, depending on whether the LSB is **1** or **0**, respectively. For example, with the brush location shown in Fig. 7-39 the position is at the boundary between the 0101 and 0110 positions. When the LSB brush indicates a **1**, the left column of brushes is selected, giving the number 0101; when it indicates a **0**, the right brushes give 0110. The erroneous reading 0111 cannot be produced at this position. The brush selection is made by external logic circuits controlled by the **0** or **1** reading in the LSB track.

7-8 DATA TRANSMISSION

Digital data, after suitable multiplexing, sampling, analog-to-digital conversion, and addition of error detection or correction bits, are processed for transmission. In message transmission the characters (letters, numerals, etc.) are represented in binary-coded form.

When data are transmitted in series some means of synchronization is required to locate the start of a word or group of words. A word is used to denote a group of binary digits—e.g., a sample quantized to seven bits could constitute a word. Synchronization can be accomplished in many ways, such as a pause in transmission, transmitting a unique word that cannot appear in any other place in the data or message, or transmitting a burst of a special frequency. Some basic digital waveforms and modulation methods are discussed below.

Binary Waveforms

Binary digits can be represented as electrical signals in several ways. The simplest is a two-valued voltage or current waveform, shown in Fig. 7-40. In each bit interval there is a condition of current or no current transmitted, representing **1**'s and **0**'s, respectively. This also is referred to in telegraph as a *mark* and *space*. The receiver is synchronized to the incoming waveform, and the data are sampled at the maximum output in the bit interval. The **1** or **0**

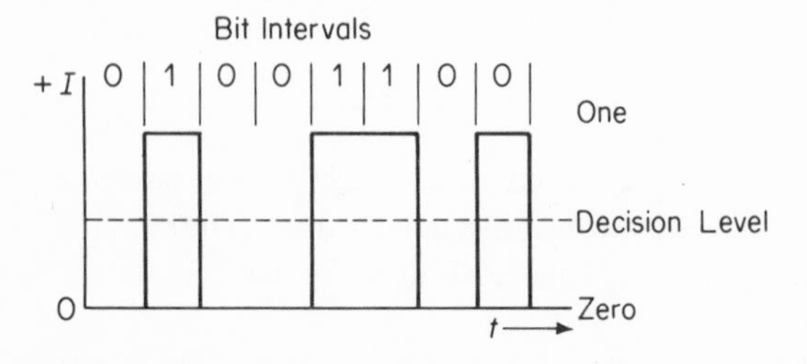

FIGURE 7-40. Two-Valued Binary Waveform, Unipolar.

lecision would be based upon a comparator circuit with a reference threshold et at one-half the amplitude level.

Instead of current, or no current, positive or negative current signals can e used, as shown in Fig. 7-41. This waveform has the advantage of having ero average current. In both cases, Figs. 7-40 and 7-41, the waveforms remain t the **1** or **0** level until a succeeding bit changes in value. This is referred to s *nonreturn to zero* (NRZ).

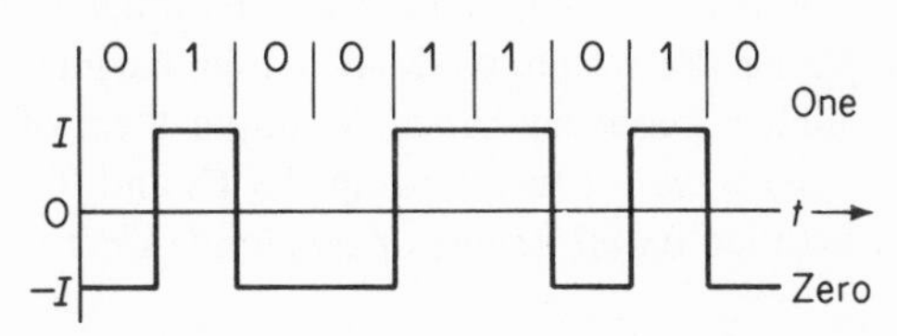

IGURE 7-41. Two-Valued Binary Vaveform with Zero Mean, Bi-olar.

A return-to-zero (RZ) signal only uses one-half of the bit interval. The wo previous waveforms are represented as return-to-zero signals in Fig. 7-42(a) nd (b), respectively. The bipolar RZ waveform, Fig. 7-42(b), contains a ransition during each bit interval that can be used to perform a clocking unction at the receiver.

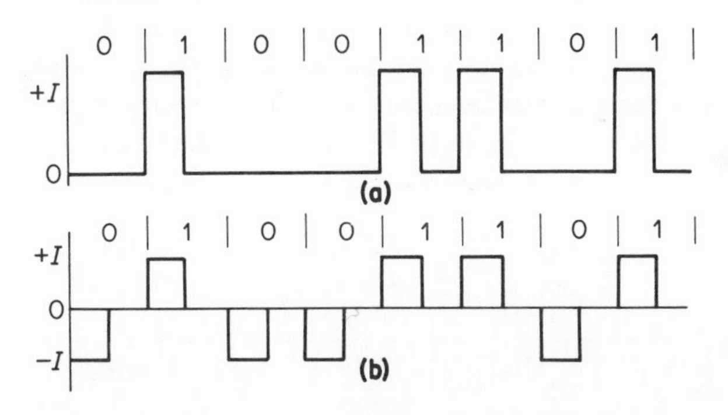

FIGURE 7-42. Return-to-Zero Signals. (a) Unipolar. (b) Bipolar.

There are other waveform representations of binary information. One other that is frequently encountered encodes the information on the basis of bit transitions. For example, there is no waveform transition for sending a **0**, and the waveform changes level for transmitting a **1** (or vice versa). This is called *differential coding*. At the receiver one signal interval is compared with

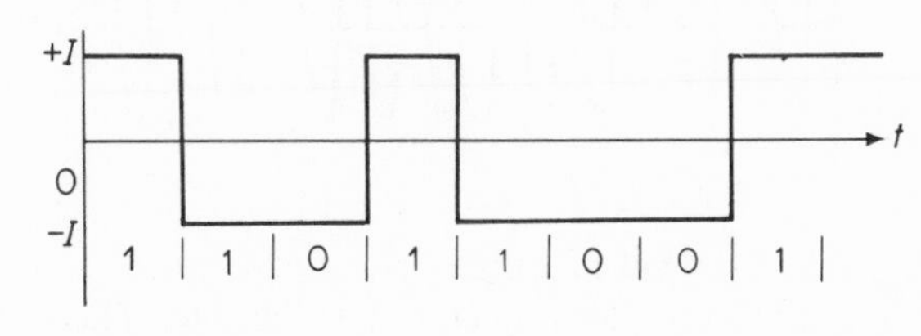

FIGURE 7-43. Differential Encod-ing.

the preceding interval to determine if a transition has taken place and therefore a **1** has been sent. This waveform-encoding method also is called a *Kineplex* system and is illustrated in Fig. 7-43.

Carrier Modulation

When the digital data are to be transmitted over a radio link the binary waveforms are used to modulate a high frequency carrier signal for transmission in a higher frequency band. A simple form of modulation is turning on and off a sine wave for transmitting the **1**'s and **0**'s. This is illustrated in Fig. 7-44 and is called *on-off keying* or *amplitude shift keying* (ASK).

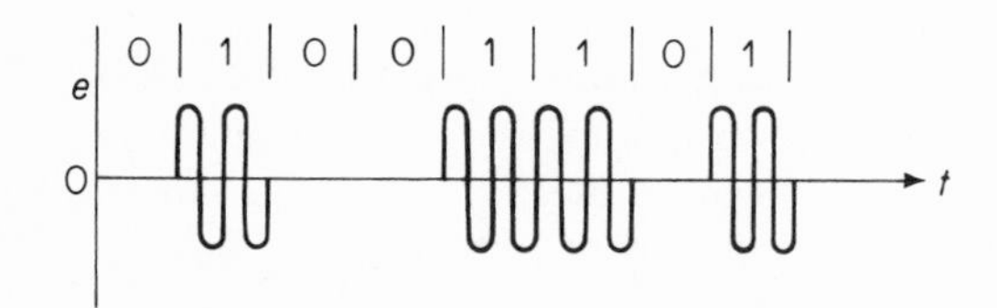

FIGURE 7-44. On-Off Keying.

Another example of modulation for binary data is to transmit one frequency for the **1**'s and another frequency for the **0**'s. This is known as *frequency shift keying* (FSK), and is shown in Fig. 7-45. In reception, the frequencies are separated by appropriate bandpass filters.

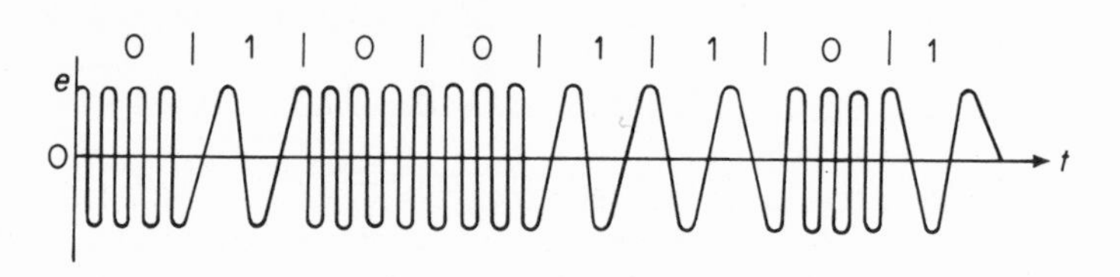

FIGURE 7-45. Frequency-Shift Keying.

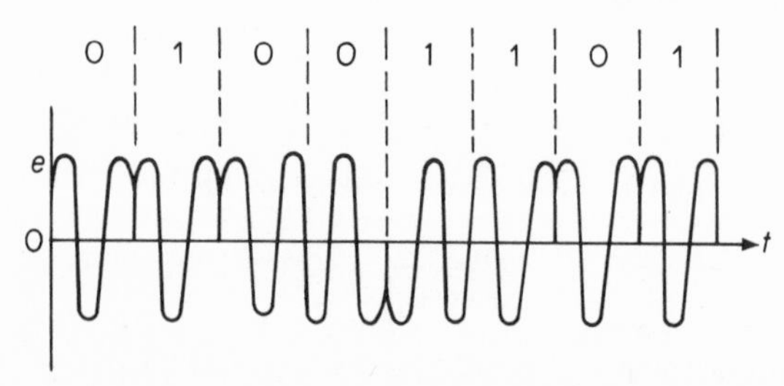

FIGURE 7-46. Phase-Shift Keying.

Instead of shifting the frequency of a carrier signal, the phase can be shifted 180° for a binary signal change. This is called *phase shift keying* (PSK) and is illustrated in Fig. 7-46. A phase detector is used in the receiver to detect the $\pm 180°$ phase shifts.

In a noisy transmission medium the probability of detection is largest for phase shift keying, followed by frequency shift keying, and is the smallest for on-off keying.[5]

REFERENCES

1. Panter, P. F., *Modulation, Noise, and Spectral Analysis—Applied to Information*, Chap. 20, McGraw-Hill Book Company, New York, 1965.
2. Hancock, J. C., *An Introduction to the Principles of Communication Theory*, McGraw-Hill Book Company, 1961.
3. Tatro, R. D., "A Microcircuit Digital-to-Analog Converter," *Int. Microelectron. Conf.* Munich, Germany, October 1966.
4. Panter, P. F., *op. cit.*, Chap. 22.
5. Bennett, W. R., and Davey, J. R., *Data Transmission*, McGraw-Hill Book Company, New York, 1965.

PROBLEMS

1. Determine the quantization signal-to-noise ratio of the average power for the triangular waveform shown in Fig. P.7-1. Calculate the average signal power by averaging the time function. Assume the waveform traverses all the levels of quantization. Compare this result with Eq. (7-11) in the text, and comment on the probability density function for the amplitude of the triangular waveform. In determing the S_0/N_0 use the statistical calculations in the text for the average noise power due to quantization.

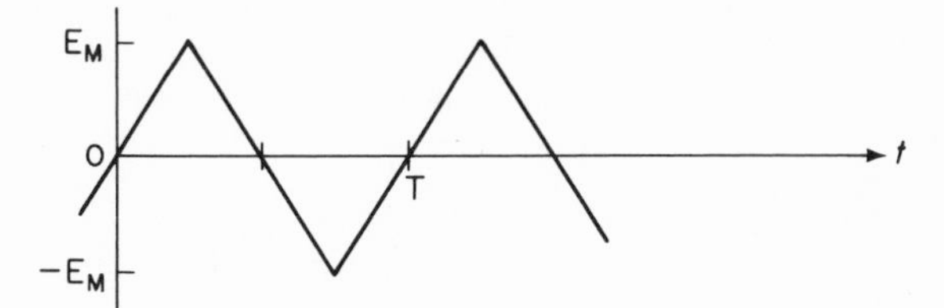

FIGURE P7-1

2. The signal dynamic range at the output of a certain receiver is 40 dB. The receiver output is fed into a 10-bit analog-to-digital converter with uniform levels of quantization. For a received sinusoidal signal, determine the average power signal-to-noise ratio and the total harmonic distortion resulting from quantization at both the high and low ends of the dynamic range.

Assume that the maximum received peak-to-peak sine wave traverses all the levels of quantization.

3. Determine the minimum sampling rate required for a 4 kHz band-limited signal with the highest frequency located at 7.5 kHz. If the spectrum had been translated such that the highest frequency were located at 8.5 kHz instead, determine the minimum sampling rate required.

4. Determine the minimum gain required for an operational amplifier for use in a 10-bit digital-to-analog converter. The analog input signal range is from 0 to 15 V maximum.

5. An operational summing amplifier is to be designed for a six-bit digital-to-analog converter. The summing resistors are R, 2R, 4R, 8R, 16R, and 32R and $R_f = R$. The gain of the operational amplifier is 5,000; assume that the input resistance is large compared to the other circuit resistances. Determine the tolerances required on the network resistors to keep the largest total conversion error less than 0.01 V. Transistor switches connect either +4.00 or 0 V to the digital inputs of the summing network. Neglect the source resistances and voltage drops associated with the transistor switches.

6. Determine the effect of the loading resistance R_L on the output analog voltage for the D/A ladder network shown in Fig. P7-6.

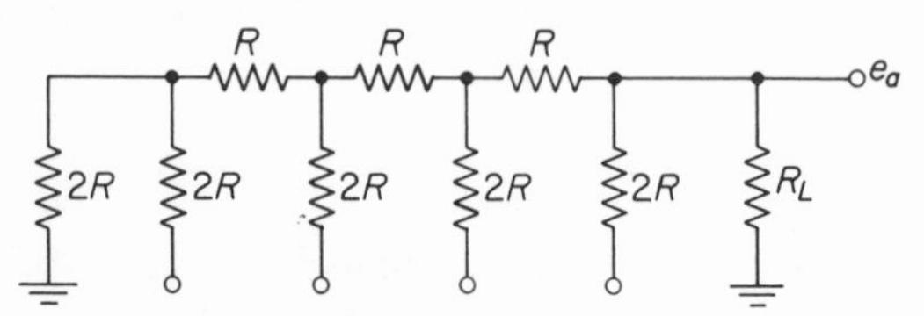

FIGURE P7-6

7. Using Fig. P7-6, determine the tolerance required for the resistors to maintain a conversion accuracy for the analog voltage of less than 25% of the contribution due to the least significant bit. Use a four-bit D/A converter and neglect the effects of the transistor driving switches and source resistance.

8. Determine the conduction states of the transistors in the D/A switching circuit Fig. P7-8, by verifying that the transistors are either cut off or in saturation for the two input voltage levels. The special output transistors

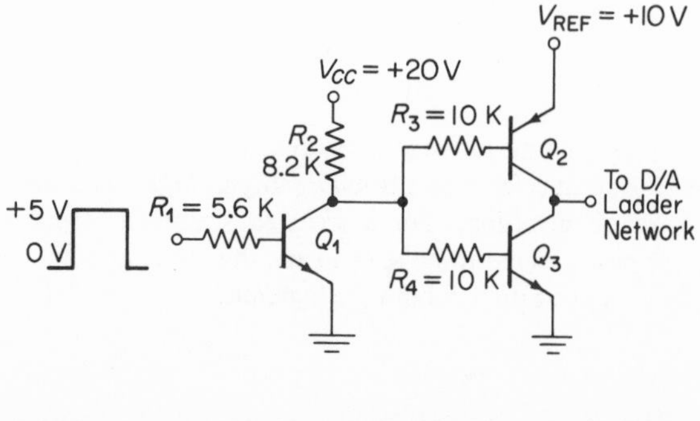

FIGURE P7-8

Q_2 and Q_3 have a mimimum h_{FE} of 50, V_{CE} (sat) $= 3$ mV max, and V_{BE} (sat) $= 20$ mV max at $I_B = 1$ mA. Transistor Q_1 has a minimum h_{FE} of 40, V_{CE} (sat) $= 0.2$ V max, and V_{BE} (sat) $= 0.5$ V max at $I_B = 1$ mA.

9. Design a two-decade decimal D/A converter by using an eight-input binary ladder network partitioned into two sections. The BCD inputs are in a 1-2-4-8 code. The analog output voltage is unipolar; i.e., the lower reference voltage is at ground potential. Determine the relationship of the positive reference voltages for each section to make the binary ladder network convert the BCD inputs.

10. Design one decade of a BCD D/A ladder network, using a single reference voltage, for a 1-2-2-4 code.

11. Using a 1-2-4 code, design a D/A ladder network for binary-coded, radix-six input signals.

12. Design a D/A ladder network for a radix-12 code. The input signals are in 1-2-4-8 binary-coded form.

13. Derive an expression for the analog output voltage e_a for the two-decade BCD D/A converter shown in Fig. P7-13. Note that

$$\frac{R_T R_S}{R_T + R_S} + R_C = R_T$$

$$\frac{1}{R_S} = \frac{1}{R} + \frac{1}{2R} + \frac{1}{4R} + \frac{1}{8R}$$

$$R_S = \tfrac{1}{9} R_T$$

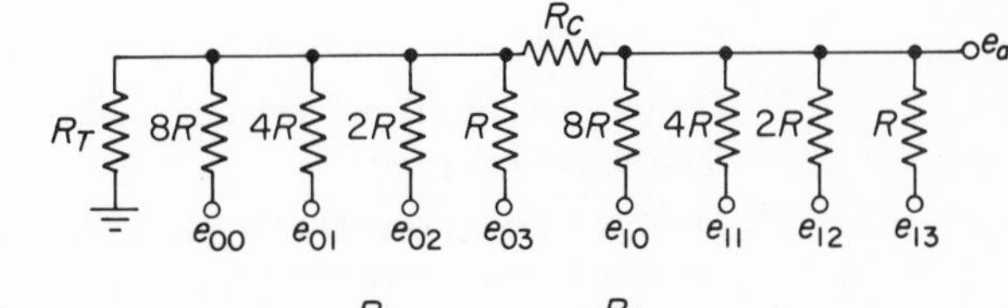

FIGURE P7-13

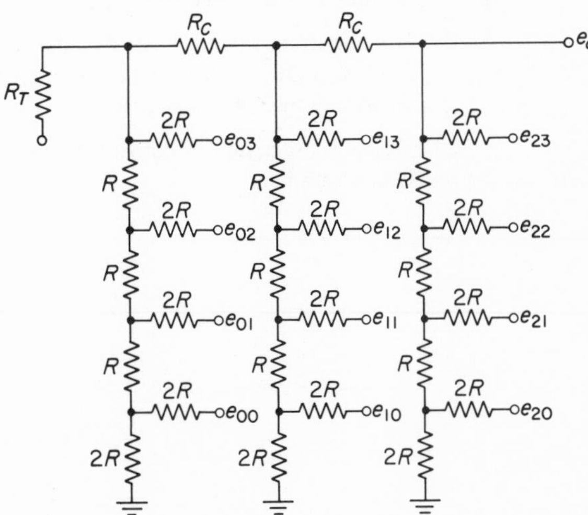

FIGURE P7-14

14. Design a BCD D/A ladder network where the decimal sections are also ladder networks. See Fig. P7-14. Find the required relationship among R, R_S, and R_T.

15. Determine the required clock frequency for a tracking A/D converter for an analog input signal with frequencies within the band of 650 to 800 Hz. The converter is to have a resolution of 127 quantization increments.

16. Derive an expression for the clock frequency required in a tracking A/D converter for the triangular waveform shown in Fig. P7-1, for a resolution of n-bits.

17. Design the logic for a successive approximation A/D converter, in block-diagram, for a resolution of seven bits. Use a sample-and-hold circuit, one voltage comparator, and the necessary switching and logic circuits for seven successive comparisons.

18. Determine the minimum value of the storage capacitor in the sample-and-hold circuit in Fig. 7-33 for a maximum analog input voltage of 15 V. The capacitor supplies 30 μA of current to the circuits connected to it. The A/D converter has a resolution of seven bits, and the voltage across the capacitor is allowed to change up to one-eighth of a quantization increment during a hold time of 50 μsec.

19. Derive Eq. (7-70), the transfer function for the modified track-and-hold circuit shown in Fig. 7-35.

20. Develop the following Gray codes:
    (a) One that is also a reflected binary code (different than the one in the text).
    (b) One that is not a reflected binary code.

21. Design a combinational network to decode in parallel the disk using the Gray code in Fig. 7-37.

22. Draw a logic schematic of a circuit that will select the proper brushes in the double-brush encoder shown in Fig. 7-39.

23. A binary number can be converted to Gray code by modulo two summing (exclusive OR) each bit with the next higher bit. Using this procedure determine the Gray codes for binary numbers 0000 through 1111.

24. Devise a procedure to convert the Gray code numbers in Problem 23 to binary numbers

CHAPTER 8

ERROR-CORRECTING CODES AND SHIFT REGISTER APPLICATIONS

8-1 ERROR DETECTION AND CORRECTION

In any digital system where messages in the form of digital words (sequences of **1**'s and **0**'s) are transmitted, there exists some probability that a certain percentage of the digits will be received in error. Only in an idealized, noise free system could absolutely error free reception of such data be expected. It is important, therefore, to devise methods for coding messages so that errors can be either detected or detected and corrected at the receiver.[1,2]

Consider a system where any one out of eight different messages is to be transmitted as a digital word. This system requires a minimum of three binary digits for encoding. The eight messages in straight binary code are given in Fig. 8-1. Note that in this system, if any of the message words has one or more digits inverted, the resultant word cannot be distinguished from another valid message. For example, if the message 001 is transmitted, and incorrectly received as 011, the receiver will assume that 011 is a correct message.

In the described system the different message words are separated from each other by a minimum of one digit position. That is, within the message set the minimum distinction, or separation, between the binary messages is a change of only one bit. This minimum separation is called the *distance* between code words. As in the illustration, when the minimum distance is *one*, errors cannot be detected. However, if four binary digits were used to encode the eight message words, then the minimum distance between words can be increased to *two*. This is illustrated in the table of Fig. 8-2.

Message Number	Binary Word		
	a_2	a_1	a_0
0	0	0	0
1	0	0	1
2	0	1	0
3	0	1	1
4	1	0	0
5	1	0	1
6	1	1	0
7	1	1	1

FIGURE 8-1. Binary Words for Eight Different Messages.

Message Number	Binary Word			
	a_3	a_2	a_1	a_0
0	0	0	0	0
1	1	0	0	1
2	1	0	1	0
3	0	0	1	1
4	1	1	0	0
5	0	1	0	1
6	0	1	1	0
7	1	1	1	1

FIGURE 8-2. Binary Code with Minimum Distance of Two.

Note that each of the eight message words in Fig. 8-2 differ in at least two bit positions. Consequently, if one digit of any message is inverted, an unused code word results. The receiver can detect this. For example, if the message is 1001, and it is incorrectly received as 1011, the receiver knows an error has occurred since 1011 is not a valid message. However, it cannot correct the error. Assuming only one digit is in error, the correct message could be 1001, 1111, 0011, or 1100. Observe also that an even number of errors (two or four) in this system would go undetected.

In order to enable a system to detect and correct single errors, the minimum distance between coded words must be increased to at least *three*. In such a system the receiver in effect *selects* the one valid message that is closest to the erroneously received message. Since the minimum distance is three, the correct message can be uniquely determined from a message containing a single error. If a double error occurs, the received message will be closer to an incorrect message than the true one. Therefore, with a minimum distance of three, either single error correction, or double error detection, but not both, can be achieved.

Code Distance	Code Capability		
	Detection		Correction
1	0		0
2	1		0
3	2	or	1
4	2	plus	1
5	2	plus	2
6	3	plus	2
7	3	plus	3

FIGURE 8-3. Summary of Detection and Correction Capability of Codes with Distances to Seven.

Further extension of the error correction and detection capabilities of coded messages requires a further increase in the minimum distance between message words. The table of Fig. 8-3 summarizes the correction and detection capabilities of codes with minimum distances of one through seven.

Increasing the minimum distance between coded messages has the basic disadvantage that the transmission rate for any given system is correspondingly reduced. Obviously the transmission time taken up by the additional bit positions (required for increasing the minimum distance) reduces the available message time. This adds what is termed *redundancy* to each message. The selection of a minimum distance is therefore a function of the parameters of a particular system. The digital systems designer must consider factors such as the probability of errors occurring in a message and the minimum error rate that is acceptable in a given system. These considerations are beyond the scope and intent of this book. In this discussion we are primarily concerned with the structure of some of the commonly used codes and the implementation of circuits that generate and detect these codes.

8-2 PARITY CODES

The *parity check code* is a simple and widely used method for increasing the minimum distance of a binary transmission system. A single parity check e.g., will increase the minimum distance of a code from one to two.

A binary code word is said to have *odd parity* if an odd number of its digits are **1**'s. For example, the number 1011010111 has seven **1** digits; therefore, the number has *odd* parity. Likewise, *even parity* is true when an even number of digits are **1**'s. Note that odd parity testing is equivalent to a modulo-2 addition of all the digits and can be determined by exclusive-OR'ing all of the digits. For an n-digit word, then, odd parity is given by

$$\text{Odd parity} = a_0 \oplus a_1 \oplus a_2 \oplus \cdots \oplus a_n \tag{8-1}$$

It also should be obvious that we could, in a similar manner, make odd and even parity checks on the number of zero digits in each word.

A single parity check code is characterized by an additional *check bit* that is added to each word to generate either odd or even parity. This was done, e.g., in the binary code given in Fig. 8-2. Note that the additional digit a_3 is simply chosen so that each word has even parity. An error in a single digit therefore is discernible since the parity check then will be odd.

In a typical system a parity generator will add the parity check bit to each word before transmission. At the receiver the digits in the original message are tested, and if the correct parity is not present the message is labeled in error. The system then will discard the faulty word or may ask that it be retransmitted. Note that only errors in an odd number of digits can be detected with a single parity check.

As an illustration, consider the odd parity check code tabulated in Fig. 8-4. A parity check digit has been added to an 8-4-2-1 BCD code to enable odd parity checking. (Negation of all the parity check digits would

Decimal Digit	4-Bit Straight Binary Code				Odd Parity Digit
	8	4	2	1	
0	0	0	0	0	1
1	0	0	0	1	0
2	0	0	1	0	0
3	0	0	1	1	1
4	0	1	0	0	0
5	0	1	0	1	1
6	0	1	1	0	1
7	0	1	1	1	0
8	1	0	0	0	0
9	1	0	0	1	1

FIGURE 8-4. BCD Code with an Odd Parity Bit Added.

generate an even parity code.) Observe that if an odd number of digits in any word are inverted, the parity will be even and the invalid word is discernible. However, if two or four digits are inverted, the parity check will remain odd and the word is assumed correct. Single parity checks therefore are effective only when the probability of two simultaneous errors occurring in a given word is negligible. In practice, odd parity codes are preferable to even parity because they preclude the transmission of all **0**'s.

Parity checking may be employed at major interfaces in a digital system. Since redundancy is added to each message, parity checks are used only when the probability of errors occurring and the risk associated with an error is high enough to warrant it.

To illustrate the application of a parity check code, consider the following example: Suppose that four-digit messages are to be transmitted at the rate of 400 bits/sec (100 words/sec). Either from test data or by suitable calculations it is determined that the probability of an error occurring in any single digit is 3.1×10^{-5}. Since each word contains four digits, the probability of an erroneous word being received is approximately 1.25×10^{-4}, or at 100 words/sec, at an average rate of once every 80 sec.

With the addition of a single parity check bit, five bits are required for each word, thereby reducing the transmission rate to 80 words/sec. Single errors can be detected and the message corrected by commanding the transmitter to repeat a faulty message. The probability of double errors occurring is computed as follows: If the five digits are A, B, C, D and E, then double errors can occur in 10 combinations:

$$\begin{aligned} &AB, AC, AD, AE \\ &BC, BD, BE \\ &CD, CE \\ &DE \end{aligned}$$

The probability of any pair occurring is $(3.1 \times 10^{-5})^2$, or 9.6×10^{-10}; therefore, the probability of a double error in a single message is equal to $10 \times 9.6 \times 10^{-10}$, or 9.6×10^{-9}. With a new transmission rate of 80 words/sec an undetected error then could be expected every 1.3×10^6 sec, or on the average of once in every 15 days. We are not considering any higher multiple errors here since triple errors are detectable and the probability of four digits being in error is negligible compared to the double error rate.

Generation of Parity Codes and Checking Circuits

Two implementations of parity bit generators now will be illustrated. If the digits are to be transmitted serially in a clocked system, the circuit shown in Fig. 8-5 will perform the function. Operation of the circuit is as follows: An n-bit binary word is entered into the storage register. Transmission is started by the *start transmission* command signal. This sets *FF*-2 and resets *FF*-1. The Q output of *FF*-2 is then a **1**, enabling gate A. This permits the next n clock pulses to read out serially the data in storage by passing through OR gate C. These pulses also are counted in the $n + 1$ counter. The $n + 1$ counter produces an output with the application of the next clock pulse. This output is shaped by the 1-shot multivibrator, producing an output pulse whose width is slightly less than the basic clock period. The 1-shot multi output resets *FF*-2 and enables gate B. If the number of **1**'s in the n-bit message is ODD, *FF*-1 will be set and the Q output is a **1**. Then the output of gate B goes HIGH and

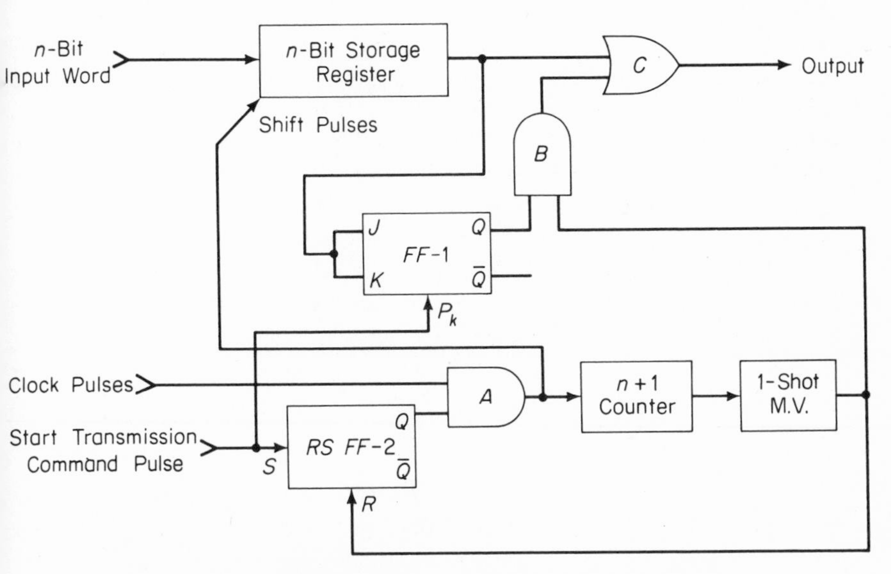

FIGURE 8-5. Even Parity Bit Generator for Serially Transmitted Data.

a **1** is transmitted at the end of the message (in the $n + 1$ bit time slot). If the number of **1**'s is EVEN, *FF*-1 will be reset and a **0** is transmitted.

At the receiver a single toggle *FF* is initially reset at the start of each message and toggled once by each **1** digit in the $n + 1$ digit message. At the end of each transmission, if the flip-flop is reset ($\bar{Q} = \mathbf{1}$), then even parity is present and the first n digits are assumed to be a correct message.

The circuit in Fig. 8-5 will generate an ODD parity code if the $\bar{Q}$ output of *FF*-1, instead of the Q output, is conected to gate *B*.

When an n-digit word is to be transmitted in parallel, a parity code bit can be generated by use of exclusive-OR gates. As an illustration, consider a five-digit word that is to be transmitted, coded with an ODD parity bit. The circuit shown in Fig. 8-6 requires four exclusive-OR circuits and one inverter to perform this function.

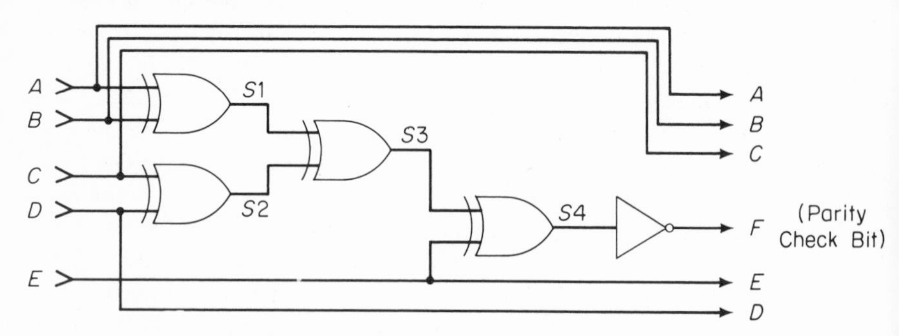

FIGURE 8-6. Odd Parity Bit Generator for Parallel Transmitted Data.

In Fig. 8-6, note that if

$$S_4 = (A \oplus B) \oplus (C \oplus D) \oplus E \tag{8-2}$$

is TRUE, then odd parity exists in digits *A*, *B*, *C*, *D*, and *E* and the output at *F* is **0** as required for odd parity in the transmitted message. If S_4 is not TRUE, then the original word has even parity and the *F* output will be a **1**. At the receiver five exclusive-OR gates determine the function.

$$S_0 = (A \oplus B) \oplus (C \oplus D) \oplus (E \oplus F) \tag{8-3}$$

to indicate if the received word has even or odd parity.

8-3 EXACT COUNT CODES

An *exact count code* is an extension of the single parity check. In the exact count code, odd or even parity is maintained, with the additional restriction that the total number of **1** digits in each word is fixed. Depending upon the application, an exact count code may require more than one additional check digit but will provide more error detection capability than the single parity check.

The *two-out-of-five* exact count code is often used to transmit decimal numbers in binary form. In this system, shown in the table of Fig. 8-7, a 7-4-2-1 weighting is used to represent the decimal numbers. The added exact count digit is then selected so that each word has exactly two **1**'s and three **0**'s. The code is sometimes referred to as a *two-out-of-five* code.

Decimal Digit	4-Bit Code – 7-4-2-1 Code				Exact Count Added Digit
	7	4	2	1	
0	1	1	0	0	0
1	0	0	0	1	1
2	0	0	1	0	1
3	0	0	1	1	0
4	0	1	0	0	1
5	0	1	0	1	0
6	0	1	1	0	0
7	1	0	0	0	1
8	1	0	0	1	0
9	1	0	1	0	0

FIGURE 8-7. Two-Out-of-Five Exact Count Code.

In the five-digit code of Fig. 8-7 the equivalent decimal word is found by using the 7-4-2-1-0 weighting factor and noting that the word 11000 represents decimal zero. Then, if any message with less than or more than two **1**'s is received, the message is known to be in error.

The exact count code enables the detection of all single errors and 40% of the double errors possible (see Prob. 8-6). Note that this error detection capability is an improvement over the single parity check of Fig. 8-4, without any increase in the required number of check digits. Both decimal word codes require five digits. This increased detection capability without added cost is deceptive. Note that, in general, four binary digits can be used to transmit up to 16 different messages. If all 16 different messages were used, then a six-digit code would be required to perform an exact count. Therefore, the added error detection capability results from the inherent redundancy in the BCD code.

8-4 GEOMETRIC CODES—BLOCK PARITY

The results of parity check codes are reasonably predictable when the corrupting signal is additive noise such as *white noise*.[3] Unfortunately, errors are caused by a variety of other sources. High frequency communications channels characteristically experience erratically spaced groups of errors due to either bursts of noise or fading. These errors can be detected more efficiently by use of block parity codes.[4]

Consider a system transmitting several messages m digits in length. If these messages are grouped into blocks of n messages each, it is possible to perform parity checks between different messages as well as on each individual message. In Fig. 8-8 a block of n messages is arranged in a rectangular pattern and coded with parity check digits in both horizontal parity (HP) and vertical parity (VP) patterns.

Horizontal Parity Bits

← m Digits →

n Words

a_1	a_2	a_3	–	–	–	a_m	HP_1
b_1	b_2	b_3	–	–	–	b_m	HP_2
c_1	c_2	c_3	–	–	–	c_m	HP_3
– – –	– – –	– – –	– – –	– – –	– – –	– – –	– – –
n_1	n_2	n_3	–	–	–	n_m	HP_n
VP_1	VP_2	VP_3	–	–	–	VP_m	HP_{n+1}

Vertical Parity Bits

FIGURE 8-8. Geometric Code with Vertical and Horizontal Parity Check.

Examination of Fig. 8-8 shows that the block parity code can detect many patterns of errors and can be used for error correction when an isolated error occurs in a given row and column. But the code cannot be used to detect errors that are *even* in number and symmetrical in two dimensions; e.g., if the block of four digits a_2, a_3, c_2, and c_3 were in error and the remainder of the digits in the a and c rows and the second and third columns were either correct or had an even number of additional errors, the parity checks all would be correct. Nevertheless, geometric codes are a powerful tool in error detection and are widely used. They are relatively inexpensive to implement and are used in telegraph transmission of alphanumeric characters or in transmitting data between computers or to and from peripheral equipment such as punched card, punched tape, and magnetic tape units.

8-5 HAMMING CODES[2,4,5]

In Sec. 8-1 we developed the concept of the minimum distance between code words and indicated that a minimum distance of three is required to enable the correction of single errors in a message word. A method of implementing this error correction is the *Hamming code.*

The Hamming code is a system of multiple-parity checks that encodes messages in a logical manner so that errors can be both detected and corrected. The total transmitted words used in the Hamming code consist of the original message and the added parity check digits. Each of the required parity checks is made upon specific bit positions of the transmitted word. The system, when properly implemented, enables the isolation of an erroneous digit, whether it is in one of the original message digits or in one of the added check digits.

The steps required for developing and using the Hamming code for a message word m digits long are as follows:

1. The minimum number of check digits k is determined. These digits are labeled $D_1, D_2, \ldots, D_k$. Each of the check digits permits the specification of a different parity test.
2. The original message is coded with the k check digits to form a new word, $m + k$ digits long. The k check digits are selected (**0** or **1**) to satisfy the necessary parity conditions.
3. The received message has the required k parity checks made.
4. If all the parity checks are correct, the message is assumed to be error free. If one or more of the checks fails the digit in error is uniquely determined by the outcome of these checks.

Number of Parity Checks

A basic consideration in the development of the Hamming code is the determination of the minimum quantity of check digits (k) required. Consider a message word m digits long. If k check digits are added, the transmitted word's total length is $m + k$. At the receiver k parity checks are performed, with the outcome of each check being either TRUE or FALSE. The results of this parity testing then may be expressed as a k digit binary word. This is capable of defining a maximum of 2^k different states. One of these states, the one indicating that all parity tests are TRUE, defines the *message-is-correct* condition. The remaining $(2^k - 1)$ states then may be used to define the position of an erroneous digit. This leads to the relationship

$$2^k - 1 \geq m + k \qquad (8\text{-}4)$$

The minimum value of k that meets the requirements of Eq. (8-4) is used. In Fig. 8-9 a table is developed from Eq. (8-4) that gives the maximum word length m of the uncoded message for different values of k up to $k = 8$.

Number of Check Digits k	Maximum Message Digits m	Transmitted Word Length $m+k$
1	0	1
2	1	3
3	4	7
4	11	15
5	26	31
6	57	63
7	120	127
8	247	255

FIGURE 8-9. Maximum Word Length for Error-Correcting Hamming Codes.

Word Format

In the conventional Hamming code word format the check digits are placed in the 1, 2, 4, 8, . . . positions. These positions are assigned to D_1 through D_k, respectively. Although the placement of check digits is somewhat arbitrary, it will be seen that the conventional positioning has the advantage of making each check digit uniquely determined by a single parity test.

To illustrate word formation, consider a system that requires the transmission of 11-bit messages. From Eq. (8-4) we determine that k must be four. This results in a 15-bit word format that is illustrated in Fig. 8-10.

Word Position	1	2	3	4	5	6	7	8	9	10	11	12	13	14	15
Check Digits	X	X		X				X							
Message Digits			X		X	X	X		X	X	X	X	X	X	X
Composite Word	D_1	D_2	m_1	D_3	m_2	m_3	m_4	D_4	m_5	m_6	m_7	m_8	m_9	m_{10}	m_{11}

FIGURE 8-10. Positioning of Check and Message Digits in the Hamming Code

Parity Checks

The k check digits are determined by performing parity checks on the $m + k$ bit composite word. In the above example, for $m + k = 15$, four even parity tests are made. These tests, labeled *A, B, C,* and *D*, are on the bit positions shown in Fig. 8-11. Note the even-parity test on word positions 1-3-5-7-9-11-13-15 determines check digit D_1 uniquely. Likewise, parity tests *B, C,* and *D* each determine check digits D_2, D_3, and D_4, respectively.

Parity Conditions	Word Position														
	1	2	3	4	5	6	7	8	9	10	11	12	13	14	15
A	X		X		X		X		X		X		X		X
B		X	X			X	X			X	X			X	X
C				X	X	X	X					X	X	X	X
D								X	X	X	X	X	X	X	X

FIGURE 8-11. Parity Check Positions.

The received message has the same parity tests (*A* through *D*) performed. If all tests are TRUE the message is assumed to be correct. If one or more of the tests fail the error position of a single digit error is identifiable. For example, if digit position 10 is inverted, tests *A* and *C* will be TRUE and tests *B* and *D* will fail. By assigning **0**'s for the TRUE outcomes and **1**'s for the FALSE outcomes, and forming the digital number *DCBA*, with *A* the least significant bit, the error position is simply the binary number *DCBA*= 1010.

To illustrate further the coding and testing of the Hamming code, consider the requirement to code the four-bit message 1001. The uncoded message is first loaded into the proper word positions, as shown in Fig. 8-12. The

Word Position	1	2	3	4	5	6	7
Digit Type	D_1	D_2	m_1	D_3	m_2	m_3	m_4
Uncoded Word	–	–	1	–	0	0	1
Check Digits	0	0	–	1	–	–	–
Coded Message	0	0	1	1	0	0	1

FIGURE 8-12. Coding of a Four-Digit Message.

check digits D_1, D_2, and D_3 can then be determined by performing even parity tests on the 1-3-5-7, 2-3-6-7, and 4-5-6-7 word positions, respectively. The required check digits are found to be $D_1 = \mathbf{0}$, $D_2 = \mathbf{0}$, and $D_3 = \mathbf{1}$. This enables us to complete the coded message, which is 0011001.

Consider now the effect of one of the seven digits being received in error. For example, if the received message is 0001001, then the parity checks on the

D_1	D_2	m_1	D_3	m_2	m_3	m_4
0	0	0	0	0	0	0
1	1	0	1	0	0	1
0	1	0	1	0	1	0
1	0	0	0	0	1	1
1	0	0	1	1	0	0
0	1	0	0	1	0	1
1	1	0	0	1	1	0
0	0	0	1	1	1	1
1	1	1	0	0	0	0
0	0	1	1	0	0	1
1	0	1	1	0	1	0
0	1	1	0	0	1	1
0	1	1	1	1	0	0
1	0	1	0	1	0	1
0	0	1	0	1	1	0
1	1	1	1	1	1	1

FIGURE 8-13. Hamming Code for Four-Digit Uncoded Messages.

1-3-5-7 and the 2-3-6-7 positions will be ODD whereas on the 4-5-6-7 positions parity is EVEN. ODD parities denote an error; therefore, the outcome of parity checks A and B are **1**'s and C is **0**. The error position (CBA) is then 011 or the third digit in the received message. The correct message is therefore 0011001.

In Fig. 8-13 the Hamming code is tabulated for all 16 messages of a system that uses four-digit message words.

8-6 SHIFT REGISTER APPLICATIONS[7,8]

Shift registers constitute one of the most important segments of the design engineer's inventory of digital devices. Their use is particularly advantageous when generating error-correcting codes and binary sequences with special properties. Random noise-like binary sequences, e.g., can be produced simply and efficiently with shift registers and exclusive-OR gates. Shift registers also are widely used for data storage, counting circuits, digital delay lines, and filters.

A second attractive feature in the use of shift registers is the ease with which they are designed and manufactured. Shift registers can be constructed with discrete components, magnetic cores, monolithic integrated circuits, and MOS-FET techniques. In recent years integrated circuit manufacturers have produced FET shift registers with several hundreds of bits on a single IC chip. These units are competitive with core memories (see Chap. 9) and show considerable advantage in operating speed, size, and power consumption. Their major limitation appears to be the number of leads that can be attached to the small chip area. Therefore, most such multibit devices are limited to sequential access.

Shift Register Sequence Generators with Direct Feedback Logic

An n-stage shift register may be configured with a network that feeds back a pattern of **1**'s or **0**'s to the input of the first stage. This is illustrated in

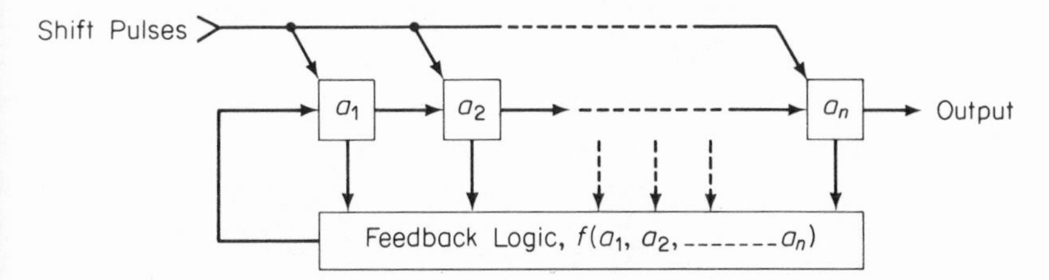

FIGURE 8-14. Shift Register of Order n with Direct Feedback Logic.

Fig. 8-14 by a generalized shift register with direct feedback logic. The feedback function is an implementation of a Boolean relationship by a combinational logic circuit whose inputs are derived from the internal states of the register.

The direct feedback shift register circuit is capable of generating a repeating sequence of length p. The sequence length is referred to as the period (p) and is bounded by

$$1 \leq p \leq 2^n \tag{8-5}$$

where n is the number of binary stages. A basic restriction on this sequence is that each group of n consecutive digits is different. For example, one of the maximum length sequences possible with $n = 4$ is the sequence 0-0-0-0-1-1-0-1-0-1-1-1-1-0-0-1. In Sec. 9-2 we shall investigate alternate methods for generating sequences where any n-digit word may repeat. However, in this example the 16 groups of four-digit words (0000, 0001, 0011, 0110, . . . , 0010, 0100) are unique as required. Observe that $n = 4$ is the maximum number of consecutive **0**'s or **1**'s permitted, if the above condition is to be satisfied. For convenience, the state that the register is in is labeled in decimal notation. The decimal equivalent of the binary states is derived by starting with a_1 as the least significant bit. Therefore, the state diagram for the 0-0-0-0-1-1-0-1-0-1-1-1-1-0-0-1 periodic sequence is drawn as follows:

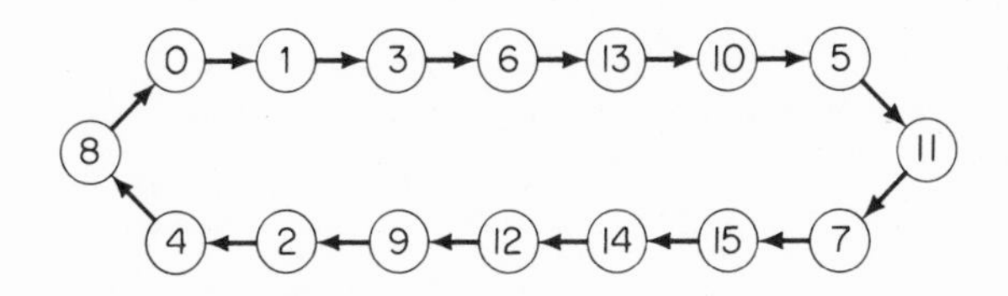

Shift Register Counters

A shift register counter may be implemented with the circuit of Fig. 8-14 by using either clocked JK or RS flip-flops. One straightforward configuration is given in Fig. 8-15. Note that the feedback inputs may be derived from both the Q and $\bar{Q}$ outputs of each stage.

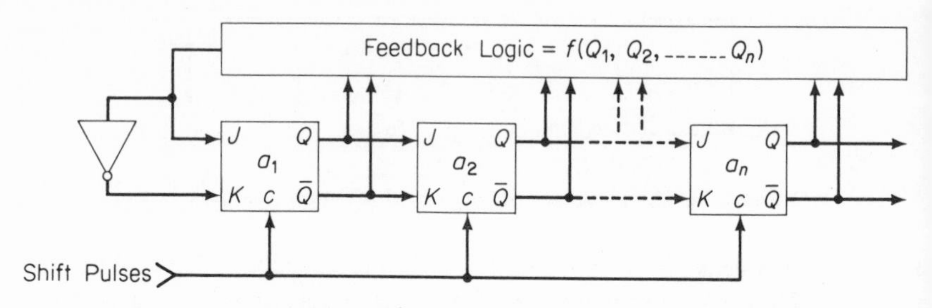

FIGURE 8-15. Shift Register Counter Implemented with JK Flip-Flops.

The output of the feedback logic is inverted to drive the K input of the first stage. Depending upon the application, the shift pulses are either derived synchronously from timing clock pulses or from asynchronous or synchronous input data. The generalized circuit can be used to generate either specific binary sequences or to function as a modulo-M counter.

To analyze the operation of the counter the input-output relationships of each stage are defined by the following difference equations:

$$Q_1^{n+1} = f(Q_1, Q_2, Q_3, \ldots)^n$$
$$Q_2^{n+1} = Q_1^n$$
$$Q_3^{n+1} = Q_2^n \qquad \text{etc.}$$

The feedback function $f(Q_1, Q_2, \ldots, Q_n)$ produces either a **1** or **0** output that in turn determines the state of Q_1 during the following bit time. Assuming that the shift register is in the 0000 state, the feedback logic will cause a transition during the next bit time to either the 0000 or 0001 states. (Obviously, feeding back a **0** when in the 0000 state is a trivial case and is included only for completeness. The counter then, is, simply a modulo-**1** counter and performs

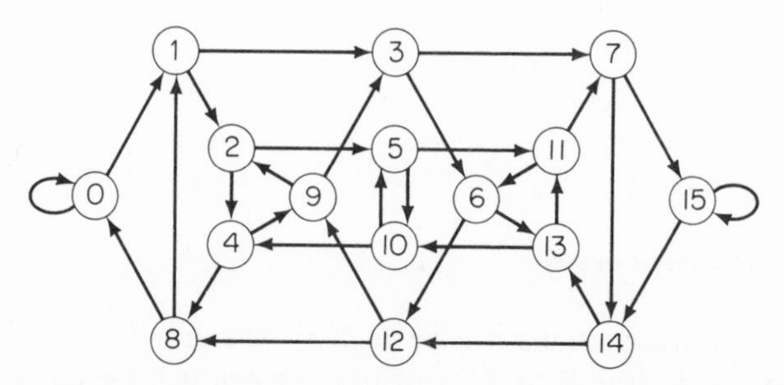

FIGURE 8-16. State Diagram for a 4-Bit Shift Register with Feedback Logic.

no useful function.) In the 0001 state the following transition is into either the 0010 or 0011 state. Likewise, the next transition is into either the 0100, 0101, 0111, or 0110 state. In this manner a universal state diagram is developed, as shown in Fig. 8-16.

The state diagram of Fig. 8-16 is a departure from the usual state diagram that defines a specific state sequence. All 16 possible internal states (0000 through 1111) are shown along with all possible transition paths. The feedback function now will determine which path will be traced. For example, careful investigation of Fig. 8-16 shows that there are six different state diagrams that can fulfill the requirements for a modulo-13 counter. (Each of these represents a different output sequence of period $p = 13$.) One of the modulo-13 sequences is given in Fig. 8-17(a).

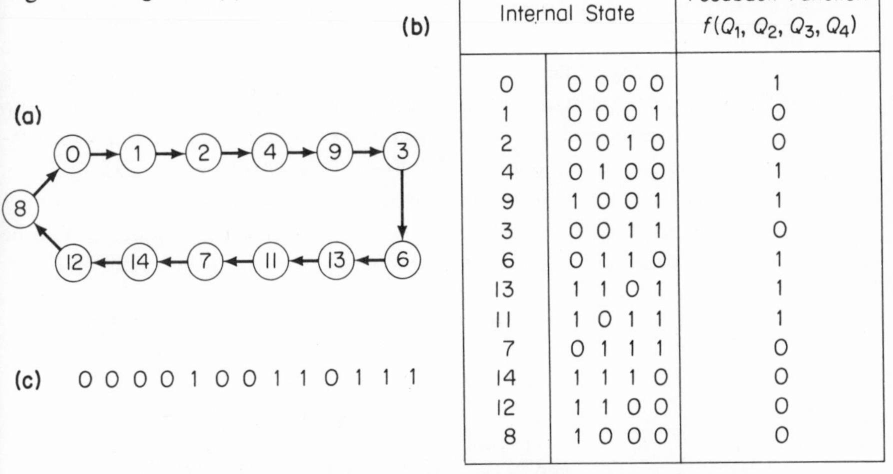

Internal State		Feedback Function $f(Q_1, Q_2, Q_3, Q_4)$
0	0 0 0 0	1
1	0 0 0 1	0
2	0 0 1 0	0
4	0 1 0 0	1
9	1 0 0 1	1
3	0 0 1 1	0
6	0 1 1 0	1
13	1 1 0 1	1
11	1 0 1 1	1
7	0 1 1 1	0
14	1 1 1 0	0
12	1 1 0 0	0
8	1 0 0 0	0

FIGURE 8-17 (a) State Diagram of a Modulo-13 Counter. (b) State Transition Table with Required Feedback Functions. (c) Resultant Output Sequence.

The internal states are tabulated in Fig. 8-17(b) with the feedback functions required to effect the necessary transitions to each successive state.

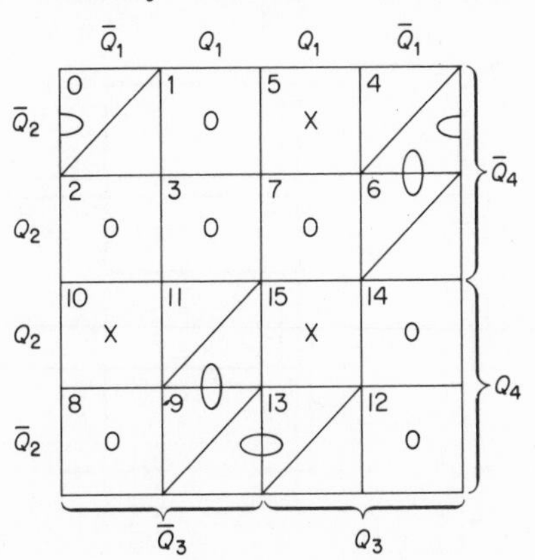

FIGURE 8-18. Map for Gating Function of Modulo-13 Counter.

The design of a suitable feedback network is now determined by use of mapping techniques. The map of Fig. 8-18 shows the required **1**'s, **0**'s and *don't care* conditions corresponding to all 16 states for the feedback function $f(Q_1, Q_2, Q_3, Q_4)$.

Observe that if the *don't care* conditions were chosen to yield a **0** output in the 5 and 10 states and a **1** output in the 15 state, the feedback function would simplify to $F = Q_1Q_4 + \bar{Q}_1\bar{Q}_4Q_3 + \bar{Q}_1\bar{Q}_4\bar{Q}_2$. However, further analysis shows that if the counter were inadvertantly placed in the 15 state a **1** fed back would cause the circuit to "hang-up" in this state. The feedback in the 15 state therefore should be a **0**, yielding the following feedback function:

$$F = Q_1\bar{Q}_3Q_4 + Q_1\bar{Q}_2Q_4 + \bar{Q}_1\bar{Q}_4Q_3 + \bar{Q}_1\bar{Q}_4\bar{Q}_2 \tag{8-6}$$

The corresponding state diagram is shown in Fig. 8-19(a) and the circuit design in Fig. 8-19(b).

The direct logic method in this manner, can be used to design any modulo-M counter (or sequence of length p, where $M = p \leq 2^n$) provided no internal state repeats itself in going through a complete cycle of the state diagram. For the four-stage counter discussed there are a total of 95 different counter sequences that can be developed. These range from two modulo-1 to eight different modulo-16 counters. Also, complementary sequences can be developed for each design by taking the output from the $\bar{Q}$ terminal of FF_4.

FIGURE 8-19. (a) State Diagram and (b) Circuit Implementation of Modulo-13 Counter with Feedback Function of Eq. (8-6).

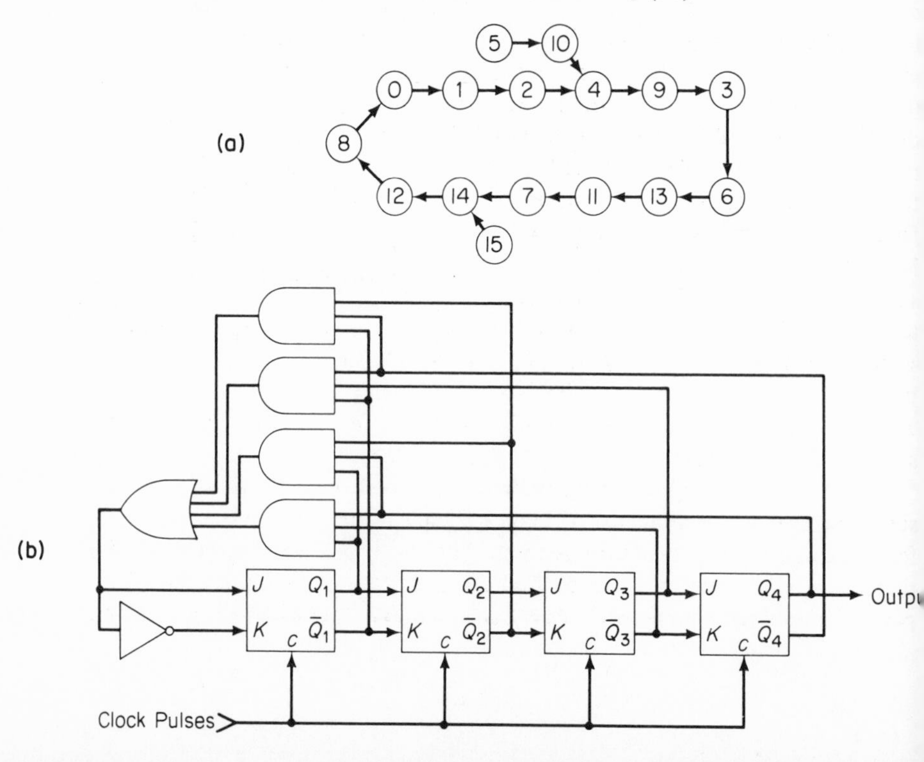

8-7 SHIFT REGISTERS SEQUENCE GENERATORS WITH OUTPUT LOGIC

By adding an output logic function, $g(a_1, a_2, \ldots, a_n)$, the direct logic counter is modified, as shown in Fig. 8-20. The direct logic sequence now may be changed to any desired output sequence of the same length. The output logic sequence generator therefore is capable of producing a sequence of length $p \leq 2^n$ with no restrictions on repeating states in the state diagram; i.e., unlike the direct logic implementation, groups of adjacent digits n bits in length

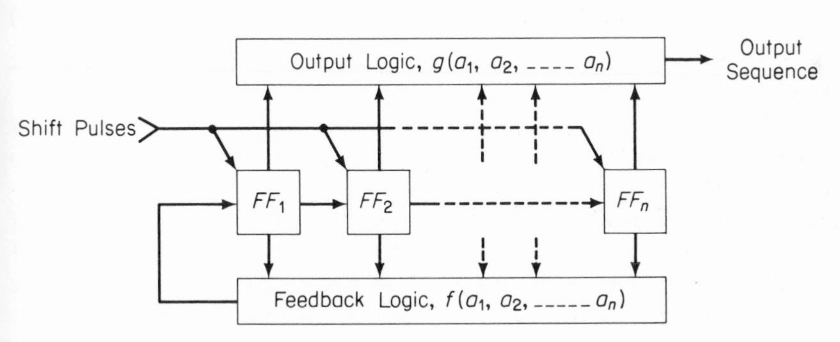

FIGURE 8-20. Output Logic Shift Register Sequence Generator.

could be repeated within the sequence. To illustrate, consider the requirement to generate the sequence 0-0-0-1-1-0-1-1 with a three-stage shift register. The eight combinations of adjacent three-digit groups are 000, 001, 011, 110, 101, 011, 110, and 100. This yields the state diagram of Fig. 8-21. Observe that the circuit cannot be designed with a three-stage shift register and direct feedback logic since an ambiguity exists when the shift register is in the six state. One method that may be used to resolve the ambiguity is to use some additional logic with a fourth bit of storage (an additional flip-flop). The two **6** states then can be *labeled* and the correct transition to the next state can be determined.

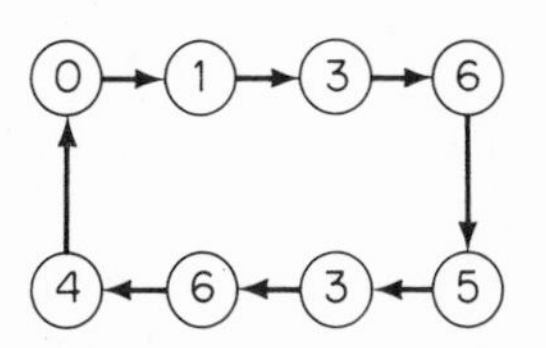

FIGURE 8-21. State Diagram of Code with Repeated State.

An alternate method, using indirect logic, will now be illustrated. As shown in Sec. 8-7, a sequence of length eight can be developed with direct feedback and a three-stage shift register. The universal state diagram for the three-stage shift counter is first derived and is drawn as in Fig. 8-22(a). Two possible modulo-8 sequences are then traced out and are listed in Fig. 8-22(b).

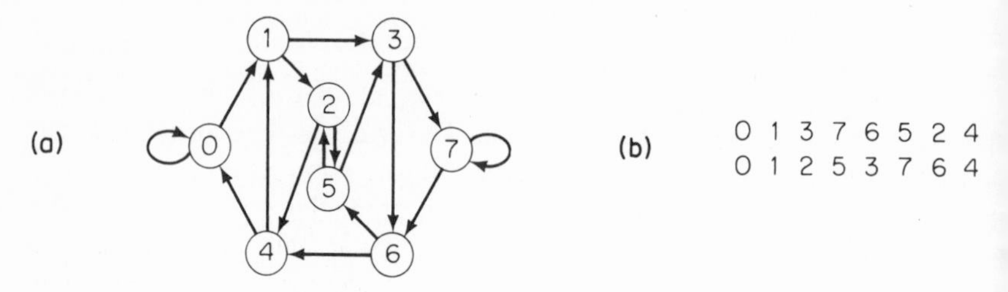

FIGURE 8-22. (a) State Diagram for Three-Stage Shift Counter and (b) Two Sequences of Length Eight.

Internal State		Feedback Function
Decimal	Binary	
0	0 0 0	1
1	0 0 1	1
3	0 1 1	1
7	1 1 1	0
6	1 1 0	1
5	1 0 1	0
2	0 1 0	0
4	1 0 0	0

(a)

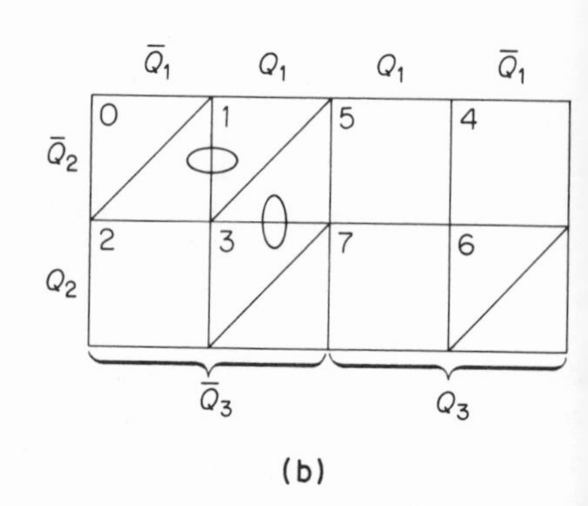

FIGURE 8-23. (a) State Table and (b) Map of Required Feedback Function.

The state table and required feedback digits for the 0-1-3-7-6-5-2-4 sequence are listed in Fig. 8-23(a). Choice of this rather than the 0-1-2-5-3-7-6-4 sequence is arbitrary. Generally the design engineer is offered more than one code for a given sequence length, and there is no way to predetermine which one will yield the *simplest* design. Depending on the application, investigation of several alternatives may be desirable. The required feedback function is therefore

$$F = \bar{Q}_2\bar{Q}_3 + Q_1\bar{Q}_3 + \bar{Q}_1Q_2Q_3 \tag{8-7}$$

A transition table is now developed listing the eight internal states of the shift register with their corresponding output states from the output logic.

The output logic circuits for G_A, G_B, and G_C are determined from the maps in Fig. 8-24, which reduce to

$$G_A = Q_1\bar{Q}_3 + Q_1\bar{Q}_2 + \bar{Q}_1Q_2Q_3 \tag{8-8}$$

$$G_B = Q_1Q_3 + Q_2\bar{Q}_3 + Q_1Q_2 \tag{8-9}$$

$$G_C = \bar{Q}_1Q_2 + Q_2Q_3 + \bar{Q}_1Q_3 \tag{8-10}$$

The complete circuit now may be designed, using Eqs. (8-7)–(8-10). If the requirement is merely to produce the sequence 0-0-0-1-1-0-1-1, however, only Eq. (8-7) and any one of Eqs. (8-8)–(8-10) need to be implemented. The shift register inherently yields the same sequence, delayed by an appropriate number of bit-times, at each binary output. Any number of different output sequences of the same length can be produced simultaneously by implementing other output logic functions.

Internal States				Output States			
Decimal	Q_3	Q_2	Q_1	Decimal	G_C	G_B	G_A
0	0	0	0	0	0	0	0
1	0	0	1	1	0	0	1
3	0	1	1	3	0	1	1
7	1	1	1	6	1	1	0
6	1	1	0	5	1	0	1
5	1	0	1	3	0	1	1
2	0	1	0	6	1	1	0
4	1	0	0	4	1	0	0

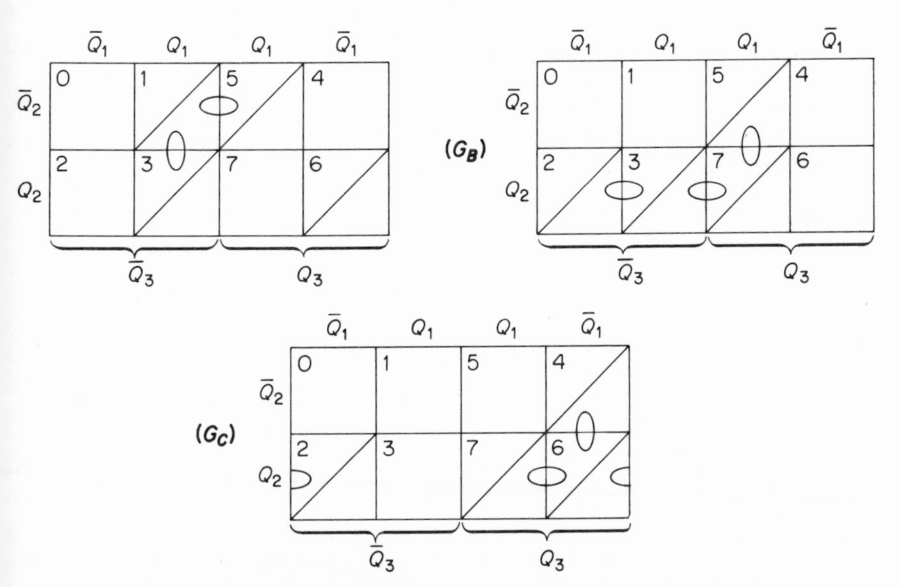

FIGURE 8-24. Transition Table for 0-1-3-6-5-3-6-4 Sequence, and Logic Maps for Required Output Functions.

8-8 SHIFT REGISTER-GENERATED LINEAR SEQUENCES[7, 8]

When the feedback logic used in the direct logic sequence generator of Fig. 8-14 consists of one or more odd parity checks on the shift register's different states, then the resultant output sequence is termed *linear*. Odd parity checking is equivalent to exclusive-OR'ing (or modulo-2 adding) the Q outputs of the required stages. The output sequence is then linear in the sense that superposition holds when performing the exclusive-OR operation. The reader should refer to Probs. 15 and 16 at the end of the chapter.

The maximum length of a linear sequence, generated by an n-stage shift register, is given by Eq. 8-11. Note that the maximum sequence length

$$p_{\max} = 2^n - 1 \tag{8-11}$$

is one less than the maximum sequence length of the general feedback logic generator. Observe that if all the stages of the shift register were in the **0** state then the feedback always would be **0** and the register would "hang-up" in this position. Therefore, the all **0**'s state must be excluded from the sequence and the maximum length is $2^n - 1$.

Not all odd parity check feedback connections result in a maximum length linear sequence. For any given register length there exists one or more feedback connections that will produce a maximum length sequence; however other connections will result in shorter (nonmaximum length) sequences.

To illustrate the linear feedback connection for a maximum length sequence, consider a three-stage shift register consisting of flip-flops a_1, a_2 and a_3. Feedback to a_1 is derived by performing an odd parity check on the outputs of a_2 and a_3. This yields the following relationships:

$$f(a_1, a_2, a_3) = a_2 \oplus a_3 \tag{8-12}$$

and

$$(a_1)^{n+1} = (a_2 \oplus a_3)^n \tag{8-13}$$

Implementation of the sequence generator in block diagram form is shown in Fig. 8-25(a). The circuit design using JK flip-flops and appropriate feedback gating is shown in Fig. 8-25(b).

The output sequence is determined by assuming the shift register is in the 100 state. This assumption is restrictive only in that it precludes the 000 state. All other states from 100 through 111 are entered in each full sequence. Since the output is periodic, starting in any other state simply results in a time shift (one or more bit-times) of the output sequence. By application of the relationship of Eq. (8-13) the state table of Fig. 8-26 is developed. The output sequence, as derived from the a_3 binary stage, is 0010111. Note that the outputs of the other stages (and $a_2 \oplus a_3$) are the same sequence shifted in time.

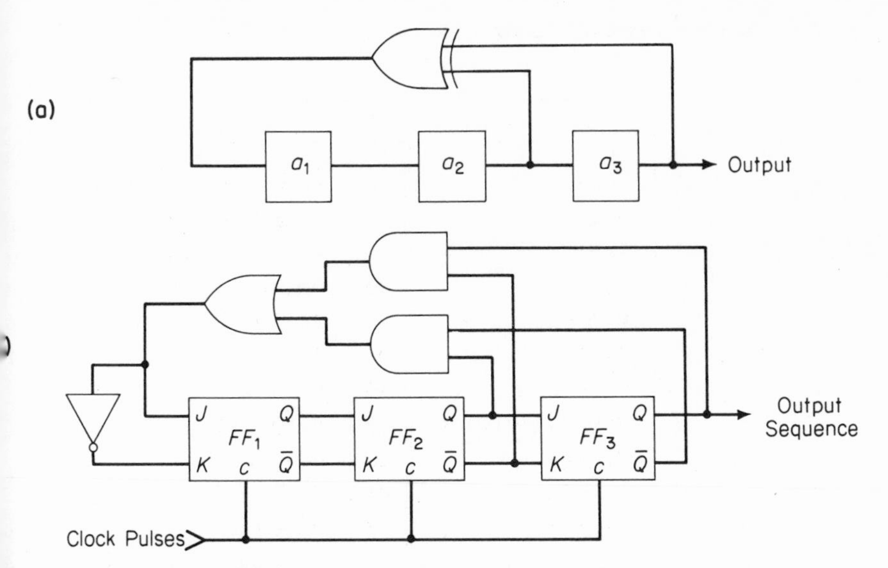

FIGURE 8-25. (a) Sequence Generator Block Diagram and (b) *JK* Flip-Flop Implementation.

Internal State				Feedback Function
Decimal	a_3	a_2	a_1	$a_2 \oplus a_3$
1	0	0	1	0
2	0	1	0	1
5	1	0	1	1
3	0	1	1	1
7	1	1	1	0
6	1	1	0	0
4	1	0	0	1

(b)

1 → 2 → 5 → 3 → 7 → 6 → 4 → 1

FIGURE 8-26. (a) State Table, Three-Stage Maximum Length Linear Sequence Generator, and (b) Output State Diagram.

Maximum Length Linear Sequences

In Sec. 8-9 we stated that a maximum length linear sequence may be generated by an *n*-stage shift register. Feedback is derived from one or more odd parity checks on its internal and output stages. One feedback tap always will occur on the output stage. Otherwise the length of the shift register is simply shortened and any stages beyond the last feedback tap are incidental. The other feedback taps, and the parity checks performed, when appropriate, will generate a sequence of length $p = 2^n - 1$. In Fig. 8-27 the required feedback functions are tabulated for sequence generators with lengths up to $n = 22$. For other traits of maximum length linear sequences see Appendix 2.

n	Logic Function	n	Logic Function
1	a_1	12	$a_2 \oplus a_{10} \oplus a_{11} \oplus a_{12}$
2	$a_1 \oplus a_2$	13	$a_1 \oplus a_{11} \oplus a_{12} \oplus a_{13}$
3	$a_2 \oplus a_3$	14	$a_2 \oplus a_{12} \oplus a_{13} \oplus a_{14}$
4	$a_3 \oplus a_4$	15	$a_{14} \oplus a_{15}$
5	$a_3 \oplus a_5$	16	$a_{11} \oplus a_{13} \oplus a_{14} \oplus a_{16}$
6	$a_5 \oplus a_6$	17	$a_{14} \oplus a_{17}$
7	$a_6 \oplus a_7$	18	$a_{11} \oplus a_{18}$
8	$a_2 \oplus a_3 \oplus a_4 \oplus a_8$	19	$a_{14} \oplus a_{17} \oplus a_{18} \oplus a_{19}$
9	$a_5 \oplus a_9$	20	$a_{17} \oplus a_{20}$
10	$a_7 \oplus a_{10}$	21	$a_{19} \oplus a_{21}$
11	$a_9 \oplus a_{11}$	22	$a_{21} \oplus a_{22}$

FIGURE 8-27. Feedback Functions for Generating Maximum Length Linear Sequences.

Generation of Nonmaximum Length Linear Sequences

An n-stage shift register that is configured with linear feedback wi generate either a maximum length or several nonmaximum length linea sequences. Except for the trivial case of remaining in the **0** state, the maximur length connection always generates the same sequence. Initial loading of th shift register merely changes the relative *phase* of the output sequence. A non maximum length connection however, will generate two or more sequence and the particular output sequence is a function of the initial state of th register. Also, for any particular feedback connection the sum of the length of all the output sequences generated is equal to 2^n.

In general the method for determining the output sequences for a par ticular nonmaximum feedback connection is empirical. A feedback connectio is assumed, and the resultant state table and state sequence are determined. W now will illustrate this procedure with an example. Consider a four-stage shif register connected with the feedback logic $a_2 \oplus a_4$. The register is initially set i the 0001 or **1** state. This results in what is referred to as the *impulse response* c the circuit and generates the longest possible sequence for that connection. Th state table and output state diagram then is developed as shown in Fig. 8-28(a) Next the register is set in the 0011 state, the lowest state not included in th first state sequence. This results in the state table and sequence of Fig. 8-28(b) Finally, the register is set in the 0110 state, and the results shown i Fig. 8-28(c) are obtained. The output sequences are summarized i Fig. 8-28(d).

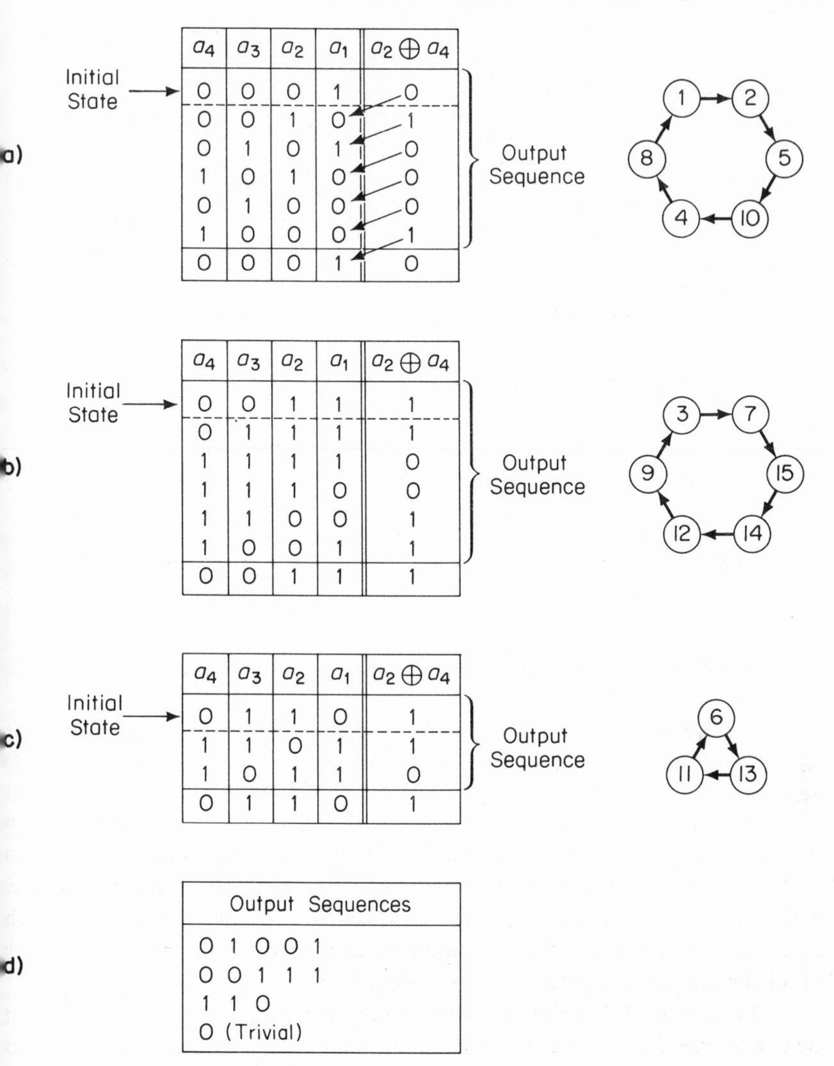

FIGURE 8-28. Sequences Generated by a Four-Stage Shift Register with $a_2 \oplus a_4$ Feedback Connection.

-9 GENERATION OF LONG NONLINEAR SEQUENCES

The procedure discussed in Sec. 8-6 may be used to generate a sequence f any desired length. However, the universal state diagram (known also as a *e Bruijn* diagram) is complex and difficult to develop when n is large. One nteresting approach, which will be given in this section, is to start with a maximum length linear sequence that is longer than the desired sequence length and then to reduce the length with a new feedback logic.

Consider an n-stage shift register that is in the S state, where

$$S = a_1(2^0) + a_2(2^1) + \cdots + a_n(2^{n-1}) \tag{8-14}$$

and the coefficients a_i are either **1**'s or **0**'s. The following state of the register then must be either S_1 or S_2, where

$$S_1 = a_1(2^1) + a_2(2^2) + \cdots + a_{n-1}(2^{n-1}) + 1 \tag{8-15}$$

$$S_2 = a_1(2^1) + a_2(2^2) + \cdots + a_{n-1}(2^{n-1}) \tag{8-16}$$

$$= S_1 - 1$$

Therefore, when in the S state of a known sequence, if the next transition is into an *odd* state (S_1), the option is available to feed back a **0** instead of a **1** and thereby go into the ($S_1 - 1$) state instead. Likewise, if the next transition is into an *even* state, we may feed back a **1** instead of a **0** and go into the S state. By searching the state diagram of a maximum length linear sequence (that can be implemented with the feedback logic given in Fig. 8-27) with the relationships Eqs. (8-15) and (8-16) as basic constraints we can shorten the sequence.

The procedure for generating a shortened sequence now will be illustrated by modifying the $p = 15$ maximum length sequence of Fig. 8-29. The maximum length sequence is first developed by applying the feedback function $a_3 \oplus a_4$. Consider now a requirement to generate a code with $p = 11$. The

(a)

a_4	a_3	a_2	a_1	$a_3 \oplus a_4$	Internal State
0	0	0	1	0	1
0	0	1	0	0	2
0	1	0	0	1	4
1	0	0	1	1	9
0	0	1	1	0	3
0	1	1	0	1	6
1	1	0	1	0	13
1	0	1	0	1	10
0	1	0	1	1	5
1	0	1	1	1	11
0	1	1	1	1	7
1	1	1	1	0	15
1	1	1	0	0	14
1	1	0	0	0	12
1	0	0	0	1	8
0	0	0	1	0	1

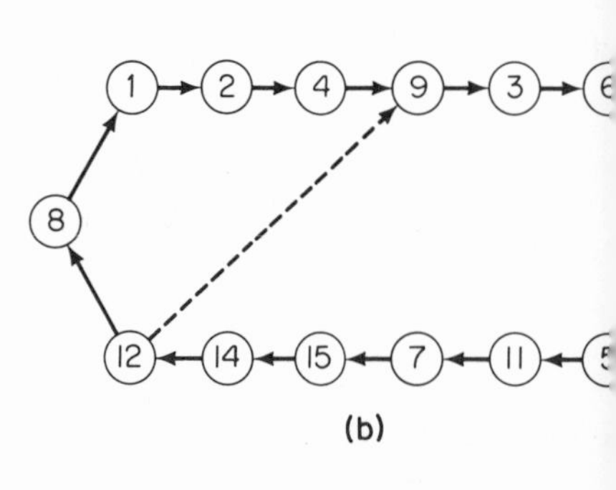

FIGURE 8-29. State Table and Diagram for $p = 15$ Linear Sequence.

problem, then, is to find four states that can be eliminated from the state diagram. A search of the state diagram of Fig. 8-29(b) shows that the **8** and **9** states are separated by three other states (**1**, **2**, and **4**). Also, the **8** state is even, so that we have the option of going from the **12** to the **9** state (shown by the dotted line in Fig. 8-29(b) by feeding back **1** instead of a **0**. This will result in a sequence with $p = 11$ as required.

(a)

Internal State	a_4	a_3	a_2	a_1	Feedback Function (F)
9	1	0	0	1	1
3	0	0	1	1	0
6	0	1	1	0	1
13	1	1	0	1	0
10	1	0	1	0	1
5	0	1	0	1	1
11	1	0	1	1	1
7	0	1	1	1	1
15	1	1	1	1	0
14	1	1	1	0	0
12	1	1	0	0	1 ← Modified Feedback

(b)

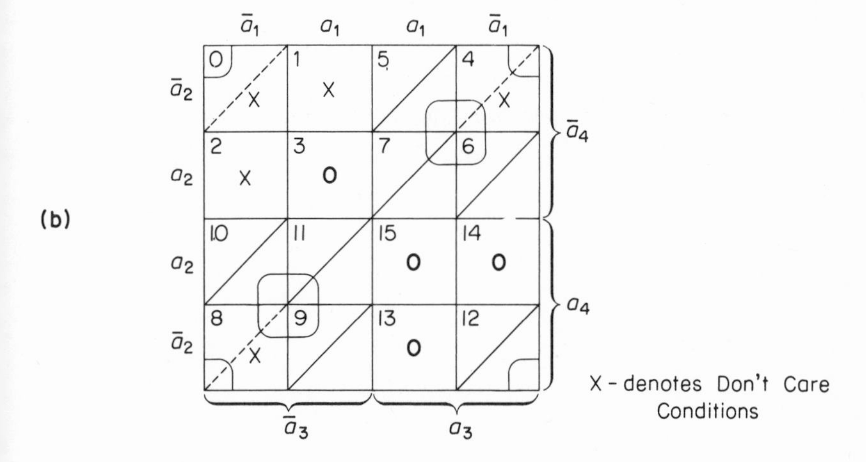

(c)

$$F = \bar{a}_3 a_4 + \bar{a}_3 a_4 + \bar{a}_1 \bar{a}_2$$

$$F = a_3 \oplus a_4 + \bar{a}_1 \bar{a}_2$$

FIGURE 8-30. (a) State Table of $p = 11$ Sequence. (b) Map for Simplifying the Feedback Logic. (c) The Required Feedback Function.

The state table for the new sequence is drawn in Fig. 8-30(a), and th required feedback in each state is determined. The feedback logic is then re duced by the map of Fig. 8-30(b). The **0**, **4**, and **8** states are don't-car conditions and are made TRUE to obtain the simplest feedback function, a shown in Fig. 8-30(c).

Other nonlinear sequences also may be developed from the same maxi mum length sequence. Those sequences, for $15 \geq p > 7$, are listed in th table of Fig. 8-31. Codes with $p \leq 7$ also can be derived but will generall be generated with a three-stage shift register. Note that all of the sequence shown in Fig. 8-31 could have been found by investigation of the state dia gram in Fig. 8-15. The basic advantage of the above approach, however, i that the universal state diagram is difficult to work with for large values of n whereas the generation of long linear sequences is relatively easy.

Sequence Length	States Omitted	Output Sequence Generated
14	15	7-14-12-8-1-2-4-9-3-6-13-10-5-12
13	10-5	1-2-4-9-3-6-13-11-7-15-14-12-8
12	2-4-9	1-3-6-13-10-5-11-7-15-14-12-8
11	8-1-2-4	9-3-6-13-10-5-11-7-15-14-12
10	6-13-10-5-11	1-2-4-9-3-7-15-14-12-8
9	4-9-3-6-13-10	1-2-5-11-7-15-14-12-8
8	13-10-5-11-7-15-14	1-2-4-9-3-6-12-8

FIGURE 8-31. Nonlinear Sequences Derived from $p = 15$ Maximum Length Linear Sequence.

REFERENCES

1. Caldwell, S. H. *Switching Circuits and Logical Design*, John Wiley & Sons, New York, 1959.
2. Hamming, R. W., "Error Detecting and Error Correcting Codes," *Bell Syst. Tech. J.*, 29: 147-150, 1950.
3. Schwartz, M., *Information Transmission, Modulation and Noise*, McGraw-Hill Book Company New York, 1959.
4. Franco, A. G., "Coding for Error-Free Communications," *Electro-Technology*, pp. 53-62, January 1968.
5. Peterson, W. W., *Error-Correcting Codes*, The MIT Press and John Wiley & Sons, New York, 1961.
6. Benice, R. J., and Frey, Jr., A. H., "Comparison of Error Control Techniques," *IEEE Trans. Commun. Tech.*, December 1964.
7. Golomb, S. W., *et. al.*, *Digital Communications with Space Applications*, Prentice-Hall, Englewood Cliffs, New Jersey, 1964.
8. Ristenbatt, M. P., and Berkowitz, R. S., ed., *Modern Radar Analysis, Evaluation, and Systems Design*, Chap. 4, John Wiley & Sons, New York, 1965.

PROBLEMS

1. One obvious method of error detection and correction is to repeat every message, i.e., transmit each message twice. If both of the received messages are the same, the message is assumed correct.
   (a) Is this method foolproof?
   (b) Why is this method not used?
2. In Fig. 8-2 a binary code is developed with an additional digit, a_3, that increases the minimum distance of the original messages from one to two. If each of a_3 digits were inverted, what would be the minimum distance?
3. Develop an even parity check code by adding a check bit to the Gray code of Fig. 7-37.
4. The code of Prob. 3 is received serially. Design a circuit, using IC gates, that performs the required even parity check on the received messages.
5. Design a circuit that implements the checking function of the exact count code shown in Fig. 8-7. Assume that the messages are received serially.
6. The simple parity code and the two-out-of-five exact count code have the same redundancy in a BCD code system. Yet the exact count code has

greater error detection capability. Explain. Devise an exact count code fc a four-digit system where all 16 messages are used.

7. A block parity code is developed for checking five-digit words in groups o four-word messages. How does the error detection capability of this cod compare to performing a simple parity check on each word?

8. The Hamming code for four-digit messages (Fig. 8-13) is used, and th messages 0110101 received. Find the error in the message and correct i Confirm this by comparing the received message with each of the 16 pos sible transmissions in Fig. 8-13. Is the received signal closest to the correc message?

9. Develop a Hamming code for six-digit messages.

10. Develop a Hamming code for four-digit messages that use the 0, 1, an 7 positions for the check digits.

11. Using the state diagram of Fig. 8-16, trace out two different modulo- counters. Using mapping techniques, minimize the feedback logic requir ed. Does one of the state sequences chosen require fewer gates than th other to implement?

12. Develop a sequence generator with a three-stage shift register and th required output logic that yields the state sequence 1-3-5-3-6-7-8.

13. Generate the same output sequence as in Prob. 12 with a four-stage shif register and direct feedback logic. (Hint: Use the 11 state for one of th three states.)

14. Develop a universal state diagram for a five-stage shift register with direc feedback logic.

15. Two four-stage shift register linear sequence generators are initially load ed in the 0001 and 0110 states respectively. The output sequences are the exclusively-OR'ed. How does the output sequence of the exclusive-OI compare to the outputs of the two generators?

16. If the sequences 0-0-0-1 and 0-1-1-0 are exclusively-OR'ed, the output i 0111. Load the sequence generator in Prob. 15 with this new initial con dition. How does the output sequence compare with that of Prob. 15 (Note that this demonstrates the linear behavior of the sequence generator.

17. Find the output sequences from a four-stage shift register configured wit the linear feedback connection $a_4 \oplus a_1$.

18. Develop a linear sequence of period $p = 31$. Find a reduction path tha will produce a sequence with $p = 17$. Design the required feedback con nection for generating this sequence. Is the new sequence linear?

CHAPTER 9

DIGITAL INFORMATION STORAGE AND CONTROL

The storage of information is a major requirement of a digital processing ystem. The stored information includes both the data to be processed and the nstructions specifying the processing steps. The memory unit in a digital sysem performs the storage function. It must provide means of access for the etrieval, or *readout*, and alteration, or *write-in*, of selected portions of the stored nformation. Therefore, the memory unit must be able to retain, identify, and etrieve digital information upon the appropriate commands.

The storage of digital information falls into two broad categories, *stationry* and *dynamic*. The storage can be both *temporary* and *permanent*. Also, ccess to the stored data falls into two classifications: *random access*, the facility o go directly to the stored location to read the data; and *sequential access*, vhere the data are scanned in a predetermined manner and access to a particlar storage location is obtained by waiting until the desired location is eached.

Permanent types of storage include punched paper tape and cards and nagnetic tape and drums. The latter are considered permanent records until hey are erased and reused. Other types of permanent storage that are used or program instructions are called *read-only* memories. These are primarily abricated by special wiring of magnetic cores or by using large arrays of semionductor gates on an integrated circuit silicon wafer (LSI read-only memory).

Temporary data storage is provided by magnetic core memories and fliplop registers. In the event of a power failure the information stored in flip-flop egisters is lost, and this is referred to as *volatile* storage. However, magnetic

core memories can provide *nonvolatile* storage; i.e., the information is retaine
during a loss of power.

An important method of dynamic storage of digital data is by use o
magnetostrictive or quartz delay lines and multistage IC shift registers. Dat
are stored by recirculating the pulses in the line, i.e., pulses at the output ar
gated back into the input driving amplifier. Dynamic storage also provides th
property of compressed time, which has an application in correlation receivers
Magnetic drums and magnetic tape loops also can be arranged to provide re
circulation of data and therefore time compressions.

9-1 DELAY-LINE STORAGE AND TIME COMPRESSION

Time compression is an important concept in the application of dynami
storage techniques. An example of time compression is the playback of a tap
recording at a higher tape speed than the original speed at which a signal wa
recorded. A chart recording of the output signal produces a waveform that ha
the same shape as the original waveform, only with a compressed scale on th
time axis. A Fourier analysis of the time-compressed waveform shows that eac
frequency component in the original waveform is multiplied by a speedup facto
η. The speedup factor is a ratio of the tape speed during playback to the tap
speed during recording. This expression is given in Eq. (9-1), where S_p is th

$$\eta = \frac{S_p}{S_r} \tag{9-1}$$

playback tape speed and S_r is the recording speed. It is important to note tha
time compression does *not* produce a frequency translation as in the familia
mixing or heterodyning process. It produces a frequency multiplication in tha
each spectral component of the signal is multiplied by the speedup factor
Similarly, time scale expansion can be achieved when the playback tape spee
is slower than the recording speed, i.e., $0 < \eta < 1$.

In the above example time compression is not accomplished in *real time*
That is, the time compression does not take place immediately following th
reception of the signal. First the signal is recorded, and then the higher-spee
playback takes place at a later time. Time compression in real time result
with multiple *looks* at the incoming signal. As an illustration, this can b
accomplished by recording a signal on a circular loop of magnetic tape with a
stationary recording head on the outer circumference of the tape; playback i
accomplished with a moving head that rotates around the inner circumferenc
of the tape. This is illustrated in Fig. 9-1. The speedup factor is expressed by
Eq. (9-2),

$$\eta = \frac{S_t + S_h}{S_t} \tag{9-2}$$

where S_t is the tape speed past the stationary recording head and S_h is the speed of the moving head around the inside of the tape loop. Note that $S_t + S_h$ is the relative speed of the tape past the moving playback head.

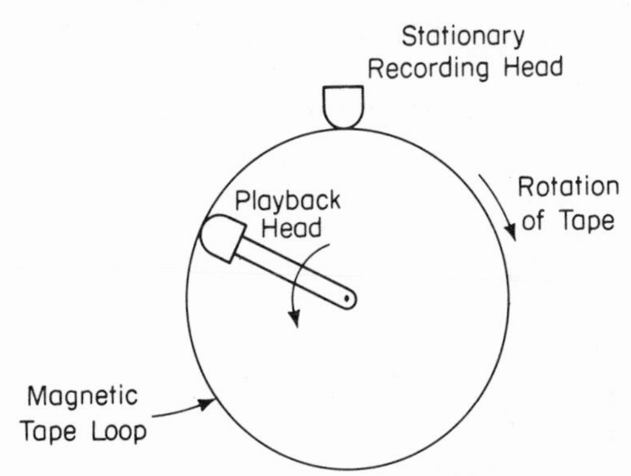

FIGURE 9-1. Time Compression in Real Time.

Each revolution of the moving head scans all the recorded data, and the output waveform is repeated. This is illustrated in Fig. 9-2, showing both the input signal and the time-compressed waveform obtained during playback. The magnetic tape loop is continuously up-dated with new input data, and the oldest data are erased just prior to recording. This is shown in Fig. 9-2(b), where each successive time-compressed waveform has shifted slightly to the left. The amount of shift is dependent upon the time compression factor and is equal to

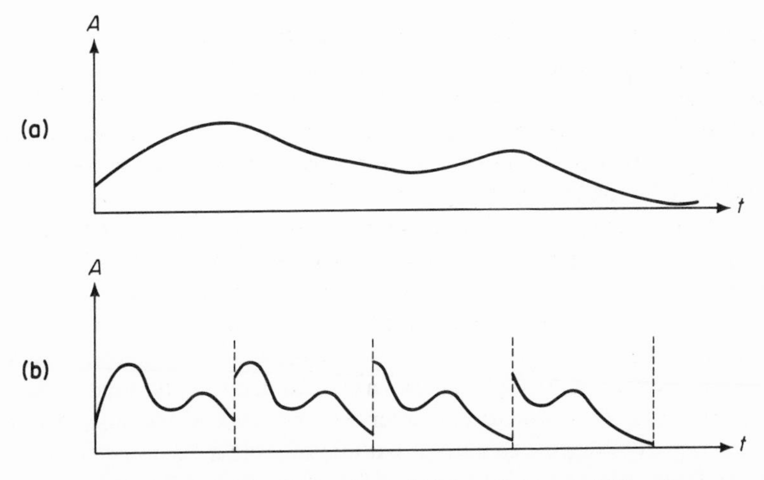

FIGURE 9-2. (a) Input Signal Waveform. (b) Time-Compressed Waveform at Playback.

the time for one revolution of the moving head. In the real time presentatio the compressed waveforms have discontinuities where the playback of the re corded data is repeated. The transient changes at these locations are not show in the illustration.

Digital Delay-Line Storage

Delay lines are used for dynamic storage of data in digital systems. A storage system using an IC, multistage, shift register as a digital delay line i shown in Fig. 9-3. This type circuit is called a *DELTIC*, an acronym for *delay line time compressor.*

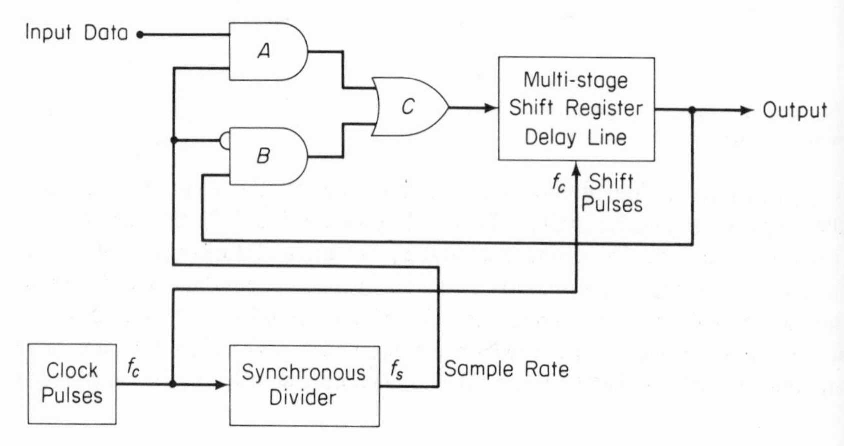

FIGURE 9-3. DELTIC Storage.

The input signal to the DELTIC is sampled and quantized and entere into the delay-line shift register at the sampling rate f_s, through gates A and C For present discussion purposes we will assume a quantization of the analo signal to 1 bit. After a time delay of τ sec., or N_t shift pulses, the first sampl pulse leaves the delay line and is reinserted back into the input through gate B and C. Immediately following this the second sample pulse is inserted int the delay line and spaced in time just one clock period behind the first sampl pulse. The total delay of the line is one clock period shorter than the samplin period. After another circulation down the line, a third pulse enters the line i position behind the first two pulses, etc. Note that the input data samples ar $1/f_s$ sec apart, but the stored data in the delay line are $1/f_c$ sec apart. Thi process produces a time compression of the sampled data.

The time compression ratio η is given by Eq. (9-3), where T_s is the perio

$$\eta = \frac{T_s}{T_c} = \frac{f_c}{f_s} \tag{9-3}$$

between input sample pulses and T_c is the period between clock pulses.

The above process is continued and each new sample is entered in the delay line after the one preceding it has traveled down the line and been

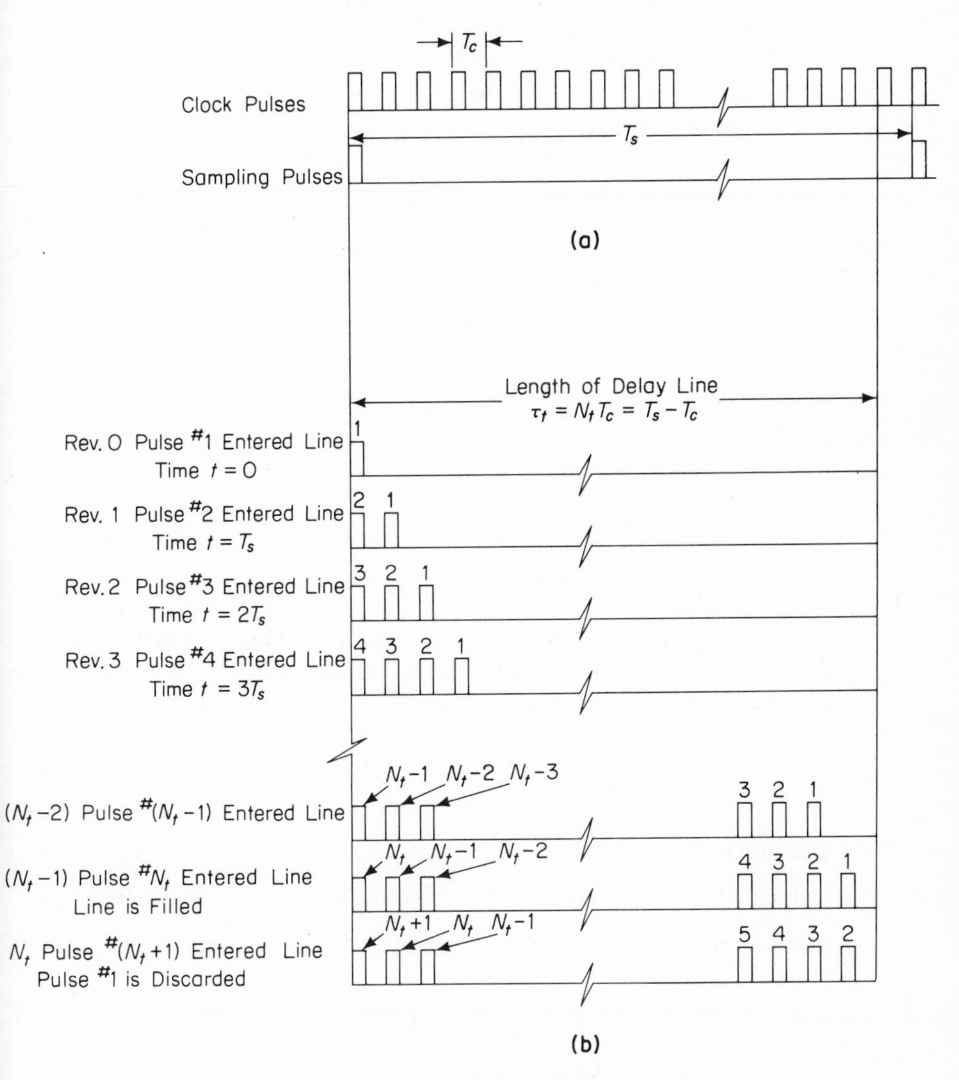

FIGURE 9-4. Timing Waveforms for the DELTIC. (a) Clock and Sampling Pulses. (b) Pulse Storage in the Delay Line for Successive Revolutions.

reinserted. The process is illustrated in the timing diagram in Fig. 9-4. The diagram in Fig. 9-4(a) shows the sampling and clock pulses in synchronization. Figure 9-4(b) shows the pulse storage in the delay line for various revolutions of the stored data at successive sampling periods. Note that the length of the delay line is 1 bit or clock period shorter than the sampling period. After N_t samples have entered the delay line, the line is filled to capacity. In this example N_t is the storage capacity of the DELTIC in number of bits. After N_t revolutions the first data pulse is discarded as a new sample enters the line. The stored samples are discarded by the inhibit input to gate B when a new sample enters the delay line. The information is continually up-dated; i.e., the oldest data pulse is replaced by the newest one.

In the above discussion of the DELTIC the stored data made one revolution or circulation in the delay line between input sample pulses. However, this is not a necessary restriction; the data could circulate twice, three times, etc. between input sample pulses. In any case, the total delay τ_t of a stored pulse must be one clock period less than the sampling period. This leads to the fundamental DELTIC relationship given in Eq. (9-4),

$$\tau_t = T_s - T_c \tag{9-4}$$

where

τ_t is the total time delay,

T_s is the input sampling period,

T_c is the clock period for the digital delay line.

Since the DELTIC is a clocked device, τ_t *must* be equal to an integral number of bit-time intervals. This relationship is expressed in Eq. (9-5), where N_t is the total number of clock periods of delay. If τ_t is achieved by X revolutions of the

$$N_t = \frac{\tau_t}{T_c} \tag{9-5}$$

data samples in a delay line of length τ_L sec, then the average delay time for each circulation is $\frac{1}{X}(\tau_t)$. However, because of the clocking requirements, τ_L also must be an integral number of bit-intervals. Therefore,

$$N_L = \frac{\tau_L}{T_c} \leq \frac{N_t}{X} \tag{9-6}$$

where

N_L is an integer equal to the bit storage capacity of the delay line

τ_L is the time delay of the digital delay line in seconds

T_c is the clock or bit time period in seconds

X is an integer equal to the number of line circulations between samples

N_t is the total number of clock periods of delay.

If N_t/X in Eq. (9-6) is not an integer, then the next smaller integral number should be used for N_L. However, in choosing this smaller number the additional delay required must be provided to make up the total N_t before the next input sample pulse. This requirement is expressed in Eq. (9-7), where N_A is the ad-

$$N_t = XN_L + N_A \tag{9-7}$$

ditional delay required to make all the variables in this equation integers to meet the requirements of a clocked system.

A logic diagram for a multiple circulation DELTIC is shown in Fig. 9-5. The stored samples normally circulate around path 1, pass through gates D and C, and reenter the delay line. After X circulations in loop 1, the pulses pass through delay line N_A and then follow path 2 through gates B and C and reenter the line. This completes the total time delay τ_t, and at the next clock pulse a new sample pulse is entered and the oldest one is discarded. The block containing the synchronous dividers generates the proper enable and inhibit gating pulses for the system.

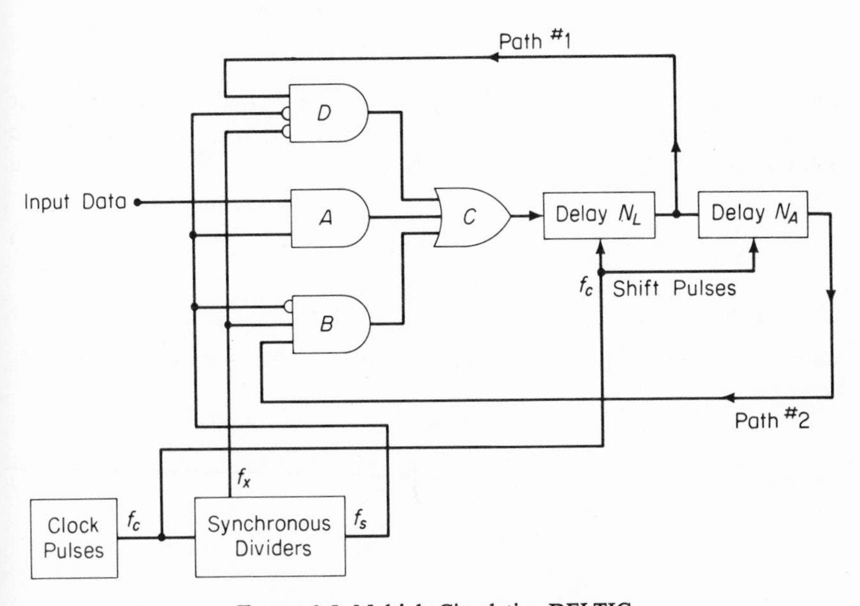

FIGURE 9-5. Multiple Circulation DELTIC.

The storage capacity of a DELTIC system is equal to the number of bits that can be stored in the delay line N_L. For the case where the data pulses circulate only once in the delay line between input samples, $N_A = 0$, and the storage capacity $N_L = N_t$, from Eq. (9-6).

As an example, assume that input data is sampled at a 1-kHz rate for storage in a DELTIC. A time compression ratio of 1,000 : 1 is required. From Eq. (9-3), the required clock pulse rate is

$$f_c = \eta f_s = (1{,}000)(1{,}000) = 1 \text{ MHz} \tag{9-8}$$

The total delay required between input samples is obtained from Eq. (9-4).

$$\tau_t = \frac{1}{1{,}000} - \frac{1}{10^6} = 999\ \mu\text{sec} \tag{9-9}$$

The storage capacity of this line is 999 bits, from Eq. (9-5), for a line using a single circulation between input pulses. Since the input signal is sampled once every millisecond, the total storage is 999 msec of data.

In the above example each sample makes 999 complete revolutions in the delay line; then, at the one-thousandth revolution the sample is discarded and at that time a new data sample is entered into the line. This DELTIC could be used for a 1-sec time delay by gating the pulse to an output register during the time of the inhibit pulse to gate B, as shown in Fig. 9-3. Using a D/A converter and filter, the original signal can be reconstructed with a 1-sec time delay. This output signal will not be time compressed.

High Speed Operation. Long shift register chains of approximately 250 bits are fabricated by medium scale integration (MSI) of semiconductor circuits on a silicon chip. These can be operated in cascade to form long delay lines. Maximum clock rates for these circuits are typically 5 MHz. However, much higher than normal speed can be achieved by operating the registers in parallel. For example, if three shift registers are in parallel, then a high speed clock can distribute the input data so that the upper register receives data bits numbers 1, 4, 7, 10, etc. The center register receives data bits numbers 2, 5, 8, 11, etc., and the bottom register receives data bits numbers 3, 6, 9, 12, etc. This distribution of input data *interleaves* the pulses in time. The bits flow through the registers at a 5-MHz rate. When the three-bit streams are combined in an output OR gate, a data rate of 15 MHz is obtained. This technique of time interleaving of data in a parallel operation of registers can be applied to other storage devices—e.g., magnetic core memories—to achieve a higher speed operation.

Parallel Operation of Shift Registers. Shift registers operating in parallel are analogous to a multihead magnetic drum recorder, where each head on a drum track corresponds to one of the parallel registers. One advantage of the parallel drum-type operation is that if the input sample is quantized to more than 1 bit, then the sample bits can be entered into the DELTIC storage unit in parallel. In general this requires less electronic circuitry than shifting parallel entry data words into a form of serial bits and synchronizing them in a single delay line.

Other Digital Delay Lines. Magnetostrictive and quartz delay lines, magnetic tape recorder loops, and magnetic drums also can provide dynamic storage in DELTIC systems. The magnetic tape loops and drums are relatively expensive, but they have the advantage of a very large storage capacity and are useful for storage requirements of greater than 200,000 bits. A disadvantage of nonshift register delay lines is that the delay is not in synchronism with the clock pulses. These delays are set by a quartz or wire line's length or a drum's rpm. The delays in quartz and magnetostrictive lines also are temperature dependent.

Because of variations in the delay time in nonshift register lines, all pulses entering the delay line must be synchronized with the clock in order to prevent timing errors. This applies especially to the pulses fed back from the output of the delay line. Any error in the time delay is cumulative with each revolution, and the operation would become hopeless if a particular pulse either advanced or retarded into an adjacent time slot. For this reason the pulses fed back are reclocked before being reinserted into the delay line. The time delay tolerance of the line is a critical design parameter. Manufacturers specify the temperature-operating range and sensitivity coefficients for the magnetostrictive and quartz delay lines. Sometimes it is necessary to place the lines in temperature-controlled ovens to provide stable operation.

Reclocking the feedback pulses in a DELTIC can be accomplished by use of a flip-flop. The delay line is adjusted to read out one-quarter clock period early. Using a magnetostrictive line in the same example as above, the delay line can be adjusted to provide a delay of 998.75 μsec, instead of the 999 μsec. The additional 0.25 μsec delay required is provided by the flip-flop. If the output pulse from the line is a **1**, it sets the flip-flop; otherwise, the flip-flop remains in the **0** or reset state. This is illustrated in Fig. 9-6, showing the pertinent waveforms. Each data pulse that is fed back is stored in the flip-flop and written into the delay line at the time of the next clock pulse, and synchronization is achieved. The clocking is accomplished by gate D in Fig. 9-6(a). This reclocking method allows some variation in the time delay of the output pulses but less than $\pm \frac{1}{4}$ clock period. This is shown as $\pm \Delta t$ in Fig. 9-6(b).

A pulse-shaping amplifier is shown in the feedback path in Fig. 9-6(a). The pulse waveforms emerging from magnetostrictive and quartz delay lines are greatly attenuated and distorted by the transmission medium in the line and the energy conversion taking place at the input and output transducers (electrical to mechanical and vice versa). The line manufacturers specify the

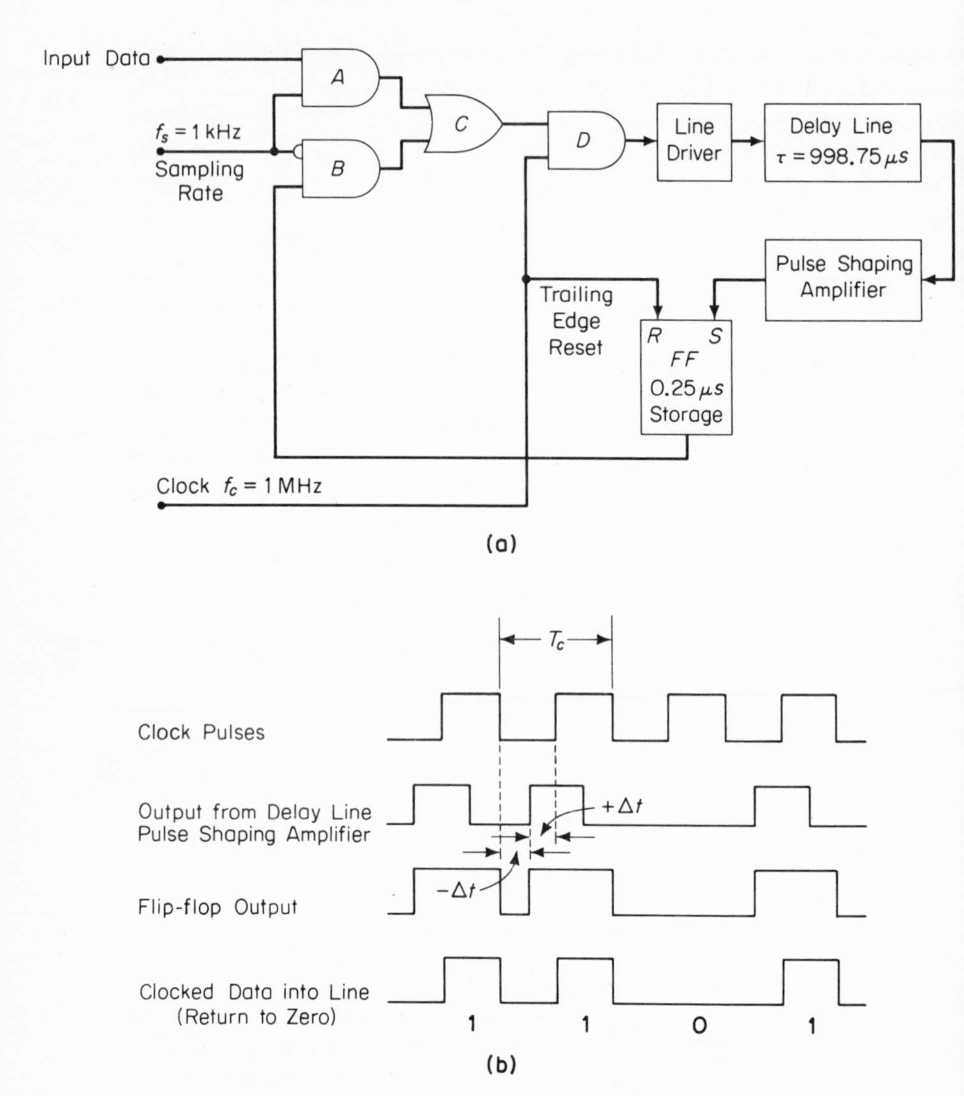

FIGURE 9-6. (a) Reclocking Feedback Pulses in a DELTIC. (b) Timing Diagram.

attenuation and other pertinent waveform data. This pulse-shaping amplifier provides the required amplification, waveform shaping, and thresholding to recover the stored **1**'s and **0**'s.

9-2 DELTIC APPLICATIONS

The DELTIC plays an important role in modern receivers used for military and space applications because of its time compression characteristic.[1] In compressed time the data are repeated over and over again with each revolution in the DELTIC. Of course, the data are up-dated by one sample each revolution and are completely up-dated in N_L revolutions. Because of these multiple *looks* at the signal it is possible to correlate a signal of unknown arrival time with a stored replica using relatively simple hardware. Digital correlation is discussed in Appendix 1.

DELTIC Correlation Receiver

The block diagram of a DELTIC correlator is shown in Fig. 9-7. In this system the analog information is quantized to one bit—i.e. two levels by a zero-crossing detector or a clipper amplifier. The reference signal may be a binary-coded signal, a CW signal, an FM sinusoidal sweep, or the like. The reference signal is written into the DELTIC at the same speed and with the same delay

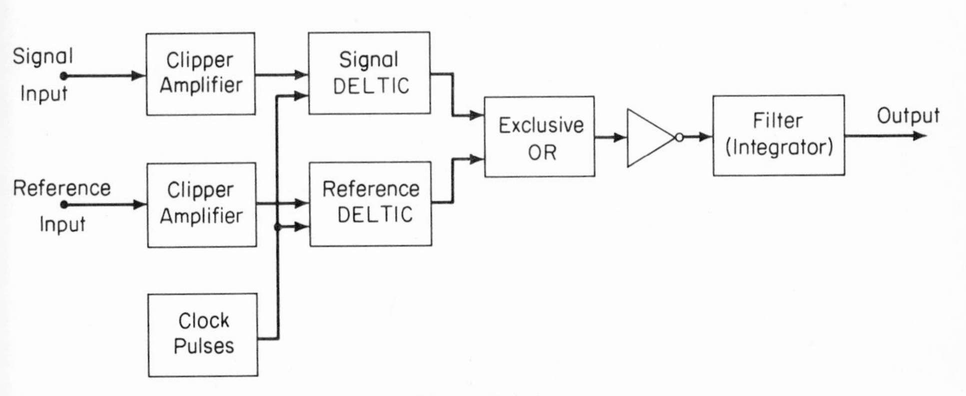

FIGURE 9-7. DELTIC Correlator.

line length and clock frequency as the signal DELTIC. In other words, the two must have the same time compression ratios. However, in correlation, the reference DELTIC must be 1 bit of delay longer than the signal DELTIC in order that the signal can precess, or slide by, with respect to the reference waveform. The precession between the reference and signal DELTICs is an important point in DELTIC correlators. The reference and signal DELTICs can be thought of as circular registers or *wheels*, with the storage bit increments distributed around the circumference. The *reference wheel* contains one more bit increment than the *signal wheel*. These two *wheels* are rotating at the same bit increment rate. Two corresponding increments on the circumferences of the two *wheels* will be one bit increment apart after one revolution of the reference *wheel*, two bit increments after two revolutions, etc. The fact that the two DELTICs are shifted at the same bit rate but that one is one bit interval longer provides the required precession. The precession between the two DELTICs is necessary because the exact arrival time of the signal waveforms is unknown in many receiver applications. The time compression in the DELTICs allows multiple looks at the signal, once each revolution. Also, during each revolution the signal and reference DELTICs are compared or correlated. They precess, and when the two waveforms line up, maximum correlation is obtained.

A reference DELTIC with additional circuitry to obtain the required precession is shown in Fig. 9-8. After the entire coded waveform is written into the reference DELTIC the input is inhibited and the stored information continues to circulate. At this point in time an extra delay of one bit is inserted into the DELTIC loop to achieve the required precession.[2]

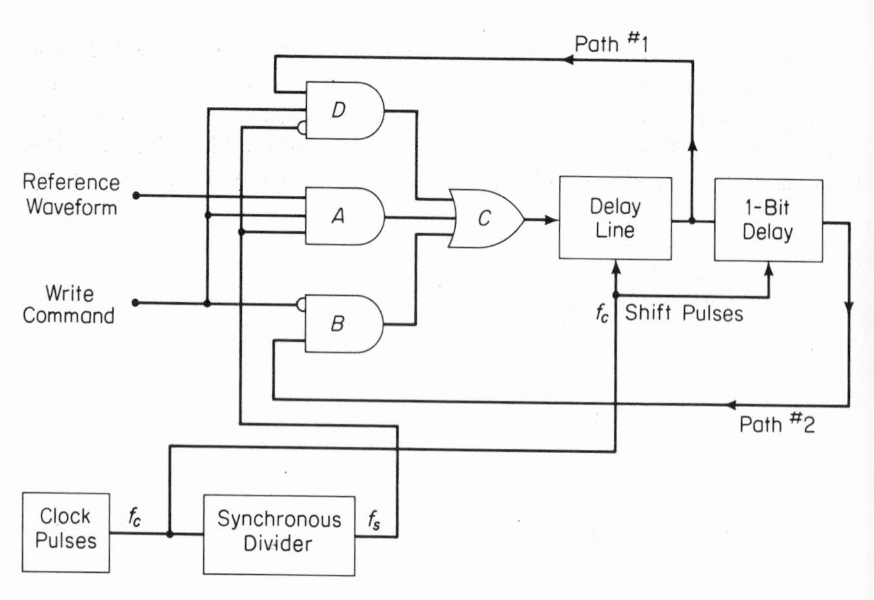

FIGURE 9-8. Precessional Reference DELTIC.

Referring to Fig. 9-8, the DELTIC is loaded with the reference waveform when the write command is a logical **1**. The command inhibits gate *B* and enables gates *A* and *D*. During write-in the data circulates in the DELTIC through path 1. When the reference waveform is completely loaded, the write command is terminated with a **0** level on the line. This inhibits gates *A* and *D*, which terminates the input data and prevents further circulation around path 1. Also, the **0** command enables gate *B* to cause the stored data to circulate through path 2, which includes the additional 1 bit of delay. No new data are written into the delay line during this mode of operation, and the stored data continuously circulate through the DELTIC. The output is correlated with the signal DELTIC, while the signal waveform precesses relative to the replica.

Another method to achieve the required precession between the two time-compressed waveforms in a correlation receiver uses a third or storage DELTIC that is longer than the reference DELTIC by the required amount. The data are transferred serially in compressed time from the reference to the storage DELTIC for correlation.

The correlator for the binary waveforms in Fig. 9-7 is comprised of the exclusive-OR circuit, followed by an amplifier and filter. The one-bit multiplication is performed by the exclusive-OR gate. The amplifier drives the filter that provides the required integration. The output would normally go to a voltage comparator threshold circuit for making a detection decision.

DELTIC Spectrum Analyzer

The time-compression characteristic of the DELTIC allows many operations to be performed in real time that otherwise would require much more additional circuitry. As an example, suppose that a received waveform is to be analyzed to determine its frequency spectrum. There are many ways that this problem can be solved. For example, a special computer can be programmed to perform a discrete Fourier transform on the input signal; the signal can be

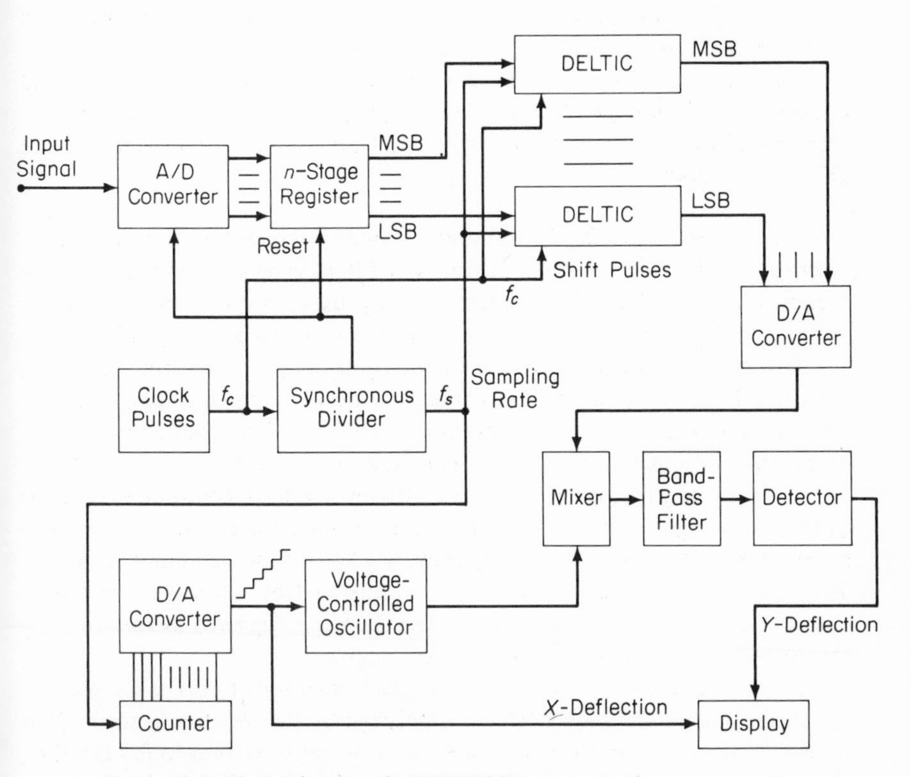

FIGURE 9-9. Block Diagram of a DELTIC Spectrum Analyzer.

applied to a contiguous bank of narrow band filters and the filter outputs scanned to determine the spectrum; or the signal can be recorded on magnetic tape or a drum and repeatedly played back through a frequency mixer and a single narrow band filter. The first two methods can be performed in real time, but relatively elaborate equipment is required. The last method requires a precision oscillator that changes frequency in incremental steps, a mixer, and a narrow band filter and has the disadvantage of nonreal time operation. The repeated playback of the data is a slow operation. With the use of a DELTIC, however, the waveform is time compressed, and this allows the last method to be performed in real time.

A block diagram of a DELTIC spectrum analyzer is shown in Fig. 9-9. The input signal is converted to digital data in an n-bit analog-to-digital converter. The number of conversion bits n depends upon the required system accuracy and therefore the allowable signal-to-noise ratio due to the quantization (see Chap. 7). If the input signal operates over a dynamic range, the use of nonlinear quantization allows the use of a smaller number of bits. The n-bit samples are clocked into the DELTICs in parallel. In this case one DELTIC is used for each bit in the digital sample. All events, clocking, conversion times, etc.—are synchronized by the master clock. The appropriate timing pulses are generated in the block entitled synchronous divider.

The digital data are stored and time compressed in the DELTICs and circulate once during each period of the sampling frequency f_s. The compressed time waveform is reconstructed with a D/A converter connected to the DELTICs' outputs.

Contiguous Filter Bank. As stated previously, the DELTIC spectrum analyzer is analogous to a contiguous bank of p filters, each filter with an effective bandwidth of Δf Hz. The system actually uses only one filter of bandwidth $\eta\Delta f$ Hz, and the compressed time signal is successively mixed with p sine waves that are separated in frequency by ηf Hz. Each frequency f_p is mixed with the compressed time signal for a duration equal to one complete circulation of the DELTICs, or for one period of f_s. The mixing frequencies, therefore, also must change sequentially at a rate equal to f_s.

The mixing frequencies are generated by a voltage-controlled oscillator (VCO). The input to the VCO is a staircase voltage waveform. Each voltage step causes the VCO to change frequency by $\eta\Delta f$, and the step duration is $T_s = 1/f_s$. The staircase voltage waveform is generated by a synchronous binary counter with a D/A converter output. The counter is triggered by the sampling frequency f_s and recycles after a count of p pulses.

The frequency mixing of the time-compressed waveforms takes place at the sped-up rate. All frequencies are multiplied by the speedup factor η. If it is desired to resolve the frequencies of the original waveform to 0.5 Hz, then the required bandwidth of the filter following the mixer is 0.5 η Hz. Also, if the original signal bandwidth is 300 Hz, then 600 frequency steps are required in the VCO, and the speedup or time compression factor $\eta = 600$. That is, in order to accomplish the 600 sequential mixing and filtering operations in real time, the original waveform must be compressed by a factor of 600 : 1. The staircase voltage to the VCO also requires 600 steps: $p = \eta = 600$. The DELTIC must have sufficient capacity to store at least the entire waveform increment that is being analyzed. This time duration, of course, should be much longer than the build-up time in the filters.

The output from the band-pass filter is amplitude detected and sent to a display or recording device. The high speed processing in the DELTIC makes the CRT a convenient display. The staircase voltage to the VCO also can be used as a horizontal sweep, and the abscissa on the CRT can be calibrated in hertz. The detected amplitudes are used for the vertical deflection to display the amplitude spectrum of the input signals.

9-3 MAGNETIC CORE MEMORIES

The most common type of high speed digital storage unit is a coincident current, ferrite core memory. Core memories can be built to provide millions of bits of storage and access times of less than 1 μsec. For storage requirements in the tens of millions of bits or greater, other storage devices are normally used, such as magnetic tapes, disks and drums. These are slow, sequential access devices and do not have the random access capability of magnetic core memories. Both types of storage are used in many digital systems: core memories for high speed operations and tapes or disks for bulk storage.

The basic storage element in a random access, magnetic core memory is a very small toroidal-shaped ferrite core. The storage in a ferrite core is represented by the direction of the magnetization in the core. If the magnetic flux is in one direction, the core is said to store a **1**; in the opposite direction, a **0**. Wires threaded through the center of the core carry current in the two directions to change the state or the direction of magnetization of the core. Each core in a large memory could have its own set of windings; however, this would become extremely bulky. A memory with a thousand cores then would have a thousand separate sets of windings. To overcome this problem, the cores are arranged in rectangular arrays in a plane and located using an *X*-*Y* coordinate system. The core planes are stacked one on top of the other to provide storage for multibit words.

The wiring of a core for the common three-dimensional (3D) memory arrangement is shown in Fig. 9-10(a). Four separate lines are threaded through the center of each core. These lines are designated *X* and *Y* for coordinate location, *Z* for an inhibit line, and *S* for a sense line.

The magnetic characteristic of the ferrite core is a nearly rectangular hysteresis loop, as shown in Fig. 9-10(b). A full positive current, $+I_m$ will reset the core to the **0** state; likewise, a full negative current $-I_m$ will set the core to the **1** state. When the cores are arranged in an X-Y plane, a coincidence of currents in both the X and Y lines is required to provide the full switching current I_m. The X and Y lines each only provide half of the total switching current, or $I_m/2$, and therefore the selected coordinate core receives the total switching current. The half-current $I_m/2$ alone is not sufficient to cause a change in the flux, and those cores receiving only $I_m/2$ do not switch. The *read* and *write* operations on the core are described below.

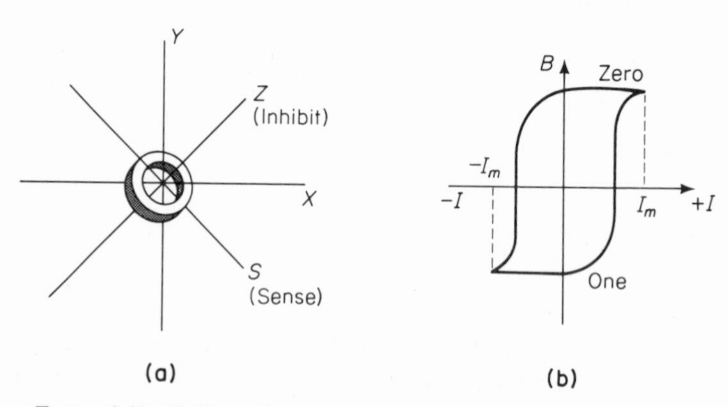

FIGURE 9-10. (a) Magnetic Core Windings. (b) Hysteresis Loop Characteristic.

During the *read* operation of the memory, positive half-currents $+I_m/2$ are applied simultaneously to both the X and Y windings so that a core receives the full switching current I_m. If the core is storing a **1** at that instant, the read half-currents will change the state of the core to a **0**. If the core was previously in the **0** state, the read half-currents will have no effect on the core state.

When the core is switched from the **1** to the **0** state, the rapid change in flux induces a voltage pulse in the corresponding sense winding. The presence of this voltage pulse indicates that a binary **1** had been stored in the core. If no voltage pulse is induced in the sense winding during the read operation, storage of a binary **0** is indicated.

During the *write* operation in the memory, negative half-currents $-I_m/2$ are applied simultaneously to both the X and Y windings. The sum of these negative currents is of sufficient magnitude to switch the core from the **0** to the **1** state and therefore results in the storage of a binary **1**. However, if it is not desired to store a **1**, then a positive half-current $+I_m/2$ is applied to the inhibit winding. This inhibit signal cancels the effect of the full switching current and holds the core in the **0** state.

The magnitude of the full switching current I_m is directly related to the core size and varies between 300 and 600 mA for 50-mil OD ferrite cores (30-mil ID). Smaller diameter cores require less current. The time required to switch a ferrite core from one state to the other is also related to the core size. Fifty-mil cores can operate at switching speeds between 0.5 and 1.5 μsec. Faster switching times can be obtained with different-type core materials, smaller sizes, and different wiring arrangements.

Coincident Current Memory Array (3D)

Figure 9-11 illustrates the manner in which the individual cores are arranged in a three-dimensional array designed for coincident current operation. Sense and inhibit lines are common to all cores in a single memory plane. The *X* and *Y* lines are wired in series through all of the planes in the array. Selection of a particular group of cores, or a complete memory word, is determined by the coincidence of half-currents in a corresponding *X* column and *Y* row.

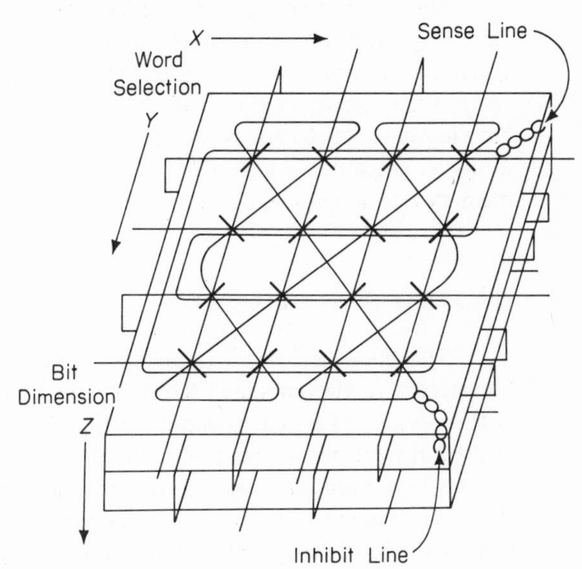

FIGURE 9-11. Coincident Current Memory 3D Array.

Each plane of cores in the array is called a bit-plane because it contains 1 bit of any given word. The bit-planes are stacked. Therefore, if a word has 10 bits, the first bit is stored in plane 1, the second bit in plane 2, etc. The 10 planes are used for storing the 10-bit word. The number of words is equal to the number of bits per plane. For example, a 2,048-word memory with 28 bits per word can be obtained with a total of twenty-eight 64 × 32 memory planes. This memory has a capacity of 57,344 bits.

In the memory read operation, the extraction of stored data leaves those cores in the selected address in the **0** state, and if it is not desired to have the word destroyed it must be written back into the cores again. In the reading process the data that had been stored may be received by an information register (storage flip-flops). Then, during the following insertion period, the data in the information register again will be stored in the cores. At the same time the data in the information register is available to the computer or other processing equipment. This type of operation is generally called a *read-and-restore* cycle or *read-write* cycle.

Typical high speed, ferrite core memory systems have read-write cycle times of 0.84 μsec. The operating speed of these cores is limited by core heating and the relaxation time of the core material. The relaxation time is a short interval immediately after each switching. At an elevated temperature or during the relaxation time, the square hysteresis characteristic of the core material is reduced considerably, which makes normal memory operation difficult. The maximum operating temperature of a core stack is approximately $+55\,^{\circ}\mathrm{C}$ for ferrite cores and $+125\,^{\circ}\mathrm{C}$ for lithium cores.

High speed ferrite core memories use small diameter cores, and it becomes increasingly difficult to thread the four windings required for the three-dimensional arrays in these minute cores. Other winding techniques have been developed that use fewer threaded leads. The smaller cores with fewer windings have less inductance and associated wiring capacitance and therefore contribute to higher speed operation. One such useful arrangement is the *linear select* memory *two-dimensional* array.

Linear Select Memory (2D)

The linear select or 2D memory organization has been used to achieve fast memory cycle times, and the core array winding is less difficult to produce. A three-wire, 2D system is shown in Fig. 9-12 using bit drive and sense lines in the X dimension and word selection lines in the Y dimension. Two-wire, 2D systems have been built using a common sense and bit drive line; but these are slower in total cycle time than a corresponding three-wire system because of sharing the bit drive and sense function on a single line.

The 2D memory is fast because a linear address selection is used for a read operation, i.e., a single drive current vs. a coincidence of two drive currents. This allows a drive current appreciably larger than full drive amplitude to be used; typically, 1.5 times full drive is used. This overdrive produces faster core switching. The 2D organization also uses shorter drive lines, which contribute less inductance and therefore permit faster drive current rise times.

The reduction in cycle time in the 2D memory is obtained at the expense of increased electronics cost. Since the core does not produce a logical AND function by summing coincident currents as in the 3D array, the address

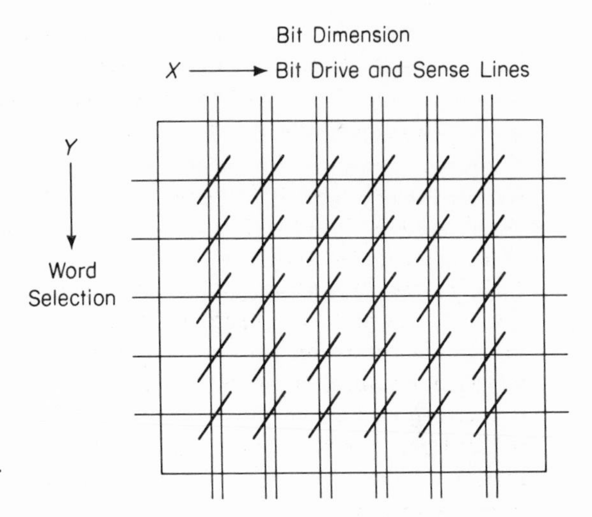

FIGURE 9-12. Linear Select Memory 2D Array.

decoding must be accomplished with an increase in word selection drivers. However, a coincident current write cycle, X and Y drive currents, is utilized. The word dimension is identified as the Y drive lines and the bit dimension as the X drive lines. This arrangement is referred to as a *word-organized* memory; whereas the coincident current memory, because of its method of core selection, is referred to as a *bit-organized* memory.

We mentioned above that the linear select memory requires more electronic components in the driving and selection circuity. For example, a 4,096-word memory would require 4,096-word selection lines in a 2D organization; however, in the 3D array only two 64-position selection switches would be required. This large requirement of electronics limits the use of the 2D organization to small capacity memories. The 2D memories are useful in high-speed *scratch pad* and *buffer* memory applications where only small quantities of data are stored.

$2\frac{1}{2}$D Memory Organization

Core memory schemes that combine some of the advantages of both the 3D and 2D memory organizations are referred to as $2\frac{1}{2}$D memories. The 2D array, of course, is fabricated in a three-dimensional package by folding it in smaller sections in an accordian-type fashion. The 2D title is given because the entire memory could be wired in a two-dimensional plane using this wiring technique. A 4,096-word, 2D memory would have very long bit drive and sense line that would be objectionable for high speed operation. The large 4;096-position switching matrix also is not practicable. A block diagram showing this 2D arrangement is illustrated in Fig 9-13(a). A 12-bit number is required to designate the desired address to the 4,096-position switching matrix. A 30-bit word length is illustrated; however, the memory can be made for longer words without serious problems being encountered.

The multiposition switch requirements can be reduced and the line length shortened appreciably by dividing the long, narrow 2D array of Fig. 9-13(a) into a number of sections and arranging them as illustrated in Fig. 9-13(b). This sectional arrangement is called a $2\frac{1}{2}$D array. These sections can be arranged in a two-dimensional plane, or groups of sections can be stacked on top of each other to form a three-dimensional array package.

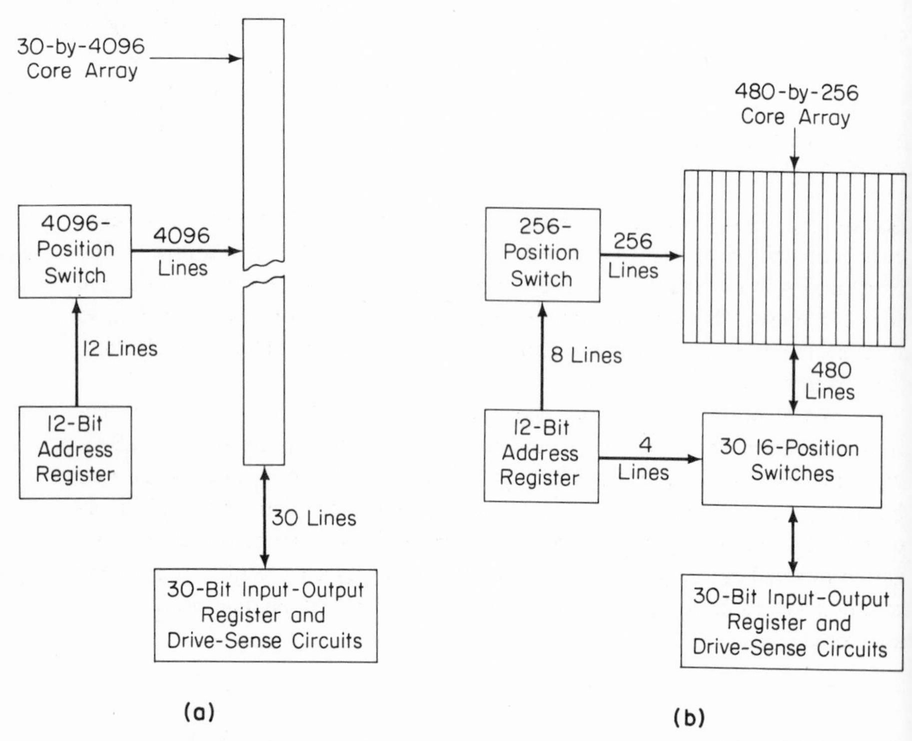

FIGURE 9-13. Illustration of a $2\frac{1}{2}$D Memory. (a) 2D Organization and (b) $2\frac{1}{2}$D Organization.

In Fig. 9-13(b) the 4,096-word memory is divided into 16 sections, each of 256-word capacity. Nine bits of the address number are required to control a 256-position switch to actuate the corresponding word drive line in each of the 16 sections. For the chosen word length of 30 bits, the total number of cores in each row is $30 \times 16 = 480$. To select a particular word eight bits of the 12-bit address word selects a corresponding word in each of the 16 sections. Finally, to select the desired word the proper one of the 16 sections must be selected, and this is accomplished by the four remaining bits of the address word. A separate 16-position switch is required for each bit in the word. Therefore, thirty 16-position switches are required for this switching matrix. The total number of matrix output lines has been reduced from 4,096 to 256 + 480 = 736. The actual reduction in cost depends upon the details of the drive and switching electronics. The original array in this example could have been split into eight 512-word sections, requiring a total of 512 + 8(30) = 752 lines from the switching matrices. Since there is not an appreciable difference in the total number of switching matrix lines in these two schemes, the complexity of the associated drive electronics would be the determining factor. The design of multiposition semiconductor switches is discussed in the next section of this chapter. In general, however, the array division is most economical when the two major switching matrices are nearly equal in size.

Other Magnetic Memory Techniques

The ferrite core memories have many variations in core shapes, winding arrangements, and drive techniques.[3] These methods include the partial switching of cores to decrease switching time, the use of two cores per bit for greater speed and noise immunity, the transposition of windings to avoid excessive interwinding coupling, and the reversal of driving current in one set of lines to double the storage capacity for a given number of drive lines.

Multiapertured Cores. A great variety of core designs have been made where two or more holes, or apertures, have been placed in each core. The objectives are to obtain a higher speed and a nondestructive readout. A nondestructive read feature results in a shorter cycle time because the rewrite operation is eliminated. Two forms of these multiaperture cores are called the *transfluxor*[4] and the *BIAX elements*.[5] The reader is referred to the technical literature for the detailed operation of these core schemes.

Batch-Fabricated Memory Arrays. While ferrite cores have been the main-stay, the development trends are to batch-fabricated memory arrays. Both magnetic film and monolithic semiconductor memories are under investigation. Advantages of monolithic memories are availability of nondestructive read feature and compatibility of packaging the driver and sense circuits with the logic. Volatility, standby power, and reliability are disadvantages. The economic crossover point with magnetic memories currently appears to be in the range of a few thousand to tens of thousands of bits, monolithic being more economical below this level. However, it appears that large MOS (metal-oxide-silicon) memories will be manufactured economically. The *plated wire* memory has emerged as an intermediate step in cost and performance between ferrite core and planar film memories.

The plated wire memory is a word-organized storage device consisting of a clothlike weave of fine copper wires. The storage medium is a magnetic film plated on the wires forming the woof of the weave. Insulated wires form the warp of the weave. The magnetic thin film is an iron-nickel alloy electroplated onto the copper wire. The plated wires are the digit sense lines for the array, and the insulated wires are the word drive lines. The details of operation are discussed in the literature. Plated wire memories have high access speeds and can be fabricated using automated techniques.[6]

9-4 MULTIPOSITION SWITCHING MATRICES

A basic part of a memory system is the electronic multiposition switches used to select the desired core location in the array. Two switches are required for the 3D memory; one large switch for the 2D memory; and multiple switches for the 2½D memory, as illustrated in Fig 9-13. The electronic switch, of course, has many other applications in digital systems. These applications include multiplexing, decoding, control logic, and generation of sequential waveforms. The semiconductor switches, or switching matrices, are generally binary actuated and have multiple output lines. An eight-position switch, e. g., is addressed by a three-digit binary number. Each binary number address energizes a corresponding output line. If an eight-position switch were sequentially addressed, the eight output lines would generate the waveforms shown in Fig. 9-14.

The previous discussion on waveform generation used combinational

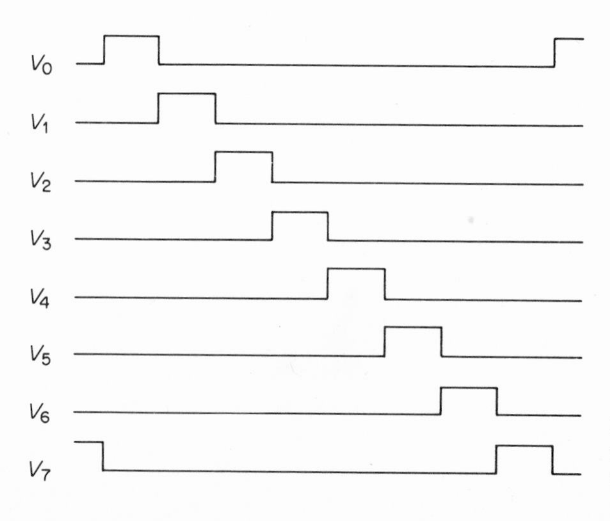

FIGURE 9-14. Waveforms Generated by an Eight-Position Electronic Switch.

tworks as the multiposition switches. We will now investigate a minim-
.tion technique that uses special groupings of the semiconductor matrices.

Diode Matrices[7]

An eight-position diode switch is shown in Fig. 9-15. The diode matrix made up of eight three-input AND gates from the three flip-flop address re- ter, A, B, and C, forming all the possible minterms ($\bar{A}\bar{B}\bar{C}$, $\bar{A}\bar{B}C$, $\bar{A}B\bar{C}$, $\bar{A}BC$, $\bar{B}\bar{C}$, $A\bar{B}C$, $AB\bar{C}$, and ABC). Rectangular diode matrices constructed in this hion require the following number of diodes: $D = n2^n$, where n is the num- r of variable or flip-flop inputs.

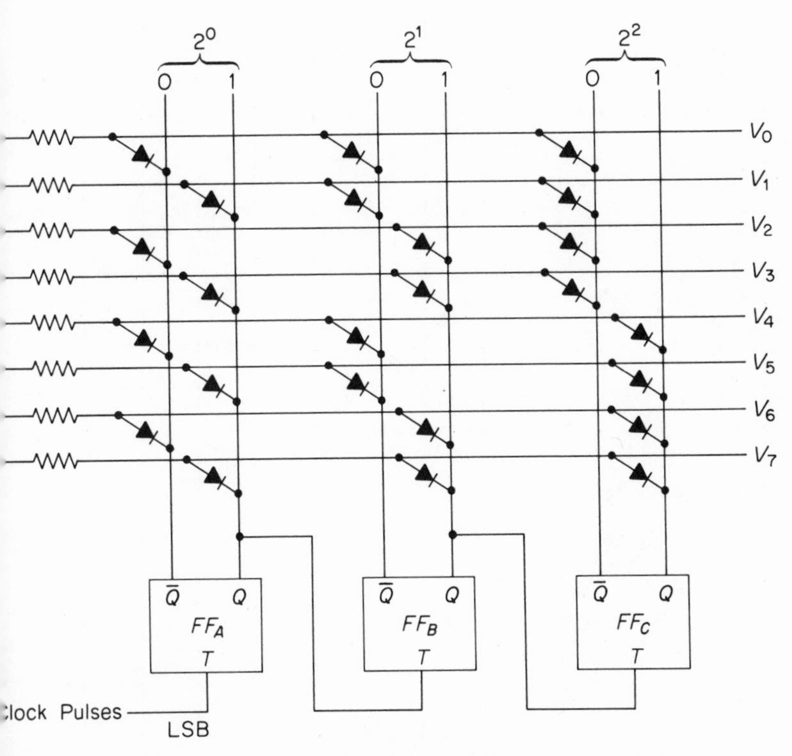

FIGURE 9-15. An Eight-Position Code-Operated Diode Matrix Switch.

Figure 9-15 shows that the locations of the diodes in a rectangular matrix rrespond to the binary numbers that the AND gates represent. This pattern diodes can be extended to any number of inputs. For example, a switch with ght input variables, and therefore 256 outputs, requires 2,048 diodes in the nary rectangular matrix. It turns out that for 16 or more outputs it is possible arrange the diodes differently such that fewer diodes are used.

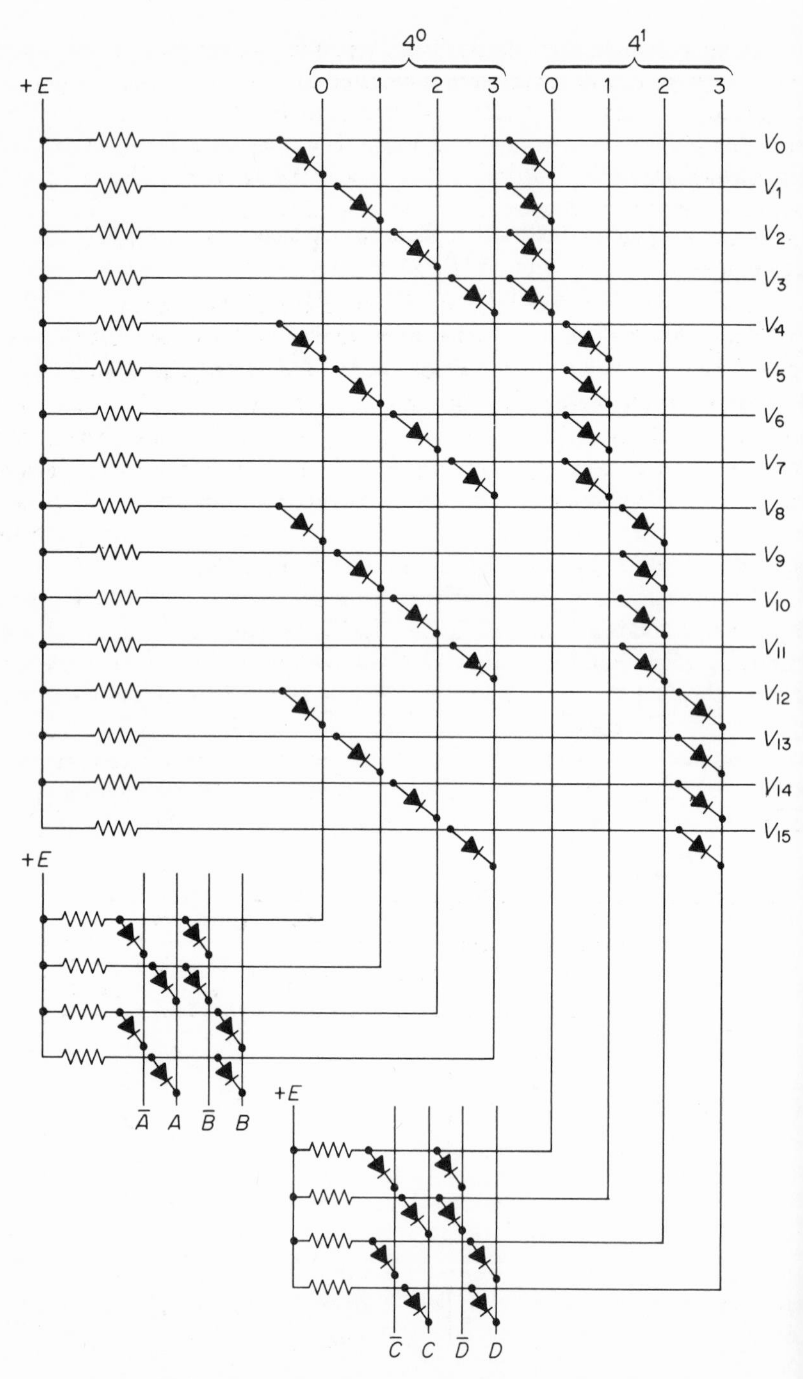

FIGURE 9-16. A 16-position Switch Using a 4 × 4 Diode Matrix. Most Economical Multilevel Matrix.

A binary rectangular matrix of a 16-position switch requires 64 diodes 4 $\times$ 2^4 = 64). This diode network is referred to as a 2 $\times$ 2 $\times$ 2 $\times$ 2 matrix. 'he diodes are arranged in a 4 $\times$ 4 matrix in Fig. 9-16. It now should be oted that the locations of the diodes in the 4 $\times$ 4 matrix correspond to numers in a radix-four number system. The 4 $\times$ 4 matrix is driven by two 2 $\times$ 2 atrices. The total diode count for the multilevel matrix is 48. This is the most conomical diode switch and represents a savings of 16 diodes over the rectanular matrix.

The above comparisons are based upon the total diode count. However, ley can be based upon the number and types of AND gates or AND gate nputs, which is the same as the diode count. The rectangular matrix of the 6-position switch contains 16 four-input AND gates. Whereas, the most conomical multilevel switch contains 24 two-input AND gates. The actual conomy depends upon the number and types of components or devices vailable for the equipment. In LSI techniques minimizing the interconections is usually more significant than the number of active components sed, and therefore the layout of the matrices is important.

The switch in Fig. 9-16 is referred to as the most economical multilevel, 6-position, diode-matrix switch. Other multilevel switches are possible. For xample, a 16-position switch can be made with an 8 $\times$ 2 matrix, where the ight inputs are driven by a 2 $\times$ 2 $\times$ 2 rectangular switch. A switch of this ype requires 56 diodes.

The most economical switch is obtained when the submatrices have *equal r nearly equal inputs.* For example, a 4 $\times$ 4 matrix is preferred over an 8 $\times$ 2 atrix.

Input Pairs n	Number of Outputs 2^n	Diode Requirements	
		Rectangular Switch	Most Economical Multilevel Switch
2	4	8	8
3	8	24	24
4	16	64	48
5	32	160	96
6	64	384	176
7	128	896	328
8	256	2048	608
9	512	4608	1168
10	1024	10240	2240

FIGURE 9-17. Comparison of Diode Requirements.

The most economical 256 position switch can be built with a 16 × 1 matrix (512 diodes), driven by two 16-position switches (96 diodes). This give a total diode count of 608 diodes, as compared with 2,048 for the rectangula matrix. A table of diode requirements for different switch arrangements is give in Fig. 9-17.

It is important to note that the AND gates can be replaced with O gates in the above switches (just turn the diodes around), and the output wave forms will be inverted. This is useful for some timing operations and for wave form generation.

Another interesting comment on these switches is that the least significa digit diodes can be replaced by the gate resistors. The eight-position switch o Fig. 9-15 is redrawn in Fig. 9-18 to shown this configuration. The output line

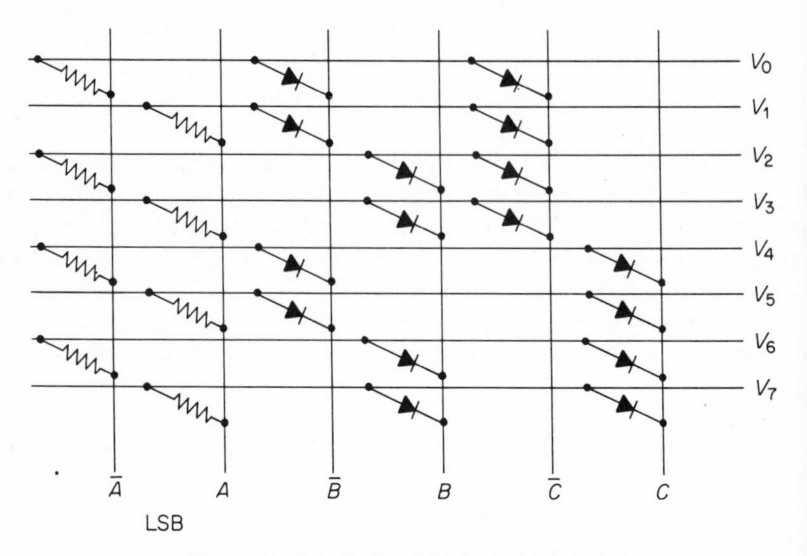

FIGURE 9-18. A Reduced Eight-Position Switch.

are still fed by three-input AND gates. However, in Fig. 9-18 one of the inpu is to a resistor and the other two are to diodes. This type gate also perform the AND function.

Transistor Tree Matrices

The rectangular diode-matrix switch was constructed by using diode ANI gates for all the minterms of the input functions. In a similar manner a tran sistor switch can be constructed using transistor AND gates. An eight-positio transistor switch is shown in Fig. 9-19. The transistors build up in a tree fashion The number of transistors required for the tree switch is

$$T = 2 + 4 + 8 + \cdots + 2^n \tag{9-6}$$

The eight-position switch, therefore, can be constructed using 14 transis-ors; this compared with 24 diodes (or 16 diodes in the reduced switch of Fig. -18).

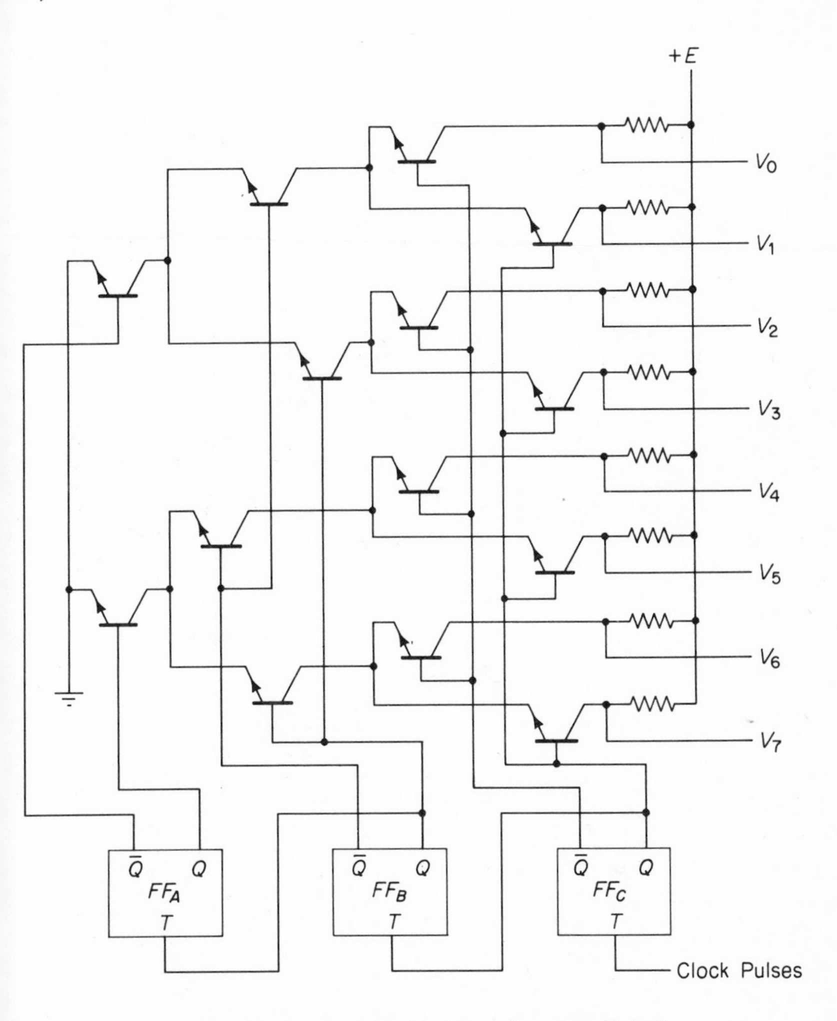

FIGURE 9-19. An Eight-Position Transistor Tree Switch.

The transistors in Fig. 9-19 are used as series contacts; therefore, a similar ree could be built using relays for very low speed applications.

Transistor trees can be expanded to any value of n, with a practical limi-ation imposed by the series voltage drops across the saturated transistors. Buffer amplifiers may be required to supply base current at the outputs of the

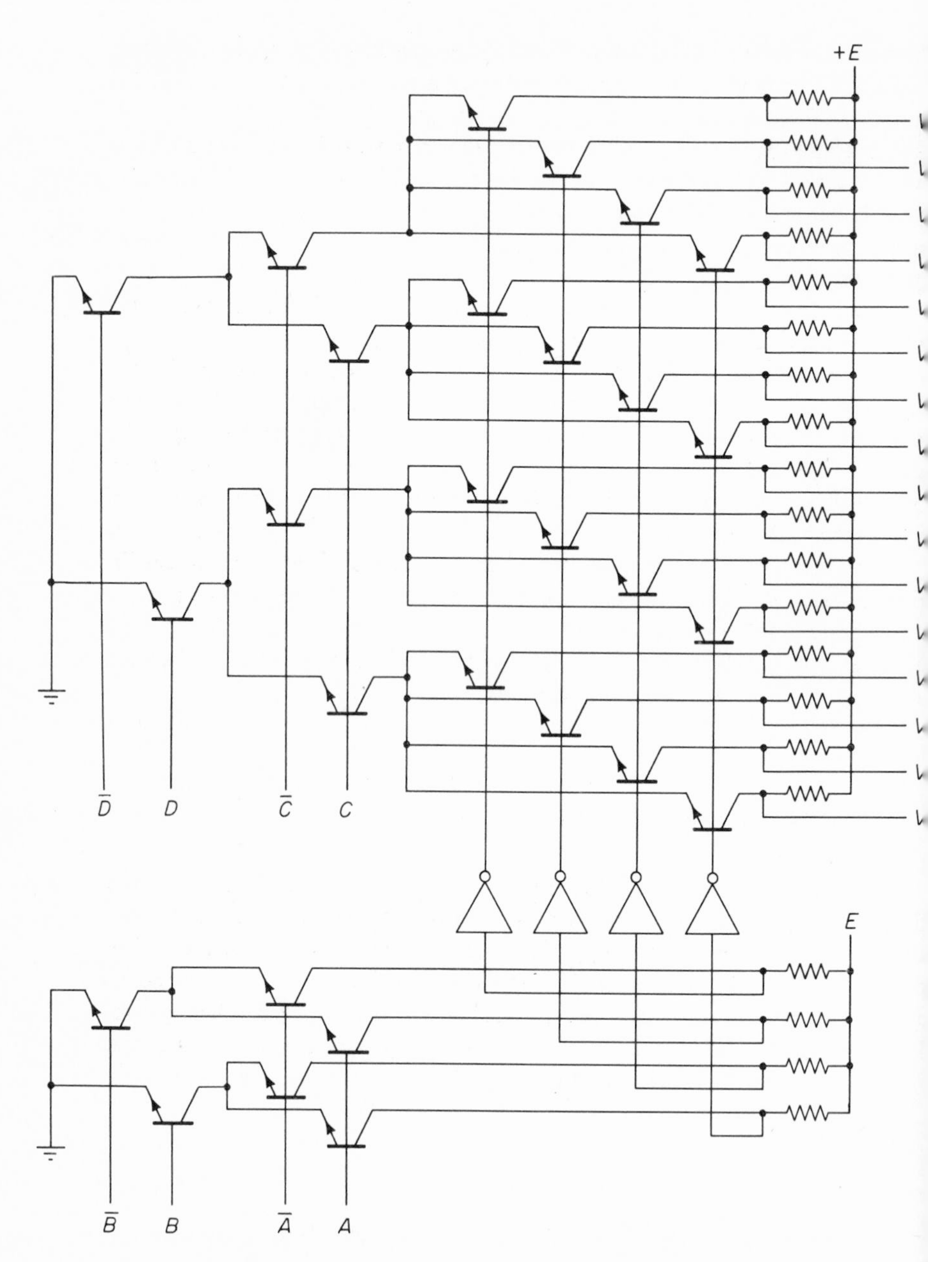

FIGURE 9-20. A 16-position Transistor Switch Using a Multilevel Tree Configuration.

ip-flops that drive many transistors. Also, considering the previous discussion n the most economical arrangement of diodes, it is natural to wonder if special roupings of transistors will result in a savings over the tree configuration. It urns out that this is also possible and that the transistors can be grouped on a nultilevel basis by associating the drives with a number system of a higher adix; however, certain problems arise because of the inverting property of the ransistors. A 16-position switch using a multilevel tree with a 4 × 4 represen- ation is shown in Fig. 9-20.

A 16-position tree requires 30 transistors; whereas the multilevel tree in 'ig. 9-20 uses 28 transistors, which is only a saving of two transistors. How- ver, the savings are greater for larger switches. A design of real significance ses a combination of diode and transistor matrices for the different levels. For xample, the lower level four-position switch (*A* and *B* controlled) in Fig. 9-20 an be replaced with a diode matrix. An alternate design is to use two four- osition transistor trees to drive a 16-position, 4 × 4 diode matrix. A reduced liode switch, similar to the technique used in Fig. 9-18, can be used for the × 4 matrix to give a very economical design. Due to the signal inversion ob- ained in transistors, care must be taken in designing multilevel transistor trees nd hybrid transistor-diode switches. The transistor switches also have a drive apability whereas diodes do not. The different combinations in a hybrid design hould be considered for each application.

)-5 INTEGRATED CIRCUIT MEMORIES

Large scale integration (LSI) of semiconductor circuits on a single chip of ilicon has made available to the digital systems designer entire functional cir- uits, e. g., arithmetic units, long shift registers, memory arrays, and other pecial repetitive logic functions. The application of LSI shift register memories n DELTIC systems was described in Secs. 9-1 and 9-2. Two other general lassifications of LSI memories have a broad application in digital systems. These are *read-only* memories and random access *scratch pad* memories.

Read-Only Memories

The term read-only memory (ROM) is given to a large scale semicon- luctor array of gates preinterconnected or programmed to perform a particular et of functions. The array storage is nonvolatile and also nonalterable. Such a levice is actually a group of combinational logic networks wired to implement given Boolean functions. Calling this device a memory is a misnomer because he stored bits cannot be altered. Nevertheless, the term does have wide usage.

Read-only memories also have been implemented with magnetic and optica storage elements.

The multiposition switches discussed in the previous section are example of read-only memories. Other read-only memory functions include programme instructions; code conversion matrices; look-up tables for mathematical func tions such as sine, cosine, tangent, logarithm, exponential, and other functions and character generation for CRT displays.

The read-only memory is built like a diode-decoding matrix. Paralle transistor AND gates also are used to perform the minterm functions. A read only memory using transistors arranged in a rectangular array is shown in Fig. 9-21. This array uses one transistor per bit. The information is writte into the array by eliminating connections between the bit lines and the wor lines. For the arrangement shown in Fig. 9-21 the transistors are not show where the connections are missing to give **0** outputs.

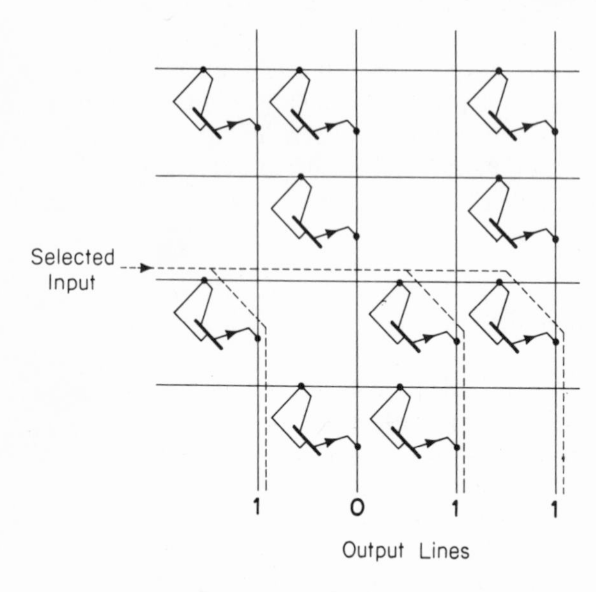

FIGURE 9-21. An LSI Read-Only Memory.

The read-only memories are manufactured according to the user's require ments. The arrays of transistors or diodes are produced and the custom inter connections are made by leaving either the emitters or gates open. There are many possible arrangements and fabrication techniques.

Scratch Pad Memories

LSI is particularly suited to the fabrication of semiconductor random access memories with a capacity ranging up to a few thousand bits and have very fast access and cycle times. These memories have a nondestructive readout

They are used primarily for intermediate storage functions associated with arithmetic or other signal-processing operations external to the main storage memory. These memories are frequently referred to as scratch pad memories. Although 1,000-bit LSI memories have been built, the principle of operation and fabrication is more easily illustrated with a smaller circuit using small or medium scale integration.

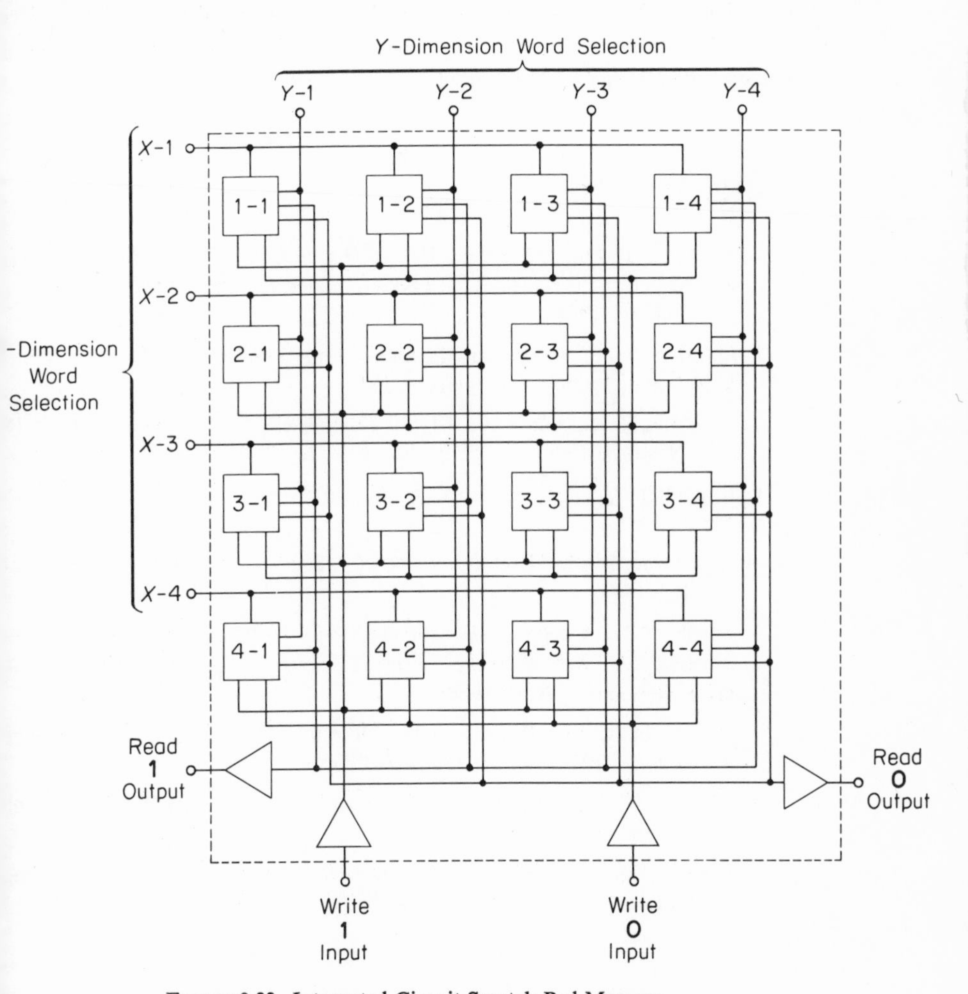

FIGURE 9-22. Integrated Circuit Scratch Pad Memory.

A logic diagram for a 16-bit random access memory is shown in Fig. 9-22. This memory consists of a single chip containing 16 set-reset flip-flops arranged to form an addressable 4 × 4 memory matrix. This structure permits non-destructive readout of all 16 bits. Addressing is through four X and Y lines that are brought out to eight external terminals. Read and write control is provided by four internal amplifiers. Each flip-flop in the 4 × 4 matrix is logically connected to its own X_m-Y_n address combination and to the sense and write amplifiers. This memory is word organized as a 16-word, 1-bit/word section. Parallel operation of a number of such elements can be employed to achieve a 16-word, multibit/word memory array. Groups of these 16-word sections can be arranged to form memories with larger word capacities.

For large memories, the decoding function for addressing can be built into the LSI chip to reduce the number of input/output leads. LSI memories have fast cycle times in the order of nanoseconds, which make them very useful as intermediate storage units for many digital system applications.

REFERENCES

1. Allen, W. B., and Westerfield, E. C., "Digital Compressed-Time Correlator and Matched Filters for Active Sonar," *J. Acoust. Soc. Amer.* 36 (1), 121-139, January 1964.
2. Sifferlen, T. P., "Precessional Delay Line Time Compression Circuit," U.S. Patent 3,488,635, January 6, 1970.
3. Richards, R. K., *Electronic Digital Components and Circuits*, Chap. 4, D. Van Nostrand Company, Princeton, New Jersey, 1967.
4. Rajchman, J. A., and Lo, A. W., "The Transfluxor," *Proc. IRE*, 44, 321-332, March 1956.
5. MacIntyre, R. M., "High-Speed BIAX Memories," *Comput. Des.*, 5(6) 54-61, June 1966.
6. Bartik, W. J., Chong, C. F., and Turczyn, A., "A 100-Megabit Random Access Plated-Wire Memory," *Proc. Intermag. Conf.*, IEEE Publ. 33-C-4, pp. 11.5-1-11.5-7, April 1965.
7. Brown, D. R., and Rochester, N., "Rectifier Networks for Multiposition Switching," *Proc. IRE*, 37; 139-147, February 1949.
8. Chu, Y., *Digital Computer Design Fundamentals*, McGraw-Hill Book Company, New York, 1962.
9. Maley, G. A., and Earle, J., *The Logic Design of Transistor Digital Computers*, Prentice-Hall, Englewood Cliffs, New Jersey, 1963.

PROBLEMS

1. Design a DELTIC storage system to store 2 sec of data. The input signal is sampled at a rate of 1,500 Hz and quantized to 1 bit. Determine the clock frequency to be used and specify the speed-up factor for the condition where the data circulate once between input sample pulses.

2. An audio signal of bandwidth 0–1 kHz is to be sampled at 25% greater than the Nyquist rate and quantized to three bits. The data bits are clocked serially into a DELTIC at a 9 MHz rate. Determine the storage capacity of the system in terms of the number of data samples, the length of the delay line in seconds, and the time-compression ratio.
3. A detected output signal from a receiver is sampled at a 1-kHz rate. The signal is quantized to four bits and fed in parallel to shift register delay lines in a DELTIC system. The data are to be circulated three times between new input data samples. A clocking frequency of 1 MHz is used. Determine the length of the delay line, the amount of input data storage, and the time speedup factor.
4. Design a DELTIC system for an input signal sampled at 1.5 kHz and quantized to one bit. The clocking frequency is 4.5 MHz, and it is desired to circulate the data three times between new input data samples. Determine the storage capacity, speedup factor, and delay line requirements.
5. Show by use of the Fourier series for a square wave that time compression in a DELTIC system results in a multiplication of each frequency component by the speedup factor.
6. Analyze a DELTIC system where the fundamental relationship is $\tau_t = T_s + T_c$. Use a clock frequency of 1 MHz and a delay line of length $\tau_t =$ 1,001 μsec. Determine the input sampling rate and the waveform speedup factor. Assume that the bits of the quantized samples are entered into parallel lines. Discuss what happens to the shape of the compressed waveform in this type of design.
7. In a DELTIC correlation receiver, explain why the reference DELTIC is longer than the signal DELTIC. Discuss the operation if it were the other way around.
8. Design a DELTIC spectrum analyzer with a frequency resolution of 1 Hz. The signal to be analyzed contains frequencies in a band from 100 to 500 Hz. The input signal is to be sampled at a rate of 2.8 times the bandwidth and quantized to five bits. The data bits are entered in parallel into the DELTIC. Determine all pertinent parameters for this system.
9. A magnetic core memory with a capacity of 4,096 words, 30 bits/word, is to be partitioned for a $2\frac{1}{2}$D array design. Determine the total number of address lines required if the array is partitioned into eight sections, each section containing 512 words.
10. Determine the minimum number of address lines required to partition an 8,192-word magnetic core memory, with 32 bits/word, into a $2\frac{1}{2}$D array organization. Draw a block diagram showing the various sections.
11. Design a 16-position diode matrix switch using an 8×2 matrix. The eight-position switch is driven by a rectangular $2 \times 2 \times 2$ matrix. Determine the total diode count. Compare with the most economical diode switch.
12. Show that the most economical 32-position diode matrix switch contains 96 diodes.

13. Determine the minimum number of transistors required to construct th following multiposition switches:
    (a) 32-position switch.
    (b) 64-position switch.
    (c) 128-position switch.
14. Design a 32-position switch using a multilevel design with transistor tre for the lower levels and a diode matrix for the higher level.

APPENDIX 1

CORRELATION FUNCTIONS

The correlation of two signals is a measure of their *likeness*, or of how much of one signal is contained in the other. Expressed mathematically, the correlation function, $R_{12}(\tau)$, is a time average of the multiple of two signals, where one signal is time delayed with respect to the other.

$$R_{12}(\tau) = \lim_{T\to\infty} \frac{1}{T} \int_{-T/2}^{+T/2} f_1(t) f_2(t - \tau)\, dt \qquad \text{(A1-1)}$$

$R_{12}(\tau)$, is obtained by multiplying the two functions and averaging their product over all time. One signal is delayed by a time factor τ, making the averaged product a function of τ. In practice, the averaging is carried out over a finite time interval. The inaccuracy caused by T being finite can be made acceptably small by choosing a reasonable computation interval.

The autocorrelation function of a signal, $f(t)$, is derived by setting $f_1(t) = f_2(t)$ in Eq. (A1-1).

$$R(\tau) = \lim_{T\to\infty} \frac{1}{T} \int_{-T/2}^{+T/2} f(t) f(t - \tau)\, dt \qquad \text{(A1-2)}$$

The autocorrelation function of both continuous and discrete waveforms may be computed from Eq. (A1-2). The discrete waveforms representing digital signals, however, are usually part of a synchronous data system where the signals change their binary states at clocked bit-time intervals. The *digital correlation function* may then be computed by shifting the signals at discrete intervals of τ, corresponding to multiples of the bit-time, rather than continuously for all values of τ. At each shift position, the signals are multiplied digitally and their products are summed. The digital autocorrelation function computed in this manner is designated $\phi(\tau)$.

To illustrate, consider the autocorrelation function of the waveform, $f(t)$, that represents the binary word 1 0 0. The amplitude and bit time are normalized for convenience, which results in the waveform in the diagram.

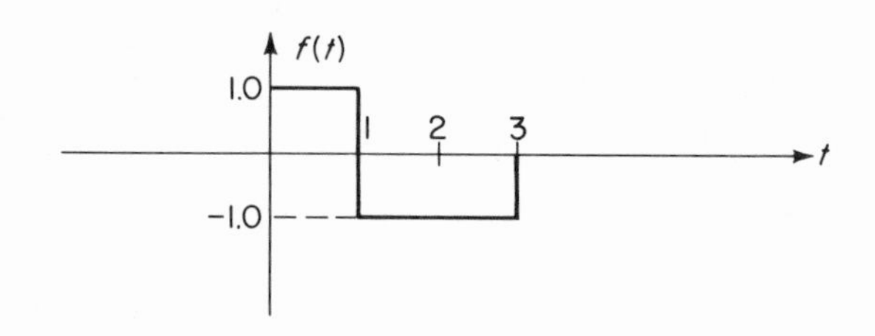

For purposes of computing the correlation functions the waveform is undefined (assumed to be zero) outside the three-bit interval shown. Also, the $1/T$ term in Eq. (A1-2) is omitted. This term is a normalization factor that scales down the magnitude of the finite-time integration, and may be omitted for computational convenience. The un-normalized autocorrelation function of $f(t)$ therefore becomes:

$$R(\tau) = \int_0^3 f(t)f(t-\tau)\,dt \tag{A1-3}$$

In Eq. (A1-3), $R(\tau)$ is zero for values of $|\tau| > 3$. Also, $f(t)f(t-\tau)$ is zero for values of $t < 0$ and $t > 3$, so that the integration is performed over the interval indicated.

It is often convenient to compute the autocorrelation of binary waveforms by graphical means. This is done, as indicated in Fig. A1-1, by drawing $f(t)$ and $f(t-\tau)$, and visualizing $f(t-\tau)$ as scanning across $f(t)$ from left to right. The term $f(t)f(t-\tau)$ is then drawn at significant shifts, and the integral which has straight line characteristics, is determined.

The autocorrelation function $R(\tau)$ of the digital word 1 0 0 is therefore as shown in the diagram.

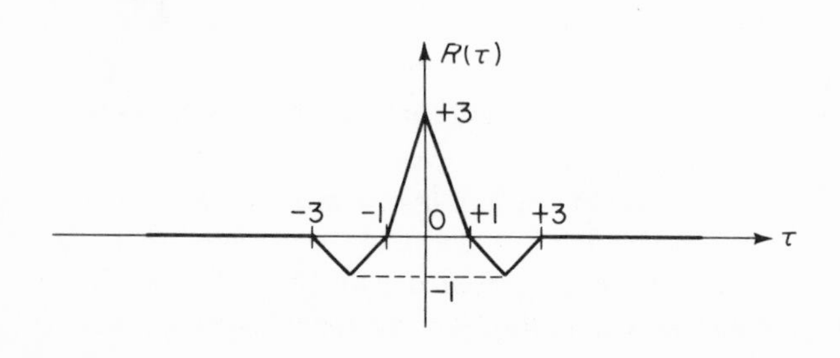

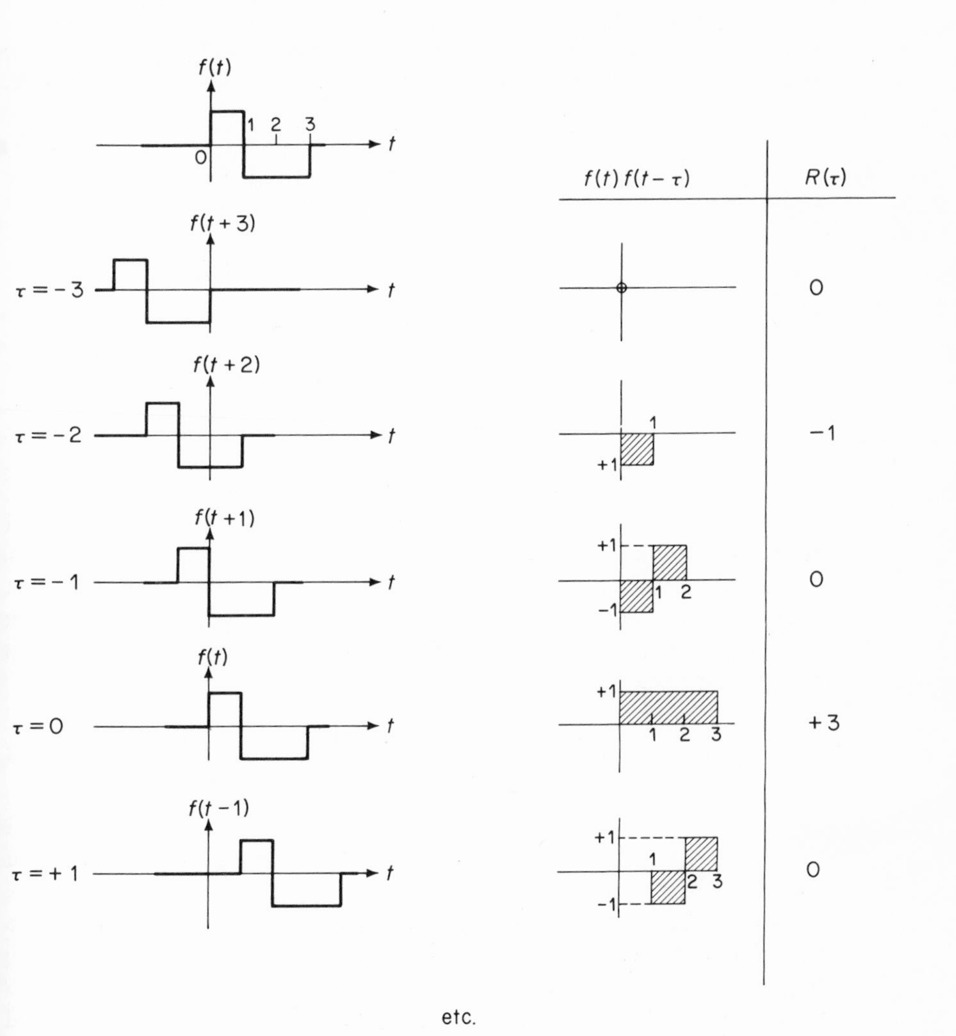

FIGURE A1-1. Graphical Computation of Autocorrelation Function of Digital Waveforms.

The digital autocorrelation function is computed by arranging the binary digits in their respective bit-time positions, and multiplying the digits in the same time slots. Mathematically, the $\phi(\tau)$ is computed as follows:

$$\phi(\tau) = \sum (A_i \cdot A_{i+\tau} + \bar{A}_i \cdot \bar{A}_{i+\tau}) \tag{A1-4}$$

where $\tau = 0, \pm 1, \pm 2$, etc., A_i is the ith digit of the binary sequence, and the summation is performed over the bit-time intervals where the A_i and $A_{i+\tau}$ digits

(**1** or **0**) are both present. A proceduce for computing the digital autocorrelation function is illustrated in Fig. A1-2 for the sequence 1 0 0.

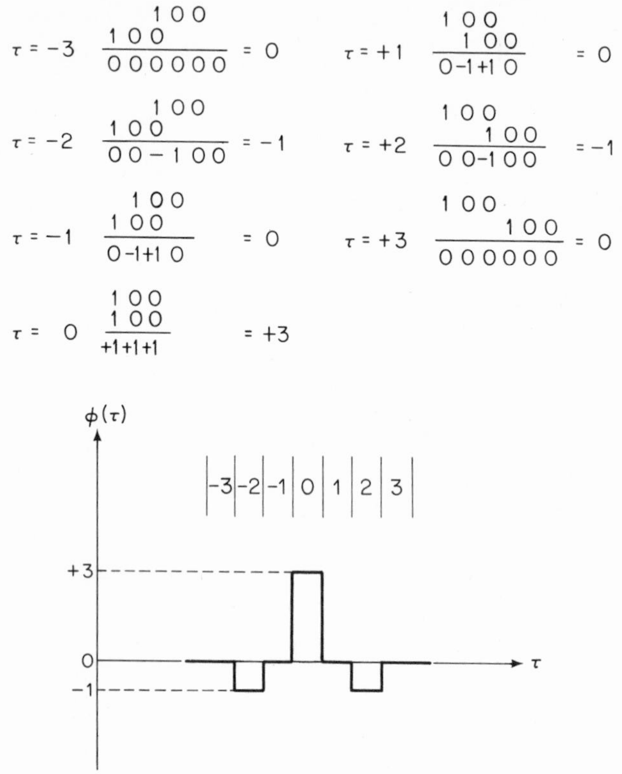

```
          1 0 0                          1 0 0
        1 0 0                              1 0 0
τ = −3  0 0 0 0 0 0 0  = 0      τ = +1   0 −1 +1 0        = 0

          1 0 0                          1 0 0
        1 0 0                                1 0 0
τ = −2  0 0 − 1 0 0    = −1     τ = +2   0 0 −1 0 0       = −1

        1 0 0                            1 0 0
        1 0 0                                  1 0 0
τ = −1  0 −1 +1 0      = 0      τ = +3   0 0 0 0 0 0      = 0

        1 0 0
        1 0 0
τ = 0   +1 +1 +1       = +3
```

$\phi(\tau)$

|−3|−2|−1|0|1|2|3|

+3

0

−1

τ

FIGURE A1-2. Computation of Digital Autocorrelation Function of Digital Word 1 0 0.

A second method for performing a digital autocorrelation calculation on a binary signal will now be illustrated. Consider the waveform corresponding to the digital word 0 1 0 0 1 1 1. The autocorrelation function is easily determined by listing the function and moving it to the right or left, one digit at a time. The central column must contain only **1**'s; therefore, when writing the numbers in any row, if a **0** appears in the center column, the complement of that row is written in. This is illustrated in Fig. A1-3(a). The columns are now summed, with **0** corresponding to -1 and **1** corresponding to $+1$. The sums are then equal to $\phi(\tau)$, as shown in Fig. A1-3(b).

$$\phi(\tau) = A - D$$

where A = number of agreements = $\Sigma\, A_i + A_{i+\tau}$,
D = number of disagreements = $\Sigma\, A_i + A_{i+\tau}$

```
0 1 0 0 1 1 1
  0 1 0 0 1 1 1
    0 1 0 0 1 1 1
      1 0 1 1 0 0 0
        1 0 1 1 0 0 0
          0 1 0 0 1 1 1
            1 0 1 1 0 0 0
-1 0 -1 0 -1 0 +7 0 -1 0 -1 0 -1
```

(a)

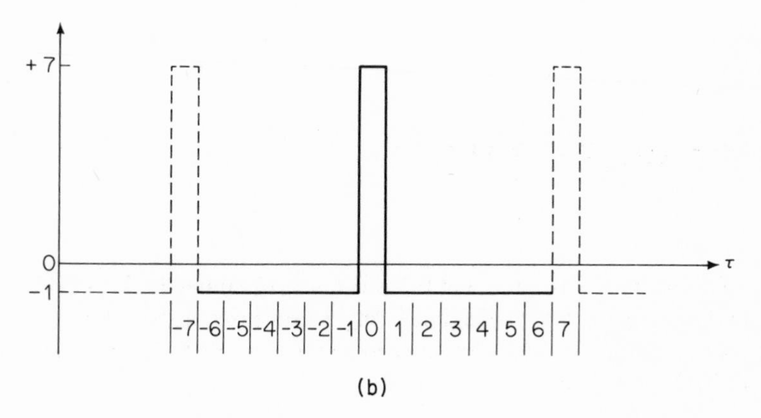

FIGURE A1-3. Digital Autocorrelation Function of the Digital Word 0 1 0 0 1 1 1.

REFERENCES

1. Hancock, J. C., *An Introduction to the Principles of Communications Theory*, McGraw Hill Book Co., Inc., New York, 1961.

2. Golomb, S. (Ed.), *Digital Communications*, Prentice-Hall, Inc., Englewood Cliffs, New Jersey, 1964.

APPENDIX 2

BARKER CODES

The binary words 1 0 0 and 0 1 0 0 1 1 1, illustrated in Appendix 1, are two of a class of non-periodic, finite-length sequences called *Barker Codes*. These constitute a class of coded sequences known as *perfect words*. The only known Barker Codes are:

1

1 1

1 0

1 1 0

1 1 1 0

1 1 0 1

1 1 1 0 1

1 1 1 0 0 1 0

1 1 1 0 0 0 1 0 0 1 0

1 1 1 1 1 0 0 1 1 0 1 0 1

Each of these sequences may also be taken in reverse order and/or complemented.

The Barker Codes are distinguished by their desirable autocorrelation functions. For a code of length N, the autocorrelation function is:

$$\phi(\tau) = N \qquad \text{for } \tau = 0 \tag{A2-1}$$

$$|\phi(\tau)| \leq 1 \qquad \text{for } \tau = \pm 1, \pm 2, \ldots \pm (N-1) \tag{A2-2}$$

$$\phi(\tau) = 0 \qquad \text{for } |\tau| \geq N \tag{A2-3}$$

Barker Codes are often used as synchronization signals in data transmission systems. The autocorrelation function of a Barker Code is advantageous in that signal processing gain is realizable without loss in timing accuracy. Consider, for example, the binary signals shown in Fig. A2-1. The autocorrelation function of the Barker Code combines the advantage of the higher autocorrelation output of longer pulses with the timing accuracy of a single narrow pulse. Note that in Fig. A2-1, the peak of $R(\tau)$ in (c) is the same as in (b), whereas the width is the same as in (a).

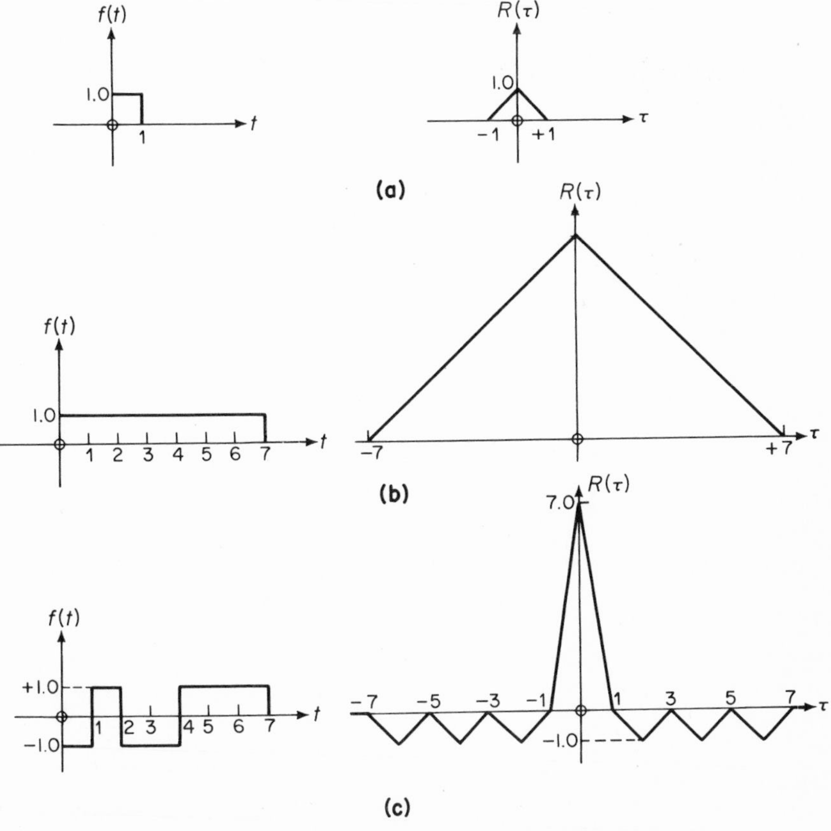

FIGURE A2-1. Three Binary Waveforms and Their Corresponding Autocorrelation Functions.

Since the longest Barker Code is for $N = 13$, alternate means are used when longer sequences are required. One alternate method is to use an N_1 bit

Barker word, each bit of which is another (N_2 bit) Barker Code. The peak $R(\tau)$ is then $N_1 \times N_2$ units for $\tau = 0$; however, for other τ, *side-lobes* appear and Eq. (A2-2) is not satisfied. Methods have been developed for calculating the magnitude of these side-lobes, and code sequences are chosen that minimize the amplitude of the side-lobes relative to the peak output at $\tau = 0$.

REFERENCES

1. Barker, R. H., "Group Synchronizing of Binary Digital Systems," in *Communication Theory*, Ed. W. Jackson, Academic Press, London, 1953.
2. Toerper, K. E., "Biphase Barker-Coded Data Transmission," *IEEE Trans.* Vol. AES-4, No. 2, pp. 278-282, March 1968.
3. Hollis, E. E., "Comparison of Combined Barker Codes for Coded Radar Use," *IEEE Trans.* Vol. AES-3, No. 1.

Appendix 3

PSEUDO-RANDOM SEQUENCES

All maximum-length linear sequences (see Chap. 8) have the characteristic of being *random noise-like* in their statistical properties. The sequences are, however, generated in a deterministic manner; therefore, they are termed *pseudo-random noise sequences.*

To qualify as a pseudo-random noise signal, a sequence must be repeatable, and possess the following characteristics of random signals.

1. The number of **1**'s and **0**'s in the sequence will differ by no more than one.
2. The occurrences of single **1**'s and **0**'s, pairs of **1**'s and **0**'s, triplets, etc., will statistically conform to the average distribution in random noise signals. For example, one-half of the consecutive runs will be single **1**'s and **0**'s, one-fourth of the runs will be pairs (1 1 or 0 0), one-eighth triplets (1 1 1, 0 0 0), etc. Also, the number of runs of equal length will be (as nearly as possible) composed equally of **1**'s and **0**'s.
3. The autocorrelation function of the sequence is two-valued. Specifically, the unnormalized digital correlation of the sequence with its replica cycle-shifted (i.e., shifted in time in a shift register whose output is recirculated into the input) will be:

$$\phi(0) = +p \tag{A3-1}$$

$$\phi(\tau) = -1 \qquad \text{for } \tau \neq 0, \pm p, \pm 2p, \text{ etc.} \tag{A3-2}$$

where p is the sequence length.

To illustrate, consider the 15-bit linear sequence 0 0 1 1 0 1 0 1 1 1 1 0 0 0 1 that is generated by a four-stage shift register and the feedback connection $a_3 \oplus a_4$. Note the following characteristics:

1. There are seven **0**'s and eight **1**'s in the sequence.
2. There are four single runs (0, 1, 0, 1), two double runs (0 0, 1 1), one triple run (0 0 0), and one quadruple run (1 1 1 1).
3. The correlation function of the sequence and its replica, for $\tau = 0$, is computed as follows:

0 0 1 1 0 1 0 1 1 1 1 0 0 0 1	sequence
0 0 1 1 0 1 0 1 1 1 1 0 0 0 1	replica
+ + + + + + + + + + + + + + +	correlation

$$\phi(0) = A - D = 15$$

where A = number of agreements (+)
D = number of disagreements (−)

If the replica is cycle-shifted one bit-time to the right, we have

0 0 1 1 0 1 0 1 1 1 1 0 0 0 1	sequence
1 0 0 1 1 0 1 0 1 1 1 1 0 0 0	shift replica
− + − + − − − − + + + − + + −	correlation

$$\phi(1) = A - D$$
$$= 7 - 8$$
$$= -1$$

The correlation function will remain -1 for all cycle shifts, right or left, except for $\tau = 0, \pm 15, \pm 30$, etc. Also, the normalized correlation function may be calculated by

$$\phi'(\tau) = \frac{A - D}{A + D} \tag{A3-3}$$

and is either $+1$ (for $\tau = 0$, etc.) or $-1/p$. The above definition and the method of computing the autocorrelation function $\phi(\tau)$ are used when the binary sequence is periodic—repeating every p digits.

As a second example, consider the maximum-length linear sequence 0 0 1 0 1 1 1. The output sequence is tested for compliance with the requirements for a pseudo-random sequence as follows:

1. We note there are four **1**'s and three **0**'s in the sequence.
2. Also, there are two singles (0, 1), one doublet (0 0), and one triplet (1 1 1) in the sequence.
3. The auto-correlation of the sequence for different cyclic shifts (starting with $\tau = 0$) is:

0 0 1 0 1 1 1	
0 0 1 0 1 1 1	
+ + + + + + +	$A - D = +7$
0 0 1 0 1 1 1	
1 0 0 1 0 1 1	
− + − − − + +	$A - D = -1$
0 0 1 0 1 1 1	
1 1 0 0 1 0 1	
− − − + + − +	$A - D = -1$

etc.

The digital autocorrelation function of this sequence is plotted in Fig. A3-1. Note the cyclic nature of $\phi(\tau)$.

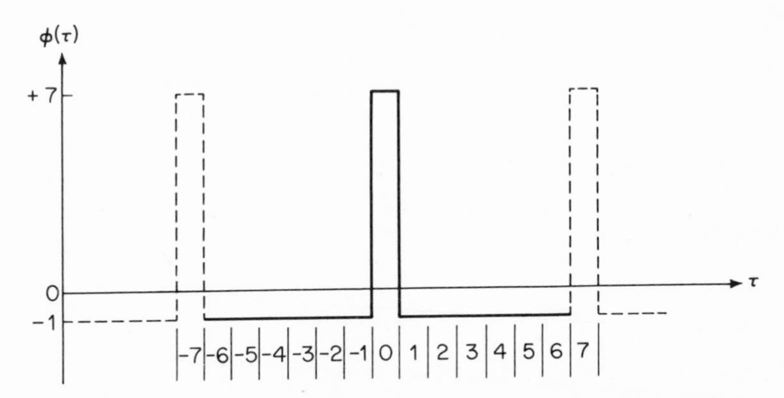

FIGURE A3-1. Autocorrelation Function of the Pseudo-Random Sequence 0 0 1 0 1 1 1.

Other interesting and useful properties of maximum-length linear sequences are:

1. Shift-and-Add Property. If a maximum-length linear sequence is exclusive-ORed with a shifted version of itself, then the output sequence will also be a shifted version of itself.

2. Sampled Sequences. To produce new sequences, maximum-length linear sequences may be sampled at fixed rates that are multiples of the basic shift rate of the register. For example, by sampling every fifth digit of the repeating waveform of Fig. A3-1, the sequence 0 1 0 0 1 1 1 is generated. If the digit sampling rate is relatively prime to the sequence length (p), then a new maximum-length sequence is produced. Sampling at odd multiples of the shift rate other than relatively prime rates will produce nonmaximum length sequences, and at even multiples of the shift rate will produce a shifted version of the original sequence.

3. Combined Sequences. Several maximum-length linear sequences may be combined in a nonlinear manner to form a new sequence whose length is equal to a multiple of the length of the original sequences. The new sequence is nonlinear; therefore, the digital autocorrelation function is less than optimum. However, the total detection time of these sequences is significantly less than it would be for an equivalent-length linear sequence.

DETECTION OF PSEUDO-RANDOM SEQUENCES

Pseudo-random sequences of length p may be detected in a straight forward manner by entering received digits into a shift register with p stages and summing the appropriate Q and $\bar{Q}$ outputs of each stage. In Fig. A3-2, the circuit implementation for the sequence 0 0 1 0 1 1 1 is shown. With this

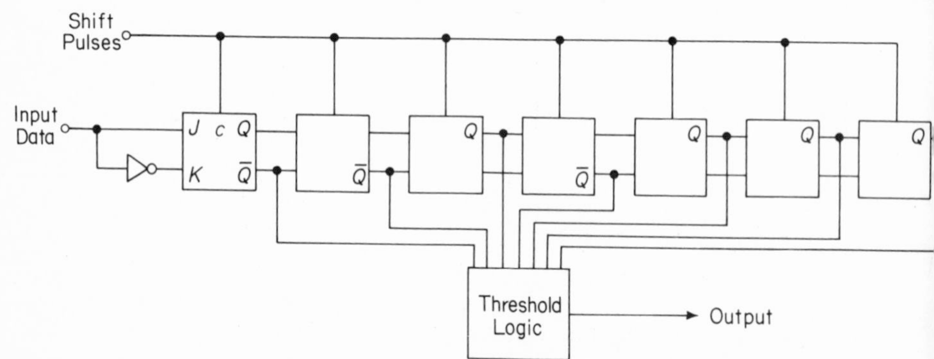

FIGURE A3-2. Pseudo-Random Code Detector.

method of detection, the input data's arrival time is assumed known, and the shift pulses are properly synchronized to the data bits.

The detection method shown becomes unwieldly when the code length is increased to 15 or higher. An alternate method is to use a replica generator that requires only an n-stage shift register rather than $2^n - 1$ stages. The incoming data are then correlated bit-by-bit with the sequence generated at the receiver, and the results are summed in a binary adder. The difficulty with this approach is that the receiver has no way of predetermining to which bit-time the replica generator should be synchronized.

In practice, one method of resolving this difficulty is to store received data and then process them in nonreal time. The stored data could, for example, be sequentially processed against a set of replicas, each generated by the same code generator. After each complete cycle, the replica generator is advanced one bit-time. In this manner, a p-bit sequence could be processed $p \times p$ bit-times. Some improvement may be realized by stopping the process when a correlation peak is detected, but the detection process is still too slow for many applications.

A second detection method, that enables the processing of data as received, (in p bit-time for a p-bit sequence), is made possible by digital time compression techniques. The receiver is thereby given multiple looks at the same data in compressed time. Digital time compression (DELTICS) is discussed in Chap. 9.

REFERENCES

1. Golomb, S., *et al.*, *Digital Communications with Space Applications*, Prentice-Hall, Inc., Englewood Cliffs, New Jersey 1964.
2. Ristenbatt, M. P., "Pseudo-Random Binary Coded Waveforms," Chap. 4 in *Modern Radar Analysis, Evaluation, and Systems Design*, Ed. R.S. Berkowitz, John Wiley and Sons, Inc., New York, 1965.
3. Burdic, W. S., *Radar Signal Analysis*, Chap. 5, Prentice-Hall, Inc., Englewood Cliffs, N. J., 1968.
4. Golomb, S. W., Welsh, L. R., and Hales, A., "On the Factorization of Trinomials Over GF(2)," JPL Memorandum 20-189, July 1959.
5. *Communication Techniques—Deep Space Range Measurement*, JPL Research Summary No. 36-1, Vol. 1, California Institute of Technology, Pasadena, California, February 1960, pp. 39–46.

INDEX

R

S

T

U

V

W